Rural-Urban Migration and Agro-Technological Change in Post-Reform China

Rural-Urban Migration and Agro-Technological Change in Post-Reform China

Lena Kaufmann

Routledge
Taylor & Francis Group
LONDON AND NEW YORK

Published with the support of the Swiss National Science Foundation

First published in 2021 by Amsterdam University Press Ltd.

Published 2025 by Routledge
4 Park Square, Milton Park, Abingdon, Oxon OX14 4RN
605 Third Avenue, New York, NY 10158

Routledge is an imprint of the Taylor & Francis Group, an informa business

ISBN: 9789463729734 (hbk)
ISBN: 9781041185673 (pbk)
ISBN: 9781003703198 (ebk)
NUR 761

Cover image: A farmer from Hunan Province, China, walking away from the village alongside a harvested rice field
Photographed and copyright by Lena Kaufmann

Cover design: Coördesign, Leiden

DOI: 10.5117/9789463729734

For Product Safety Concerns and Information please contact our EU representative: GPSR@taylorandfrancis.com
Taylor & Francis Verlag GmbH, Kaufingerstraße 24, 80331 München, Germany

To LFT and the other Chinese rural women
who are in a similar situation

Table of Contents

Acknowledgements 11

Introduction 15

Arguments and aims of the book 22

Agriculture and migration 25

From 'migrant worlds' to 'community of practice' worlds 28

Knowledge, repertoire, and agency 35

Accessing the rural-urban community of practice 39

Structure of the book 47

1 **How the Predicament Arose** 61

Modern agriculture in Anren County 65

De-collectivization and marketization 87

Abolition of the collective welfare system 93

The new urban economy and increased migration 94

2 **Rice Knowledge Systems in Transition** 105

Transformation of agricultural knowledge transmission 108

Transformation of the repertoire of knowledge 113

Agricultural deskilling and extended knowledge repertoires 136

3 **Reference Models for Transmitting Knowledge** 145

Transmitting farming knowledge through proverbs 148

Educating the masses 159

Textualizing vernacular knowledge 160

Negotiating knowledge and farmer-state relationships 162

4 **Technological Choice in the Wake of Migration** 167

Tilling with power ploughs and oxen 171

Harvesting with sickles and combine harvesters 173

Choosing harvesting technologies 177

Technological choice from a repertoire perspective 181

5 **Land-Use Strategies** 187

Sustaining intensive rice farming 191

De-intensifying rice farming 202

More than linear, more than technical 219

Conclusion: A Skill Perspective on Migration 229
Agency beyond resistance 232
Decision making beyond economic reasoning 233
Technology beyond linear progress 234
Migration beyond dichotomies 235

Appendix 239
I Glossary 239
II Solar terms 244
III Song of the 24 Solar Terms 244
IV Examples of proverbs and encoded knowledge 245

References 273

Index 299

List of Figures and Tables

Figures

Figure 1 Map of mainland China 40
Figure 2 Map of Hunan Province 42
Figure 3 Throwing the bundles of seedlings and transplanting the seedlings 123
Figure 4 A proverb painted on a wall: 'People shouldn't relax in winter, and the fields shouldn't waste in winter' 146
Figure 5 Fields harvested with a combine harvester (left) and a sickle (right) 168
Figure 6 Field preparation with a power plough 172
Figure 7 A hand-made sickle (above) and an industrially-produced sickle (below) 174
Figure 8 The only combine harvester in Green Water 176
Figure 9 An abandoned paddy field in Longshi Township 205
Figure 10 The foundations of a house under construction on a former paddy field 207
Figure 11 Zhou Wenlu and Mrs. Luo water their dry fields (former wet fields) 213
Figure 12 Lamp rush growing in a wet field 215
Figure 13 Granny Li peels the dried lamp rush 218

Tables

Table 1 Simplified overview of the changing Chinese system of rice knowledge transmission 109
Table 2 Overview of the villagers' land-use strategies 190
Table 3 The 24 solar terms and equivalent dates in the Gregorian calendar 244

Chinese Measurements

dan	50 kg
fen	0.1 *mu* (about 66.67 m^2)
jin	500 g
li	500 m
luo	25 kg
mu	1/15 ha (about 666.67 m^2)

Acknowledgements

十年操出个文秀才，
十年操不出一个田秀才。

Shi nian caochu ge wen xiucai,
shi nian cao bu chu yi ge tian xiucai.
(XT 1988, 206)

'It takes ten years to make a literary scholar, but you cannot make a "field scholar" in ten years'.

This farmers' saying from Hunan acknowledges the rich knowledge and practice that is needed to be a good rice farmer. I fully agree. In writing this book, I have attempted the challenging task of documenting such often-tacit knowledge. I have endeavoured to investigate the transformation of this knowledge held by Chinese rice farmers between home villages and domestic migration. In the course of this journey, moving between Berlin, Shanghai, Beijing, the Chinese countryside, and Zurich, my own knowledge has also grown and been transformed. This is thanks to the numerous people and institutions who have supported my project in one way or another.

I am most grateful to the interlocutors who have contributed to this book, all of whose names I have changed to protect their privacy: my interviewees from Anhui, the villagers from and around Green Water, and the Zhao couple from Shanghai. They have all given me precious insights. I especially thank the Wu siblings, who invited and hosted me in their respective homes in Anhui, and Yuemei and her family who hosted me in Hunan. I am greatly indebted to them for their incredible hospitality and for sharing their knowledge, as well as the joys and sorrows of their translocal lives. Yuemei was particularly helpful in assisting my research, always ready to help me find material and answers to my numerous questions, even after I had left China. She kindly allowed me to use her photos.

My first long-term stay in China in 2006-2008 was made possible through a grant from the German Academic Exchange Service (DAAD) and the Chinese Scholarship Council. Although at that time I was working on a different project on rural migrants' skills in urban China, many of the insights gathered during that time have made their way into this book and inspired this book's topic. For my field research in 2010-2011, I received funding from the Ethnographic Museum of the University of Zurich. In fact, I would never have ended up in Zurich and written this book if it wasn't for my academic mentor Mareile Flitsch, the museum's director, who encouraged me to do so. Her detailed critical feedback on earlier versions of the manuscript and her belief in my project were invaluable. I took from her how we can attribute individuals and societies with the dignity they deserve only if we pay close attention to their often-overlooked knowledge and skills – a perspective I believe is highly beneficial in terms of mutual understanding and appreciation, way beyond this book.

The Swiss National Science Foundation generously awarded me a two-year Marie Heim-Vögtlin grant, in addition to supporting the publication of this book. I would like to express my gratitude to Gonçalo D. Santos and Francesca Bray for their support with the grant application for that. Gonçalo has offered many useful thoughts and, moreover, convinced me to write in English. Thanks to Helen Rana's professional copy editing, I am confident that it has become readable, although all the mistakes in this book remain, of course, my own. Francesca Bray's thoughtful critique of my proposed research framework was particularly beneficial in helping me to clarify the framework and thread of my argument. More than she may realize, her seminal insights into Asian rice farming economies provided an essential intellectual foundation for me to build upon. Here, I would also like to thank Dagmar Schäfer, who hosted me as a short-term visiting fellow at the Max Planck Institute for the History of Science in Berlin in 2014, supported by the Gender Equality Commission of the University of Zurich. In Berlin, I had the opportunity to benefit greatly from the inspiring discussions in Dagmar's research group, of which Francesca Bray was also a member.

In Zurich, the Ethnographic Museum and the Department of Social Anthropology and Cultural Studies (ISEK) at the University of Zurich have provided the ideal environment to pursue my project. I could not imagine a better working atmosphere and colleagues, many of whom have become friends. I would especially like to thank my second academic mentor Annuska Derks. Her insightful comments on my research and the presentation of my argument were extremely valuable and much appreciated. I am also very grateful to my Writing Group, consisting of Rehema Bavuma, Francesca Rickli, and Maria-Theres Schuler, who closely read and commented on an early draft of this book. In addition, more colleagues at the University of Zurich than I can name here gave me helpful comments on parts of this book. Among these are the late Ingo Nentwig, Esther Leemann, Stefan Leins, and Juliane Neuhaus at the ISEK. In addition, Katharina Woodhouse, Martina Wernsdörfer, Maike Powroznik, Michèle Dick, Rebekka Sutter, Andreas Isler, Alexis Malefakis, and others all shared office life in the Ethnographic Museum with me, and provided inspiration. The expertise of the respective library teams, especially Jörg Schlatter, helped me save a great deal of time and expense in my research. In addition, Wu Jing and Hu Junli assisted me in purchasing much-needed literature and carrying it all the way to Zurich.

I received valuable input and feedback at numerous conferences and workshops, which helped me to sharpen my arguments. One very productive occasion was a workshop on gynocentric technologies, organized by Jacob Eyferth, Suzanne Z. Gottschang, and Gonçalo D. Santos at Smith College, Northampton in 2018. I benefitted immensely from the remarks of my paper's discussant Jacob, as well as from feedback from all the participants. Moreover, throughout the past five years, the China Regional Group of the German Association of Anthropology has been a particularly beneficial source of scholarly exchange and support. I especially thank Madlen Kobi, Robert LaFleur, and Christof Lammer for their astute comments on parts of my writing.

The final stages of this book have been completed at the Department of History of the University of Zurich, amongst a new group of stimulating colleagues. I especially thank Monika Dommann for being so generous with giving me so much time to dedicate to finishing this book and, more generally, for engaging me in fruitful methodological reflections on the cross-fertilization of history, science and technology studies, and anthropology.

I am deeply indebted to the two reviewers at Amsterdam University Press for their careful reading and for sharing their rich expertise on the topic in their highly constructive comments. They have helped greatly to improve this book. Moreover, I thank editor Saskia Gieling, as well as the New Mobilities in Asia series editor Pál Nyíri, for their helpful suggestions and support. Thanks also to Victoria Blud for giving the manuscript a final meticulous look, and to Jutta Turner for drawing the maps.

There are numerous people who I owe my gratitude to, since they have supported my everyday life in Switzerland. I thank the team at the Arche for their flexibility, for always assuring me that my children were in good hands, and for providing me with the necessary space to concentrate on my research. Simone and Felix Balke, as well as Paula and Thomas Grossmann Rodriguez, were always ready to assist, making me feel welcome in Switzerland through their friendship.

It is not possible to put into words the depth of my gratitude for the unfailing support of my family, especially my parents Ulrike and Heinz, my husband Jean Pierre and my grandmother Helga. This book would not exist without them. Finishing this book amidst the Covid-19 pandemic has been a special challenge, all of a sudden removing the boundaries between working and private life. I would like to thank Noah and Mika for their patience, for always keeping my feet on the ground, and for making me smile every single day throughout this book project.

Introduction

Abstract

This introduction introduces the basic predicament being faced by rice farmers in post-reform China: the conflicting pressures to both migrate into cities and yet preserve their family land resources in the countryside. It posits that paddy fields play a crucial role in shaping farmers' migration strategies. More generally, it proposes that socio-technical resources and related skills are key factors in understanding migration flows and migrant-home relations. Furthermore, the chapter proposes a socio-technical approach to investigating this paddy field predicament and explains how this approach contributes to existing literature at the intersection of the literature on agriculture, migration, and skill. Finally, it introduces the main field site, a rice-farming village in southern China, and briefly discusses the data and sources.

Keywords: China, materialities of migration, agriculture-migration nexus, socio-technical knowledge and skills, rural-urban farming community of practice, migrant-home relations

Mr. Wu and his family never mentioned the necessity of maintaining their rice fields. Instead, when speaking about home, they talked about house construction, food, and especially their children, who they had left behind and missed dearly. They called them once a week from a nearby telephone booth, meeting them only once a year during the Spring Festival, the Chinese New Year celebrations. Mr. Wu's small restaurant selling spicy noodle soup first attracted my attention when, in spring 2007, I was looking for a place to eat on the outskirts of the former French Concession in downtown Shanghai. It was located in one of the last blocks of two-storey houses not yet replaced by the high-rise glass facades of shopping malls, hotel restaurants and hospitals. I saw the bustling queue of lunchtime customers, escaped the loud honking of buses and motorcycles, and snuck inside. The crammed and windowless interior, with diners loudly slurping hot soup and wiping sweat from their

Kaufmann, Lena, *Rural-Urban Migration and Agro-Technological Change in Post-Reform China*. Taylor & Francis Group 2021
DOI: 10.5117/9789463729734_INTRO

brows, made it easy to fall into conversation with the other customers as well as Mr. Wu and his family, who ran the restaurant, and who were all eager to make sense of me, a foreigner: 'Where do you come from? What are you doing in China? How old are you? Are you married? What do you eat at home?'[1] This first encounter led on to numerous regular, longer visits. Gradually I learned that Mr. Wu and his family were originally rice farmers from rural Anhui Province, a day's bus ride from Shanghai. They were part of the one fifth of the entire Chinese population, or more than one third of Chinese farmers who had become migrants since the 1980s (NBSC 2019, sec. 2-3). Eight years ago, having tried out various informal jobs in different provinces, they had followed a group of fellow villagers to run a noodle shop in Shanghai.

When I joined the family on their annual trip home for the Chinese New Year in 2008, it struck me that they were maintaining their rice fields. I followed Mr. Wu's wife Li Cuiping from the main road, where the overland bus had dropped us off, far away from any township or even bus stop. We continued our way on foot, balancing one after another along the narrow ridges between the rice fields. As we approached the village, Li Cuiping pointed at a neatly cultivated and harvested field to her right: 'This is ours'. Rather than simply letting the fields lay fallow during their years away, the family tried to sustain rice cultivation. Obviously, these fields were of central importance. Nevertheless, the necessity of maintaining the fields seemed so self-evident to Mr. Wu and other migrants I met that they hardly ever mentioned it. As Mr. Wu's niece Caixia later explained: 'You don't talk about your bathroom either. There is no need to talk about it'. She went on to explain that fields were something everybody had, similar to a garden, which made it unnecessary to talk about (video conversation, 5 September 2017).

During the course of my research, however, it became clear that rice fields are not a trivial aspect of migration at all. In fact, a lot of strategic efforts are made to maintain this valuable resource, regardless of migration. The fields play a crucial role, not only for those left behind, but also, and perhaps especially, for the migrants. For those staying behind they provide subsistence. For migrants, this farmland is an asset that provides seed capital and an important economic safety net for their often highly precarious city life. Indeed, some of the migrants I interviewed inferred that their fields were so central to their social and economic security that they had specifically left close family members behind to look after them. Preserving wet rice fields is a real challenge, especially where skilled people have migrated,

1 Unless stated otherwise, all the translations of written and oral Chinese sources, as well as the quotes from French and German secondary literature in this book, are the author's.

so are unavailable to cultivate them. There are certain socio-technical particularities about farming rice fields.

First, each step of wet rice cultivation requires considerable skill, and many cannot be mechanized, making it more labour- and skill-intensive than most other Chinese crops (Bray 1994). Even where it is possible to mechanize certain stages, few farmers can afford to do so. Therefore, it is crucial that a sufficient number of skilled people are around to carry out the necessary tasks to ensure successful rice cultivation. Second, in order to retain their worth and yield, wet rice fields need to be cultivated with rice continuously. In contrast to dry fields, they actually increase in value if they are cultivated regularly over a long time (Bray 1984; 1994). Not cultivating the fields or transforming them into dry fields therefore means significantly decreasing their value. This is tangibly related to the particular soil characteristics and the requirements of wet rice itself.

Wet rice, or paddy fields, have specific soil characteristics, and fallowing or switching crops alters these characteristics in both the short and the long term. Heavy rainfall in south China normally leaches the soil and makes it acidic. The continuous long-term cultivation of wet rice reverses this unwanted process, producing soils that are particularly favourable for wet rice cultivation. These are characterized by an upper layer of fine, grey, low-acid silt, and a lower layer that is hard and impermeable (Bray 2004, 17). Consequently, fallowing fields would expose the soil to leaching, degrading the soil quality needed for wet rice farming.

This also implies that it is not easy for farmers to turn wet fields into dry fields, or to change transformed fields back into wet fields, and there are consequences of doing so. As agronomists and geographers note, the creation of paddy soil is a long-term transformation of the soil. Therefore, it is not feasible to successfully cultivate other crops such as vegetables by simply planting them in drained paddies. Similarly, it is difficult to switch from planting non-rice crops to wet rice. Once non-rice crops such as beans have been cultivated in paddy fields, they deplete the soil's nitrogen fertility, creating a new soil condition which is not tolerated by conventional rice varieties. Changing a wet field into a dry field, or the other way round, therefore takes many years, so it is not a decision that can be taken lightly (Kleinhenz, Schnitzler, and Midmore 1996; McKay 2005).

Weeds that quickly populate fallow fields have a similar effect. According to my interlocutors, weeds are *the* major issue when fallowing fields. They 'eat up all the fertilizer' and nutrients in the soil. In addition, once they are there, weeds such as the tenacious barnyard grass (*Echinochloa crus-galli* Beauv.) are persistent and almost impossible to get rid of. This weed invasion is precisely what happens, however, if paddy fields lay fallow, in an unwanted condition called *huang* (waste, desolate).[2] If this happens, the fields are 'no longer good to cultivate' (personal interviews, 2011). In short, when confronted with off-farm migration, it might at first sight appear that mechanization, fallowing fields or switching to less labour-intensive crops would be easy ways to compensate for the missing skilled labour. However, the constraints described above show that none of these are actually straightforward possibilities.

This places Chinese rice farmers in a tricky situation, because staying home to ensure constant rice cultivation is not an appealing option either. The pressure to migrate is enormous, as the following two accounts from Green Water Village in Hunan Province demonstrate. According to my interviews with several Green Water villagers, most migrants from the village move to neighbouring Guangdong Province. There, many women work in textile factories, while many men work in mining and become excavator operators.

The two labour migrants Zhou Wenbao and Zhou Wenlu, however, are not among these men. When I met them in 2011 during the Spring Festival, they were in their forties and fifties respectively and had just come home from another year of migrant work. As the first two syllables of their names suggest, they belong to the same lineage and generation. Having turned their backs on rice farming, they were now working in construction, moving to different provinces each year. Their boss was a local man, too, recruiting workers from his immediate surroundings. In the past year, both men had worked in Beijing, whereas in the following year the company was going to operate in Gansu Province. Zhou Wenbao and Zhou Wenlu had both specialized in steel and iron – 'you do what you know' – in contrast to other workers who laid tiles, cement, did plastering or carpentry. As Zhou Wenbao stated, 'it is very hard (*xinku*)'.

When asked why they had migrated, they explained that it was mainly for financial reasons, like the other migrants I interviewed. However, some other factors were also involved. These included gaining higher social standing, attracting potential future spouses by constructing a new house, or financing their children's education. The younger of the two men, Zhou Wenbao, had only ceased rice farming five years earlier. He described his personal family situation:

> My wife, Wu Guizhen, also works (*dagong*) outside the village, in a textile factory in Zhongshan City in Guangdong. Only my parents and

2 I use the official Chinese system of *pinyin* for phonetic transcriptions.

> my paternal grandmother live at home and plant rice. My four siblings have also migrated. My grandmother was born in the 1920s. She is over 87 years old and can hardly walk. I am the oldest son, so I have to take care of her and my parents.
>
> My two daughters, Lanxiang and Lanying, are in their early twenties [born in 1990 and 1991 respectively]. They are studying in Changsha [the provincial capital]. Lanxiang is in the last year of her bachelor studies in automotive insurance. Lanying did not pass the university entrance examination. She attends a vocational college and will become a primary school teacher. Lanxiang has already been recruited to an automotive insurance company in Shenzhen [one of Guangdong's major cities] as soon as she finishes her degree. Lanying will probably become a teacher in one of the primary schools here. I don't think they will ever work as farmers. But [because there are no sons] they will inherit the house and the fields. For us [me and my wife] it is very hard (*xinku*)! We have to send two children to university! And it cost us 200,000 Yuan [about 28,250 USD][3] to build this house – other people even spend 300,000 or 400,000.

Zhou Wenbao continued with the following calculation:

> From rice farming alone, you [i.e. a household] can earn about 10,000 Yuan [about 1400 USD] per year by cultivating eight to ten *mu* [just over half a hectare].[4] From this you have to subtract 2000 Yuan of capital input for pesticides, harvesting, and fertilizer. Harvesting alone costs 80 Yuan per field. You cannot send your children to university with these few thousand Yuan per year!
>
> But with a middle school degree, you can earn between 1000 and 2000 [about 140-280 USD] per month, as a construction worker [i.e. up to three times as much as a rice farming household]. (Interview, 28 January 2011, from fieldnotes.)

Zhou Wenbao's fellow villager and colleague Zhou Wenlu had migrated for similar reasons. His family hosted me during my stay. As his elder daughter Yuemei explained:

> There are three of us children, two sisters and one younger brother. When I went to primary school [in the late 1980s and early 1990s], school fees were

3 10 Yuan Renminbi equates to about 1.41 US Dollars (as at 25 June 2020).

4 One *mu* equals one fifteenth of a hectare, i.e. about 0.067 hectares.

> still very high. Therefore, ever since then my dad had to work outside the village. Now my siblings and I have also left, only my mother remains. My brother is 22 now and will have to marry soon. But he works in mining, so it is difficult for him to get to know a woman there. He needs a piece of land and to build a house [here], otherwise it will be hard to find a wife. (Interview, 4 February 2011, from fieldnotes.)

Regarding her own decision to go to Beijing, where she had recently graduated from university, she stated: 'I have always liked studying. I studied hard, because I wanted to get out of this cycle [of hardship and of being bound to the countryside]' (ibid.). While most of my interlocutors were absolutely certain that they would return to their ancestral home (*laojia*) once they became too old to work in the city, Yuemei clearly did not see her future in farming. Nevertheless, she was sending money to her mother, which ensured that rice cultivation could continue. As these accounts reveal, the pressure to migrate is strong, not only because rice farming barely provides subsistence-level incomes, but also due to social pressures. Therefore, all of my interlocutors felt it was imperative to migrate, although for some of them leaving was difficult due to their current circumstances. This put them in a difficult situation, which challenged them to find suitable ways to simultaneously migrate, whilst still ensuring the sustained cultivation of their farmland.

I define this situation of conflicting pressures to both migrate into cities and preserve their resources in the countryside as a predicament. Tom Shakespeare in the field of disability studies suggests that, to 'call something a predicament is to understand it as a difficulty, and as a challenge, and as something which we might want to minimize but which we cannot ultimately avoid'. Yet, while such difficulties 'make life harder, [...] this hardship can be overcome' (Shakespeare 2006, 63). Notably, Shakespeare's concept evokes an active, problem-solving subject rather than victimhood. Accordingly, I see the farmers I studied not as victims, but as actors who are capable of finding workable solutions despite the complications they are in.

To be more specific, Chinese rice farmers are undoubtedly in a difficult situation, one that constantly requires making new decisions that take into account long-term needs and ambitions, but also short- or mid-term adjustments in line with changing household constellations and potential future circumstances. These include, for instance, the death of a parent, the out-marriage of a daughter, the birth of a baby, youngsters' migration, the return of a sick migrant, or a child starting their formal education. Furthermore, even where a solution is temporarily identified and decided upon, it might not be an easy or satisfying choice. Migration might be an opportunity, but it is also a burden. For example, migrant worker Xiao Chen felt deeply hurt when her small children in Anhui no longer recognized her upon her return, calling another woman 'mother', because a fellow villager had jokingly told them to do so. Similarly, after Mr. Wu's daughter Guilan got married, she and her husband opened their own noodle restaurant, leaving their little boy behind with her mother-in-law. She was upset: 'When I have a video conversation with him [my son], he does not even care about talking to me. He is close to his grandmother, not to us' (video conversation, 12 November 2017).

Migration-affected households face several pressures at once. There is, for example, the double burden on those who migrate to provide financially for the children and elderly relatives left behind. Alternatively, migrants need to care for children in the city while earning a living for themselves there, as well as looking after the elderly in the village. For the old people, the burden commonly consists of having to tend the fields whilst looking after grandchildren. In addition, they are often left unsupported by the state if they become ill, due to insufficient insurance coverage. Moreover, migrants experience homesickness and miss their family members, even though their decisions are always made in the hope of finding a solution that will lead to a better future. At the core of all these quandaries lies concern for their major asset, land. People are – and remain – paddy field bound, even if they migrate.

Thus, understanding this situation as a predicament means acknowledging that rice farmers are actors struggling to find suitable solutions. To better adapt the concept of predicament to the context of Chinese migration, it is useful to sharpen the term against a related one to underline the agency of rice farmers, within the limits of their predicament. I thus propose the following working definition of 'predicament', which draws upon anthropologist Susan R. Whyte's reflections on 'uncertainty'. Highlighting its social dimension, she defines uncertainty as 'a lack of protection from danger, weakness in the social arrangements that provide some kind of safety net when adversity strikes' (Whyte 2009, 214). Chinese farmers aim to avoid uncertainty by drawing on the large array of possible social arrangements that can provide protection for their paddy land and continue their family line to prevent adversity, and this often comes at the cost of what an individual would consider the good life. The predicament moment of decision making within a migration setting challenges and compels the actors to evaluate and define a solution, thereby accommodating constraints and making multiple concessions. This occurs within social arrangements that provide some kind of safety net for young and old age, or for future potential hardships. This predicament and farmers' strategic responses to it form the centre of this book.

Arguments and aims of the book

The conflicting pressures on farmers around either migrating into cities to work or staying home to cultivate and preserve their fields as a safety net is a major predicament of contemporary rural China. In this book, I provide a comprehensive analysis of this situation. I describe how this predicament emerged, what it entails, what socio-technical resources farmers have available to cope with it, and how they strategically do so. On a practical, ethnographic level, I explore how Chinese rice farmer households preserve their land resources when confronted with migration pressures. I discuss what land-use and land-arrangement decisions they take, in view of their circumstances and the resources at their disposal. I elaborate on their strategic, social and agrarian land-use decisions, which they take as conscious actors. These include their repertoire of knowledge, labour, social networks, financial resources, and farming technologies. I pursue three main arguments.

First, I argue that paddy fields play a key role in shaping farmers' everyday strategies. Scholars from various disciplines have repeatedly stressed that fields play a crucial role in, and for, migration.[5] Yet, the specific socio-technical challenges in preserving this key asset and the knowledge needed to do so remain largely unexplored. In this book, I scrutinize these challenges in more depth, proposing the need to look at the repertoires of knowledge that both staying and migrating farmers revert to.

Related to this, second, I argue that ostensibly technical farming decisions are always also social decisions that are closely interlinked with migration decisions. In taking seemingly operational decisions, farmers are actually pursuing various long-term and short-term projects that best match their current, fluctuating household situation. What looks like simple technical ability is, in fact, multi-dimensional reasoning for potentially manifold purposes. Applying skills practically and economically always includes simultaneously performing social responsibilities. This means that farming decisions also take into consideration aspects like educational, career, or

5 For Asian contexts see, e.g., Fan and Wang (2008, 228); van der Ploeg and Ye (2016); Ye (2018); and Rigg (2019).

marriage aspirations, child or elderly care, long-term engagements and future responsibilities and, more generally, the social and economic reproduction of the household and the patriline.

This brings me to my third, more general argument, namely that we need to pay more attention to the material world of migration and the related knowledge and skills. I argue that socio-technical resources are key factors in understanding migration flows and the characteristics of migrant-home relations. Importantly, 'resources' here are understood broadly as being socio-technical, reaching far beyond their mere economic value. Such resources are, I suggest, material interfaces. They are an objectification and materialization of the transformation of migration-affected rural Chinese society. In the case of China, for example, a focus on such resources helps to explain why there are so many divided households, why migration is often circular, why relationships with home remain important, and why most migrants envision returning to rural areas in the future.

In following these arguments, I aim to contribute to the migration literature both empirically and theoretically. On an empirical level, rather than focusing on the well-studied phenomenon of migrants in their places of destination, I provide a rare study of migrants' origins and, in particular, the rural side of Chinese migration. More generally, I aim to provide a qualitative analysis of Chinese internal migration that adds valuable ethnographic insights to standard quantitative analyses. Since the reform policies of the 1980s, Chinese mobility has sharply increased, both domestically and transnationally (Pieke et al. 2004; Oakes and Schein 2005; Chu 2010; Nyíri 2010; and Xiang 2016). In view of this augmented mobility, it is my objective to provide new socio-material insights relevant to understanding the most widespread pattern of migration within contemporary China: rural-urban migration from the inner provinces to the large cities of the east coast, which often results in households whose members reside separately in different locations (Lu and Xia 2016; Chen and Fan 2018). Although China's inner migrants are increasingly migrating westwards, choosing closer destinations or moving with their entire households, to date, the split-household arrangement is still the dominant migration pattern in China (NBSC 2019; Wang and Chen 2019; Fan and Li 2019). Focusing on the role of farmland in migration, this book contributes a new perspective on why this pattern remains so common. This entails comprehensively examining both those who stay and those who migrate, and acknowledging that both are part of a rural-urban farming 'community of practice' (Lave and Wenger 1991). The members of this community of practice are connected through circular migration, embodied farming skills and joint efforts to preserve home resources. This approach innovatively complements studies arguing for an integrated view of the Asian countryside (van der Ploeg and Ye 2016; Rigg 2019).

Moreover, perceiving migration in this way lets us rethink the implications of China's *hukou* system of household registration, which has strictly divided the population into either rural or urban, agricultural or non-agricultural since the 1950s (Cheng and Selden 1994). This system has long prevented rural Chinese from gaining permanent settlement rights or any entitlement to the welfare, pension and education system available to registered urban-dwellers. The recent reform of China's *hukou* system in 2014 increasingly allows rural people to move and obtain an urban registration. In this regard, the book is part of a new strand of scholarship that discusses not only the obvious constraints, but also the advantages of being registered as 'rural' (Andreas and Zhan 2016; Chen and Fan 2016). Highlighting the central role of land and land entitlement, it contributes to understanding why many rural inhabitants refuse to change their status into 'urban' citizens despite having lived in cities for years, and why the peasant smallholder model remains important, despite massive urbanization.

On a theoretical level, I integrate insights from three distinct bodies of literature – the anthropology of agriculture, migration studies, and the study of skilled practice. My objective is to contribute especially to a recently-established subfield of migration studies, materialities of migration.[6] I contribute to the material turn in migration studies a perspective on things that stay – paddy fields – and the related embodied skills. The latter are important socio-technical aspects of migration that, nevertheless, generally escape our attention because they usually remain tacit. I intend to show the value of a socio-technical perspective for studying migration phenomena, as a way to offer new understandings of migrant-home relations and dynamics.

With these ethnographic and theoretical aims in mind, it is, moreover, my goal to challenge prevailing narratives about backwardness and progress. I wish to contribute to a better understanding of the particularities of Chinese modernity, disputing the notion of linear technological progress. Challenging public discourse which portrays Chinese peasants as passive and backward (Murphy 2006; Day 2013; Schneider 2015), I want to show that farmers are, in fact, forward-looking decision-making agents who are actively shaping China's modernity.

6 See, particularly, Basu and Coleman (2008) and Wang (2016).

Agriculture and migration

Looking at the rural side of migration and the role of farmers' socio-technical resources in migration requires us to inquire into the practical details of farming in more depth. Preserving paddy fields, whether at home or as a safety net in migration, requires knowledge and skill. So far, the investigation of farmers' knowledge and skill has mostly been overshadowed by two strands of research, however. On the one hand, peasant studies have traditionally been more concerned with peasants' politics and economic decision-making, rather than with related technical details (e.g. Wolf 1966; Scott 1977). On the other hand, the long-standing agricultural intensification debate has extensively discussed the relationship between agricultural technology, the size of the farming population, and the degree of intensification.[7]

For a better understanding of farmers' knowledge and skill, it is therefore more useful to turn to the field of a more technically informed agro-anthropology. Forerunners such as Paul Richards (1985) highlight that farmers' knowledge is not only crucial, but also highly scientific. The French agro-anthropologists, in turn, draw our attention to the importance of studying techniques and the interrelation of technical and socio-cultural aspects.[8] André Leroi-Gourhan (1964) contributed the influential tool of a *chaîne opératoire* or an 'operational sequence' for the systematic analysis of farming processes. This notion is useful for analyzing the technical, organizational, ritual, and various other elements that constitute farmers' techniques. Taking these diverse facets into consideration helps us to understand the complex issue of 'technological choice' (Lemonnier 1993) – a topic that gains new importance at the intersection of technological choices and migration decisions.

The possible choices are very specific in relation to rice farming in China. Rice economies follow their own logics, as rice historian Francesca Bray shows (Bray 1984; 1994; Bray et al. 2015). Her model of Asian wet rice economies is particularly helpful in explaining this, since it shows that rice economies follow their own distinctive trajectory of technological progress and cannot be compared to Western agricultural experiences. Such rice economies are commonly characterized by scarce land, high population density, enduring smallholdings, and high requirements for skilled labour

7 The debate goes back to Malthus (1798), and has continued throughout the twentieth and twenty-first centuries, e.g. Boserup (1965); Geertz (1963); Stone (2001); and Bray et al. (2015).

8 See, in particular, Haudricourt and Delamarre (1955); Leroi-Gourhan (1964); Sigaut (1994); Lemonnier (1993); and the contributions in van Gijn, Whittaker, and Anderson (2014).

input. This makes them evolve in particular ways that cannot be explained through Western capitalist approaches, which equate progress with the efficiency of economies of scale, mechanization, and the substitution of capital for labour (Bray 1994). As I will show, these particularities have important implications when it comes to interlinkages between rice farming and off-farm migration, and to understanding farmers' choices of particular technologies in this context.

Most of the studies in the field of agro-anthropology have two things in common. First, they mainly deal with non-industrial agriculture. For understanding the complex processes of knowledge transformation and how these translate into a context of off-farm migration, it is therefore useful to explore the effects of new technologies. This includes those that were developed during the Chinese Green Revolution, and genetically modified crops, as well as the issue of agricultural deskilling (Stone 2007; Schmalzer 2016).

Second, the majority of studies, especially the older ones, hardly acknowledge how rural, urban and global worlds are tightly interconnected. It is now widely recognized that migration from farms strongly influences agricultural practice and land use. Economic anthropologists were already considering the effect of off-farm labour on production decisions back in the 1960s and 1970s (e.g. Hanks 1972). Scholars have mainly explained the situation in economic terms and with regard to the availability or lack of household labour resources (see Barlett 1980, 557). Still, anthropology is generally rather hesitant about reviewing the simultaneous pressures of migration and resource preservation, and farmers' strategies to achieve this. Only a few, more nuanced ethnographic and human geographic studies even acknowledge the local complexities at play (Murphy 2002; Linares 2003; Gaibazzi 2015; Wu 2016; Rigg 2019).

With regard to this book's central problem, it is interesting that commentators from various disciplines, including geography, economics, development and area studies, study how off-farm migration affects a range of spheres, such as agrarian transition, rural restructuring, and the rural environment (Kelly 2013; Qin and Liao 2016). Furthermore, particularly large volumes of research exist about the impact of migration on rural development, livelihoods, and agricultural production. The migration-development literature has been discussing the value of migration for development and livelihood diversification for many years.[9] Thanks to these studies, it is now

9 The body of literature concerned with the migration-development nexus is impressively large. For some recent studies in this field see, e.g., Scoones (2009); Manivong, Cramb, and Newby (2014); and Hickey (2016).

widely recognized that agriculture and migrant work are complementary elements of Asian rural employment, and that land plays an important role in securing these livelihoods.

Meanwhile, studies into the impact of migration on agricultural production include long-standing debates about whether migration leads to an intensification or de-intensification of farming, and whether migrant remittances foster or inhibit the adoption of new technologies.[10] Unsurprisingly, these studies produce varied findings, reflecting the complexity of the issue, but they all show that agricultural technology plays a central role in rural-urban migration. So far, the issue has mainly been studied in terms of labour power and measured in economic terms. I suggest, however, that preserving resources such as paddy fields is first and foremost a socio-material matter. It is a question of soil quality, cultivation, agricultural techniques and technology, knowledge, and skills.

With regard to the nexus of Chinese wet rice farming and migration, the existing research provides information about a range of strategies that left-behind people and, to some extent, migrants, employ to manage their fields. The majority of these studies only investigate particular strategies at a general level (see Chapter 6). Most of them perceive land-use strategies as part of an overall household strategy that effectively combines farm work and migrant work in order to reduce risks to people's livelihoods. Although less formalized and with more differentiated results, this perspective bears some similarity to the New Economics of Labour Migration (NELM) theory, which understands migration as a household strategy that minimizes risks and raises incomes through economic diversification (Stark and Bloom 1985; Stark 1991).

While acknowledging that risk reduction and income generation are certainly important migration motivations for my interviewees, I do not focus on migration strategies or the reasons for migration. Rather, I look at the strategies used to protect land resources *despite* migration. These involve the land-use and land-arrangement strategies of both migrants and those left behind. This approach is much less evident in the literature.[11] Moreover, existing studies do not describe the wider range of strategies employed, because they tend not to perceive these strategies as part of an overall repertoire of knowledge and solutions used to deal with

10 This body of literature is equally expansive and interdisciplinary. See, e.g., Müller and Sikor (2006); Hull (2007); Gray (2009); and Chen et al. (2014).

11 Exceptions on China are He and Ye (2014); van der Ploeg and Ye (2016); Xie and Jiang (2016); and Xu et al. (2017).

paddy fields. In contrast, this book delves into the strategies utilized in precisely this comprehensive way. On a more abstract level, the book shows that this means we need to understand migration phenomena more comprehensively.

From 'migrant worlds' to 'community of practice' worlds

Anthropological migration studies from the late twentieth century onwards highlight the complexity of migration. They show that migration is not just about individuals, but about households and social networks that span different locations. Moreover, it is also about the interrelated movement of people, capital, technology, information, images, and objects.[12] As Wang (2016) notes, the earlier studies in particular observed an abstract and generalized fluidity of movements, but it has now become widely accepted that movements are more diverse and grounded in everyday life. In line with this, a critical reassessment is currently emerging, which seeks to overcome a whole range of dichotomies, such as between internal and international migration, skilled and non-skilled migrants, mobility and immobility, transnationalism and emplacement, migrant experiences and ideals, and people and things (ibid., 2). My interest in thinking beyond such binary oppositions lies in making visible the intersectional and agentive aspects of migration, and the ways in which it is materialized and objectified.

While this reassessment is relatively recent, the resilience of earlier binary visions seems to have obscured our understanding of the study of four important realms of migration: internal migration, migrants' places of origin, those left behind, and the material aspect of migration. Studying migrants' places of arrival – generally big cities in China – was my own point of departure, when I first set out to explore the experiences of rural migrants in Shanghai in 2007. Nevertheless, I soon recognized that the places of origin and 'nonmovers' in general play a crucial role in migration decisions and processes (Cohen and Sirkeci 2011, 87). It is now widely acknowledged that migrant households in China and elsewhere frequently span different locations (Fan 2016). This also implies, however, that we need to pay more attention to the agency of not only migrants, but also of those who stay behind (Resurreccion and Van Khanh 2007; Jacka 2014; Ye 2018).

12 For some prominent contributions see, e.g., Glick Schiller, Basch, and Szanton Blanc (1992); Hannerz (1996); Appadurai (1999); Ong (1999); Sheller and Urry (2006); Brettell (2008); Vertovec (2009); Castles, de Haas, and Miller (2014); Hoang and Yeoh (2015); and Salazar (2017).

Regarding the material aspects of migration, as far back as the 1970s and 1980s, groundbreaking works enlightened us about the processes of internalization (Bourdieu 1977) and objectification (Miller 1987).[13] These significantly enhanced our understanding of objects in relation to society and social differentiation, and prompted the material turn in the social sciences (Appadurai 1986b). Nevertheless, despite the fact that migration involves things as well as people, and the realization that these things matter, migration studies are only reluctantly beginning to explore the material element.

Only a few seminal texts, including Basu and Coleman (2008) and Wang (2016), tackle the material side of migration theoretically, making important steps towards conceptualizing the materialities of migration.[14] They show that material culture plays a central role in migration processes and provide a viable conceptual lens for understanding migration in more nuanced ways. Basu and Coleman (2008) propose the notion of 'migrant worlds' rather than 'migration', since this suggests that a 'world' can itself be mobile. Moreover, it captures the materiality of migration itself, the material effects of migration, and the 'inter-relatedness of the movements of people and things' (Basu and Coleman 2008, 313).

Building on these insights, Wang and his colleagues extend the notion of 'migrant worlds', stressing the temporal, embodied, and methodological dimensions of studying interactions between migration and materiality. They do this from a multidisciplinary and philological approach that allows them to resolve a number of dichotomies, notably that between migrant people and things. With regard to the embodiment of migration, they draw on a phenomenological approach to material culture (Ingold 2000). This implies studying 'how people make place and construct identities through situated multidimensional sensuous and corporeal engagement (through sight, sound, touch, smell, taste) with the material world' and drawing attention to the objectification, articulation and extension of migrants' emotions and desires through things (Wang 2016, 5).

13 'Objectification' is a concept that tries to overcome the dualism between subjects and objects. Instead it acknowledges that '[t]hrough making, using, exchanging, consuming, interacting and living with things people make themselves in the process' (Tilley 2006b, 61).

14 The concept of materiality is, itself, ambiguous and heterogeneous (see Tilley 2006a, 5). I draw on Basu and Coleman who 'use the term "materiality" straightforwardly to refer to physical objects and worlds, but also to evoke more varied – multiple – forms of experience and sensation that are both embodied and constituted through the interactions of subjects and objects' (Basu and Coleman 2008, 317; see also Wang 2016). For me, these latter include migrant and left-behind skills.

I agree with Basu, Coleman and Wang that 'migrant worlds' is a useful concept for approaching the material aspect of migration and its interlinkages with the migration process. In addition, I explicitly include in this notion the world that migrants leave behind, one that shapes and continues to be shaped by migration, as in the example of paddy fields. Recent case studies on the material turn in migration studies conceptualize the material in various promising ways, e.g. in the role of constituting home, belonging, identity, memories, suffering and, more generally, mobility. They highlight the role that objects play in linking migrants and non-migrants, and show how mobility is enabled by transportation technology (e.g. Tolia-Kelly 2004; Frykman 2009; Chu 2010; Burrell 2011; Abranches 2013). Reflecting wider trends in migration research, most of these current studies on the material focus on mobile objects that are taken with, or sent to, the migrant. These include, especially, things related to consumption rather than production, such as food, or monetary and other remittances, which are sent through specific material infrastructures. While this is important, I assert that we definitely also need to pay more attention to how migrants and non-migrants jointly collaborate to preserve resources in their places of origin.

Furthermore, I seek to develop the dimension of the embodiment of migration in relationship with the material world that Wang proposes. I propose to do so by bringing the perspective of knowledge and skill into migration studies. Skills are an integral part of migrants' material culture, and these skills play a role in migration processes. A common-sense notion of skilled migration, whether academic or public, mainly equates skill with formal educational achievements. I suggest, however, integrating the two fields of migration studies and the study of skilled practice to understand migrants' skills as a form of tacit, often embodied knowledge. In this regard, studies in the field of skilled practice have much to offer. They greatly enhance our knowledge about skill, its transmission, formation, and transformation.[15] In particular, they discern the centrality of the whole range of bodily senses and related skills that are needed to engage with our environment, including tactile, visual, or auditory skills (Ingold 2006; Grasseni 2009; Rice 2010). Thus, they draw our attention to the everyday aspects of learning that often remain unspoken and have, therefore, escaped the attention of many academics, including migration scholars. Importantly, a skill perspective opens up a

15 It is worth noting that, in the field of studying skilled practice, scholars generally elaborate their research around crafts, through what has, for example, become known as the 'apprenticeship debate', spanning between Coy (1989) and Marchand (2010). See Flitsch (2008) and Eyferth (2009) on rural China.

view of migrants and those left behind as being knowledgeable actors. By engaging skilfully with their socio-material surroundings, these actors processually craft 'migrant worlds' through their sensuous involvement with these worlds.

This is essentially a social process. As Lave and Wenger show in their pioneering work in the field of cognitive anthropology, learning is not an activity that takes place exclusively in individual minds, but is primarily social. Accordingly, learning is situated within a 'community of practice' (Lave and Wenger 1991). The concept of a 'community of practice' has implications for how we learn, including how ethnographers learn in the field. It also endorses the fact that learning is intimately connected to an individual's identity and positioning in the social order of a community (Wenger 1998).

In prioritizing the social learning environment, however, Lave and Wenger pay less attention to how skill as a very specific type of knowledge is internalized and embodied in practice. This aspect is tackled by more phenomenologically and technically-informed scholars. From a sentient ecology perspective, Ingold (2000; 2006) sheds light on enskilment as complex learning processes that comprise the intimate interaction of the body with materials and tools, the natural as well as the social environment. This makes a substantial contribution towards resolving major Cartesian dichotomies, e.g. between body and mind, people, and things. In other words, it is 'close to the realities of lived experience' (Ingold 2000, 1).

The agronomist and anthropologist Sigaut's more technical perspective points out that the spheres of social and technical activity cannot be investigated separately from each other when studying skilled practice. One of his many contributions to the field is the explicit differentiation between knowledge and skill, his assertion that skills have to be acquired gradually through a learning process, in which knowledge is turned into skills. This implies that knowledge 'fades' in the process of being embodied or incorporated, since it becomes 'embodied in the very process of action' (Sigaut 1994, 438). Siguat's assertion that this transition occurs within a 'skill-producing group' is similar to the concept of a 'community of practice'. The former refers to a group which defines its identity through common abilities, which Sigaut sees as the basic social unit in all societies, claiming that social life can only proceed normally when everyone acquires a sufficient number of materially and socially effective practices, as well as skills that support these practices. Importantly, every social group requires a certain number of skilled members to be effective and function well. If a group is too small or too large, skills cannot be transmitted properly (ibid., 447).

Sigaut's reflections on the proper functioning of social life in relation to skill and the optimum group size are very relevant to the migration-affected farming households studied in this book. They point towards the question of what will become of the skill-producing group of Chinese rice farmers as more and more members migrate early and for the long-term, and as post-Green Revolution technologies transform embodied farming skills. Nevertheless, Sigaut's concept does not go into as much detail about the practical social arrangements of learning as Lave and Wenger's (1991). It is therefore useful to integrate Lave and Wenger's community-focused approach with Sigaut's more technically-informed approach to skill, to highlight both the social and the technical sides of skill.

In this sense, I propose that we should rethink 'migrant worlds' as 'community of practice worlds'. In the Chinese context, such worlds comprise both the migrants and the people left behind in a rice farming community of practice. This is reflected in my terminology. I use the term 'farmers' to not only draw attention to the actual practice of farming, but also to refer to both migrant and non-migrant household members. This is because the borders between farm work and migrant work are fluid in practice, with people often fluctuating between the cities and the countryside. Moreover, most migrants grew up in a farming environment. Even the younger ones, who tend to have received more formal schooling and migrated early, have spent most of their first two decades in a farming background. In addition, using the term 'farmers' for migrant workers is closely aligned to Chinese perceptions of rural migrants. Even after migrating, they generally continue to be registered by the state as rural residents with agricultural *hukou*, in addition to being considered by the public – and by themselves – as *nongmin*.[16] Accordingly, rural migrants themselves and the populace more generally use the term *nongmin gong* ('peasant workers').

Even though the binary division between migrants and the people left behind should be discarded in order to better understand Chinese rice farmers as part of a community of practice, sometimes it is still useful to retain the dichotomy for analytical purposes, for example when looking at the strategic actions of individual household members. In such cases, I distinguish between 'migrants' and 'those left behind'. The latter is related

16 See, e.g., Fan and Wang (2008, 221). As far back as ancient China people were ideologically classified into 'four classes of people' (*simin*), i.e. gentry/scholars (*shimin*), farmers (*nongmin*), artisans (*gongmin*), and merchants (*shangmin*), according to their occupation and perceived contribution to the state (Huang 1995, 26). In the twentieth century, Chinese intellectual debates about modernization have contrasted farmers/peasants (*nongmin*) against citizens (*gongmin* or *shimin*, literally referring to urban citizens) (Day 2013, 50).

to the Chinese term *liushou* ('stay behind', 'stay to take care of'), from *liu* ('remain', 'stay') and *shou* ('guard', 'conserve', 'protect'). In the rural-urban migration context, it is commonly used in composites such as *liushou ertong* ('left-behind children'). In contrast to the English connotation of the word, which implies that people are initially left behind and expected to join the migrants later on, in Chinese the term implies instead that a person is staying to take care of the farm, and that migrants will return to them. As well as this, the term *liushou* also has more general implications. As Xiang argues, 'many rural communities as a whole have been left behind economically and socially' (Xiang 2007, 179).

The categories of migrants and left-behind people infer that these two groups of people belong together. They are not lone individuals, but – below the level of the community of practice – members of a household. It is only through an additional focus on the household that it is possible to take into account the full picture of Chinese internal migration. However, what is meant by a 'household' is not clear-cut, as there are overlaps between local concepts of family (*jia*) and household (*hu*). Moreover, in my case households span different locations, 'incorporating multiple members in diverse places who remain part of the income-pooling unit directly, or who continue to exercise influence over household dynamics' (Lawson 1998, 43, cited in Fan, Sun, and Zheng 2011, 2166).

This is another key point in this book, which does not focus on the household *per se*, but on the strategies that households employ to manage their fields. These field preservation strategies can be seen as part of 'householding', i.e. the ongoing, dynamic social processes through which rural households create and reproduce themselves (Douglass 2006, 423; Jacka 2012, 2). As Tamara Jacka emphasizes, 'householding is not just a matter of maintaining livelihoods, but also of caring for dependants, sustaining household members' health and wellbeing, and maintaining the patriline' (Jacka 2012, 11). In this regard, I emphasize commonly overlooked technical aspects of householding. These are not only deeply intertwined with the social aspects, but also fundamental to the process of householding. I focus mostly on outcomes of household decision-making processes, rather than the decision-making process itself. It is clear that such a perspective obscures power differentials and individual agency below the household level, which is a criticism that has previously been levelled at household strategy approaches (e.g. Wolf 1992, 12-23; Toyota, Yeoh, and Nguyen 2007, 157). Whilst bearing in mind the point that decisions are often the products of complex evaluation processes that may be challenging or even painful, a household approach still makes sense for two reasons.

First, paddy fields and their use rights are *de facto* a resource that belongs to the household, so its preservation usually involves the whole household. Second, when it comes to Chinese internal migration, there are undeniably certain household patterns regarding who migrates and who stays. Migrants are usually young or middle-aged, while children, old and sick people, and women at particular life stages, such as new mothers, typically stay behind. While precise numbers are lacking, it is now widely accepted that left-behind children, women, and elderly people in the Chinese countryside constitute about 61 million, 47 million, and 50 million respectively (Ye 2019, 21). It is common for paternal grandparents to take care of their grandchildren in the absence of their migrated sons and daughters-in-law. These 'left-behind children' currently constitute almost one quarter of all Chinese children and nearly one third of the nation's rural children (ACWF 2013; Santos 2017, 93).

Thus, only through this approach, taking into account all the household members – both those who stay and those that migrate – as part of a wider community of practice world, spanning not only people but also things, can we grasp the Chinese internal migration phenomenon in its full socio-material complexity. The paddy field problem faced by Chinese rice farming households in a migration context is both a social and a material issue. If we want to understand this particular situation, how farmers as agents devise strategies and figure out solutions, we need first of all to understand what knowledge and options they have at their disposal to deal with it. At the core of such a migrant world are two things – knowledge and skill. Chinese migration-affected rice farmer households form a community of practice that is centred on the question of how to preserve paddy fields as a safety net today and a long-term resource for the patrilineal family in future. There is, therefore, much more at stake than just the technical skills needed to preserve this asset. Maintaining their paddy fields, which depends on knowledge and skills, crucially influences the constellations in which people migrate.

A skill turn within 'the material turn of migration studies' is therefore long overdue, to better understand migration phenomena in general, and the relationships and actions between migrants, their places of origin, and the people they leave behind in particular. Viewing it in this way offers resolutions to many of the prevailing dichotomies, not only between migrants and those left behind but, notably, also between people and things. This means, in practice, that we can understand and thus investigate farmers and land as one, shedding light onto the materialization and objectification of the Chinese farmers' predicament. This will not only provide a more complete picture of migration but eventually, will also open up a way to conceive migrants as active agents rather than victims – as people who, despite immense costs and pressures, are capable of dealing with specific challenges, of planning and finding their own solutions.

Knowledge, repertoire, and agency

This integrated household strategy and community of practice world approach is useful for showing how both staying and migrated household members deal with their home resources. However, with regard to the actual farming strategies employed, we need some fine-tuning in order to render visible the actors, their strategic agency, and their knowledge and skills. In this regard, I propose to follow a knowledge-strategic, socio-material, and actor-centred framework. This approach is holistic, seeing rice farming as a dynamic knowledge system. At the same time, it opens up a view onto how individual migrant and left-behind rice farmers pursue their own endeavours by engaging with their social and material world through their repertoire of knowledge. This framework is achieved by drawing on a triage of three methodological-theoretical approaches: first, Barth's (2002a) model of knowledge transmission, second, Schippers' (2014a; b; c; d) approach to the farmers' repertoire of knowledge in an agro-system and, third, Ortner's (2006) concept of agency, enriched by Farquhar's (2006) reflections on agency, embedded in visions of the good life.

I am inspired by Fredrik Barth's idea of putting knowledge at the centre of investigation. Knowledge, especially as it translates into action, proves a valuable lens for analyzing the socio-technical transformations and dynamics of Chinese rice farming over recent decades, including its intersections with migration processes. In Barth's model, knowledge refers to 'all the ways of understanding that we use to make up our experienced, grasped reality' (Barth 2002a, 1). This includes feelings, attitudes, information, embodied skills, verbal taxonomies and concepts. Knowledge not only structures how people understand the world, but also how they act in it. Importantly, knowledge is distributed in society, rather than diffusely shared (ibid., 3). The key is to focus on (human or social) action (Barth 2002b, 35).

At the core of Barth's model are the three 'faces' or aspects of knowledge: corpus, communicative medium, and social organization. The corpus of knowledge includes 'substantive assertions and ideas about aspects of the world' (Barth 2002a, 3). In my Chinese case, this includes, for instance, knowledge about specific cultivation techniques. According to Barth, the media in which this corpus of knowledge is represented and communicated comprise words, symbols, gestures, and actions. I see these media of knowledge representation, among others, in Chinese farmers' bodies, rituals, and farm tools. Finally, Barth claims that knowledge is distributed, employed, and transmitted in the aspect of social organization, reflected in specific collective or household divisions of labour (ibid.). In the Chinese case, this is relevant to the issue of labour depleted through migration. These three aspects are closely related, mutually determine each other, and interconnect in specific ways in different traditions of knowledge (ibid.).

By looking at the interplay of these three faces of knowledge, we can see the dynamics of any given knowledge system: how people attribute validity to certain knowledge, how knowledge is transmitted or not transmitted under specific local conditions and constraints, and which trajectory a particular system of knowledge takes under these conditions. In Chapter 2, I trace the interplay of the three faces and model the trajectory and transformation of the local Chinese system of knowledge surrounding paddy fields as a resource, from the strongly regulated collective system of the 1980s to a more diversified and migration-affected household farming system in the 2010s. This was the period when the predicament of migration pressure versus resource protection emerged, so comprises a crucial moment in the realignment of the knowledge system. Here, the model is also useful for highlighting where various aspects of the system no longer seem to fit, and where contradictions and challenges for the people involved have occurred, have had to be dealt with and resolved.

When applied to the socio-material and technical aspects of agriculture, it is useful to think of Barth's model together with the actual agricultural practices and repertoires in which skills play a crucial role. Thomas Schippers' agro-technological approach (2014a; b; c; d) is particularly inspiring here. Three notions are central to his approach: (1) the agro-system, (2) repertoire, and (3) agricultural practice. The first notion, agro-system or agricultural system, finds parallels in the 'socio-technical system' approach outlined by Bryan Pfaffenberger (1992), which views the social and technical aspects of any (agricultural or non-agricultural) system as being closely integrated and inseparable from each other. I adopt this notion in the form of a general lens through which I view the socio-technical world of Chinese rice farming.

Schippers' second concept, which is most central to my own approach, is that of a repertoire (Schippers 2014b). Referring to farmers' repertoires is, first of all, not unique to Schippers. Nevertheless, it is his idea of farmers' capacities to deal with varying, sometimes unforeseen circumstances that I find particularly inspiring, since it transcends a narrow understanding of the notion of technology as merely a technical set of knowledge and skills. Starting from here, I aim to develop the notion of the farmers' repertoire one step further, to render it more explicit by placing it at the centre of my analysis, alongside a general focus on knowledge systems. The repertoire idea is especially valuable with regard to studying a concrete case of farmers' socio-technical resources. In my field site, Green Water Village, this practical repertoire of knowledge consists of elements including the corpora of knowledge on different aspects of farming such as soil, water, farming technologies, agricultural practices, and embodied techniques. Moreover, it comprises climate and time knowledge, as represented and transmitted in the farmers' calendar and in proverbs. Finally, it also includes the knowledge of how to organize farm work efficiently, for instance, along gendered lines. In fact, much of this knowledge is skill, requiring learning and 'constant renewal in the course of practical action' (Sigaut 1994, 445). The repertoire notion therefore proves particularly useful for grasping the local Chinese rice farmers' resource pool as a basis for understanding the possibilities and capabilities, but also the constraints of their actions (e.g. in terms of available technology).

Schippers' third notion is that of agricultural practice, stimulated by leading figures from French anthropological academia, such as Haudricourt. Agricultural practices are 'specific ensembles of knowledge and skills brought into play to domesticate certain plants and/or animals in order to satisfy human nutritional, material or immaterial needs' (Schippers 2014a, 339). This perspective on agricultural practices is useful for analyzing aspects of change and stability in the local Chinese system, which has undergone significant transformations such as mechanization. It allows us to examine both the socio-technical aspects of knowledge and skill, and the performative aspect of agriculture. I believe that agricultural practices should also be seen as part of the repertoire of farmers in a certain system. Hence, drawing on the notions of the agro-system, the repertoire, and agricultural practices opens up a view onto particular technological choices under specific, changing social or environmental conditions.

In order to extend this perspective to include issues of power and intention, it is useful to draw on the concept of agency. Agency is a highly influential concept that has been conceived in different ways by various disciplines and schools.[17] From a practice theorist's point of view, agency has been broadly defined as 'the capacity to affect things' (see Ortner 2006, 137). At the core of practice theorists' debates about agency is the dialectic relationship between an overlying social structure and (collective or individual) human agency, and the way and the degree to which the two influence each other (Ahearn 2001, 54). Adding this agency perspective

17 For concise overviews, see Farquhar (2006) and Postill (2010).

helps to highlight how farmers still continue to pursue their own projects, despite the unfavourable structural conditions and being located at the lower end of the power continuum in Chinese society.

In a recent theoretical contribution, Ortner (2006) argues that agency is always closely connected to power, and that in agency there is always an intention. She therefore defines agency with regard to two fields of closely related meaning: firstly, as the exercise of or against 'power'; and secondly as the pursuit of 'projects' (Ortner 2006, 134-149). In the case of Chinese rice farmers, I see Ortner's 'agency-as-power' most clearly in the situation of farmers *vis-à-vis* the Chinese government's rural policy complex. Ortner's 'agency-as-projects', in turn, becomes most obvious when looking at actual household decisions around the resource of paddy fields. Even though these are closely interrelated, it is the more subtle issue of intention that informs my analysis, rather than power relations themselves.

For the purpose of this research, however, which centres on paddy fields as an important part of the rural material world, it is helpful to enhance Ortner's concept with a material culture perspective. Here I refer less to leading academics in this field, who have contributed greatly to studying the agency of things (notably Latour 1988; 1999; Gell 1998). Rather, I find Farquhar's (2006) reflections on agency useful, because she focuses on the interaction of people's bodies and things as a form of craftwork. Farquhar proposes the notion of 'the crafting of a good life' – which I understand as being similar to Ortner's concept of 'projects'. I view the 'imagination' (Appadurai 1999) of the good life as part of the modernity that farmers individually strive for and which drives their actions. In post-reform China, this takes the form of an imagined ideal of middle-class standards of wellbeing and material prosperity, often associated with urban life (e.g. Chen 2001, 167; Zavoretti 2017, 5). Focusing on how the good life is crafted, Farquhar bases her notion of agency on the example of food practices in everyday life, suggesting that 'agency in everyday life is a form of craftwork involving intimate collaborations among embodied humans and material objects like food'. Moreover, she shows that 'the crafting of a good life is an improvisational project in which a great deal goes without saying' (Farquhar 2006, 146). Here, the notion of Ortner's agency-as-project gains momentum from an embodiment and material culture perspective, because it draws attention to a much more subtle and unspoken agency. This agency, I suggest, also becomes visible when farmers engage with their fields, for instance, in the everyday practice of planting a certain crop, or applying a specific type of manure, while striving towards the ideal of the good life.

Taken together, this triage of concepts – comprising the knowledge system, the repertoire and the concept of agency that is driven by projects informed by the idealized imagination of a good life – is extremely productive. It allows for an analysis of the Chinese migrant world as a community of practice world, thereby taking into consideration the distributed knowledge and skills that underlie the actions of both staying and migrating farmers. This lets us explore how farmers as agents cope with their specific socio-material situation of being paddy field bound.

Accessing the rural-urban community of practice

My methodological approach to the rural-urban community of practice is through ethnographic fieldwork, proverbs and written qualitative and historical sources. The ethnographic fieldwork for this research was conducted during nineteen months' research in the People's Republic of China (PRC) in 2007-2008 and 2010-2011, as well as through follow-up correspondence and video conversations conducted between 2012 and 2017. The book focuses on rural Hunan Province in the 2010s. Additional data were collected from Anhui migrants in urban Shanghai and rural Anhui Province (see Figure 1).

I chose these sites for practical reasons, but also as part of my theoretical-methodological approach of studying a migrant world as a community of practice that comprises migrants' places of origin as well as their destinations. Paddy fields are assets – or artefacts – that remain in their location, in contrast to mobile objects that migrants can take with them. Therefore, my methodological aim is not to 'follow the things' (Appadurai 1986a, 5). Instead, the general emphasis within the migrant world discussed here is on the rural side of migration, because that is where the paddy fields are located. I look at how both the people who stay and those who move away manage this artefact that stays behind. Accordingly, data were obtained mainly through direct and participant observation, as well as semi-structured, open interviews, and informal conversations in standard Chinese (*putong hua*). My interlocutors were both rural-to-urban migrants and those who had stayed in the countryside. I met some of the migrants, both in their villages and in the city where they worked as migrants. Throughout my time in China I talked to numerous people who are relevant for this research, the most central of which were twelve households interviewed in Hunan, and five in Shanghai. Only one household was from Shanghai and of urban origin, and I included them because of their experience of being sent to the countryside during the 1960s and 1970s to engage in rice farming.

Regarding the origin of most of my interlocutors, when it comes to farming and migration, Anhui and Hunan have some similarities. Both are

Figure 1 Map of mainland China

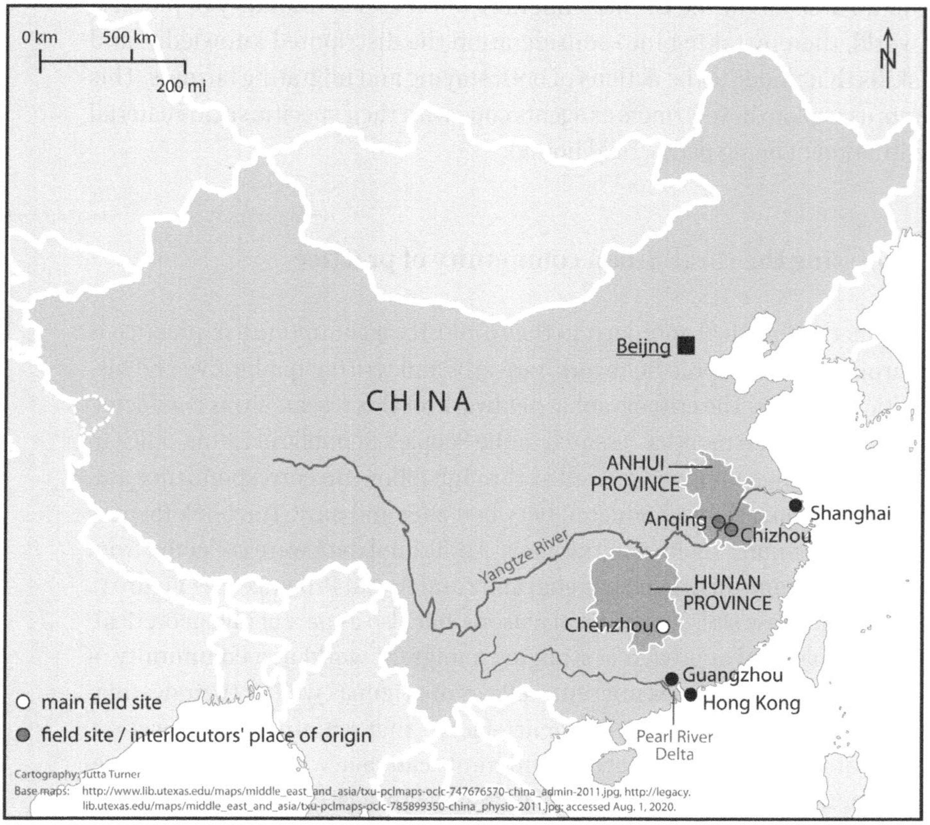

Cartography by Jutta Turner

among China's main rice-producing provinces. Moreover, both are densely populated, land-locked provinces near the Yangtze River, rather remote from China's big coastal cities and mainly inhabited by Han Chinese. They are the country's second and third major sending areas respectively for internal migrants after Sichuan (Lu and Xia 2016, 593). This has been attributed to economic regional disparities and inequalities (Naughton 2007, 26; Fan 2008). Similar to Anhui Province, but in contrast to the coastal provinces of southern China such as Guangdong, Hunan does not have a marked history of overseas migration. This implies a lack of investment into the province by overseas Chinese (Wang 2003, 319). This, in turn, gives Hunan a regional disadvantage compared to other provinces which earn significant income from abroad. This situation contributes to confining Hunan to the group of migrant-sending provinces, and favours domestic migration in the light of a lack of overseas networks.

Despite these broad similarities, however, there are also differences in the farming conditions between the two provinces. Hunan's agricultural economy, which is located in the Middle Yangtze region, is specialized and primarily based on rice production. In contrast, the Lower Yangtze region has a more diversified economy (Naughton 2007, 26). My Anhui interlocutors came from villages under the administration of the two neighbouring cities Anqing and Chizhou, located in central south Anhui. There, farmers planted one crop of rice, followed by a crop of cotton and one of rape seed. Moreover, despite being located near the Yangtze River and several lakes, the land is flatter and water resources are scarcer there, compared to my field site in Hunan. Farmers needed to use pumps to irrigate their fields, and in wintertime it was difficult to find enough water to fill up our hot-water bottles.

Nevertheless, in view of the more general constraints outlined above, the key challenge that rice farming households from Anhui and Hunan faced – the problem of protecting their field resources at home while simultaneously feeling prompted to migrate – was the same. As a result, although their specific crops and farming implements differed slightly, the way my interlocutors from both provinces dealt with the problem was still similar. This is still the main predicament faced by many people migrating from farms to cities in many parts of China today, who attempt to earn income from their urban jobs, yet retain some security in the form of their paddy fields back home. Nevertheless, this book focuses mainly on Hunan Province because of the more suitable research conditions I encountered there: being able to move around independently, in addition to accessing the local written sources described below.

Hunan Province (see Figure 2) is particularly apt for investigating questions at the nexus of agriculture and migration. On the one hand, parts of its topography make it especially suitable for wet rice cultivation. While most of the province is mountainous and hilly, it lies south of the middle reaches of the Yangtze River and south of Dongting Lake, which gives the province its name, literarily 'south of the lake'. Rice cultivation benefits from the lake and river crossings, as well as the subtropical climate. Accordingly, the province holds one of the world's longest histories of rice production and still maintains a local economy that is based mainly on rice. Today, Hunan produces more than 12 percent of the PRC's entire rice output on only 3 percent of the country's area of cultivated land (NBSC 2019, secs. 8-21, 12-10).[18] This facilitates surplus grain production

18 All national rankings and statistics in this book refer to mainland China, including its 31 provinces, autonomous regions and municipalities, but excluding Hong Kong, Macao, Taiwan, and overseas Chinese areas. I use 'China' to denote this administrative unit of mainland China.

Figure 2 Map of Hunan Province

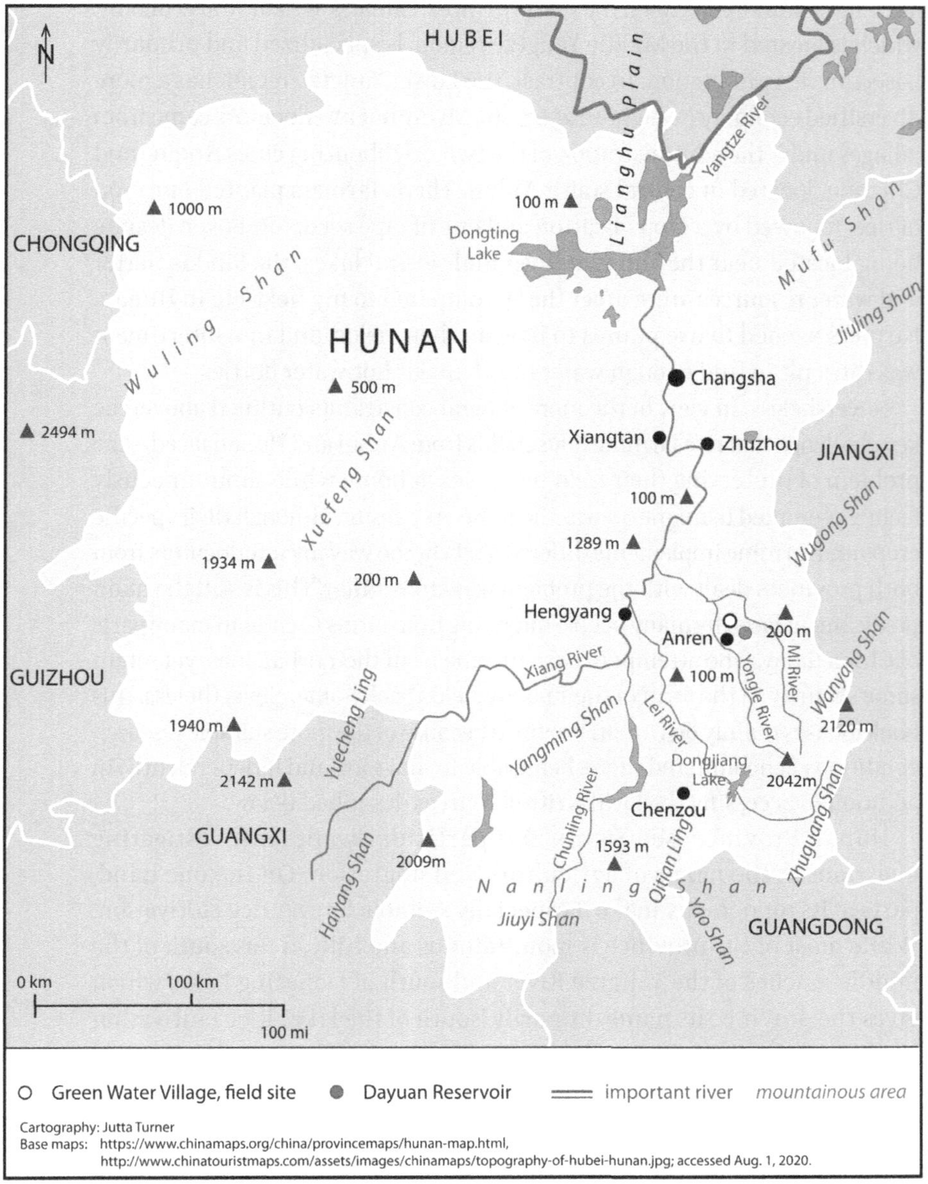

Cartography by Jutta Turner

and export to other provinces. On the other hand, as mentioned above, Hunan Province is among China's major sending provinces of internal migrants. Although Hunan's population structure is close to the national average, it is more densely populated than the national average (HPBS

2014, sec. 1-2). This population pressure adds to migration pressure. In fact, about 5 percent of China's population lives in Hunan, which only comprises about 2 percent of the country's land mass (Hunan Government 2015). In 2011, 65.96 million people lived in Hunan (HPBS 2012b), similar to the entire land area and population of the United Kingdom, but this had grown to 73 million inhabitants by 2018 (HPBS 2019, sec. 1-2). Much of the land is mountainous and not suitable for farming, however. The closest major metropolis is Guangzhou in the Pearl River Delta, about 500 km away. Most of Hunan's migrants move to this area in nearby Guangdong Province.

The area of my field site in the province is fairly representative of the provincial average. The prefectural-level city of Chenzhou, which is located in the far southeast of the province, is neither a very poor and remote mountain area, nor does it belong to the rich urban areas in the northeast. Apart from rice, the major agricultural products are tubers, tobacco, bamboo shoots, mutton, and pigs (HPBS 2014, secs. 19-30, 19-33). Moreover, the prefecture produces some mining products, energy, and building materials (Hunan Government 2015). At the county level, in 2019 my field site, Anren County, had a resident population of 464,800 people, and a total area sown to grain (mainly rice) of 44,100 hectares (ACBS 2020).

In Chenzhou I focused on six villages in Longshi Township in Anren County and, among these, in particular a rice farming village I call Green Water, one of ten administrative villages in Longshi Township. The township is reported to have originated in the Song Dynasty (about 960-1279). At that time it was famous for producing oil and paper, as well as being the location of an imperial academy. According to the township gazetteer, each of the ten administrative villages has around eight natural villages, subdivided into 13 village groups. In 2010, there were an average of 1,470 registered inhabitants per village, arranged into 370 households (Wu 2010, 4, 278).[19] Temporary migrants are included in these figures.

I mainly collected data in two adjacent village groups, which comprise about 230 people in total and constitute one natural village.[20] People here see themselves as belonging to the same patrilineage, which is the customary form of Han Chinese social organization (Santos and Harrell 2017). Virilocal marriage practices mean that brothers and agnates are usually neighbours as well as parties of mutual aid with regard to the organization of agricultural

19 For reasons of privacy, the exact village data are not provided here.

20 Often, several natural villages constitute one administrative village. See Wu (2016) for the differences and dynamics between natural and administrative villages in China.

labour. Similar to other Han Chinese lineages, in their seven-volume family record (*jiapu*) the lineage constructs a millennia-old family history, linking itself genealogically to the legendary Yan Emperor. The family record also contains a village map which depicts geomantically significant 'dragon lines' (*longxian*) and documents the position of hills, houses, tombs, fields, and irrigation ponds, reflecting the centrality of wet rice farming to the local economy.

Land suitable for house construction or farming is perceived as scarce. Therefore, houses – whether the old flat clay houses, or the new multi-storey brick and concrete ones – often have no courtyards, as is common in North China. Instead, the walls of a house are commonly shared with those of the neighbouring houses. There is also insufficient space to build every house with its entrance facing south, as is geomantically preferable (see Feuchtwang 2002). Generally, the area is so densely populated that villages are located in close walking distance from one another, sometimes only divided by a short stretch of paddy fields. The area counts as hilly land and there are many mountains and few fields. Mountain forest makes up almost 87 percent of Longshi Township's land resources, compared to only about 13 percent of farmland (Wu 2010, 4). Moreover, not all of the farmland is good quality or suitable for rice farming.

It only became possible to access all the villages in the township by car in 2001, when paved roads were constructed with governmental support (Wu 2010, 245). From Longshi Township it takes about half an hour on the bus to reach the county seat, and from there it is about a four-hour bus ride on the highway to the provincial capital, Changsha. Hence, Green Water Village is quite remote from major cities, which also has implications in terms of migration distances and duration. This affects the strategies of field resource protection, since commuting is not a common option.

My research in Hunan in 2011 was the final part of almost four years that I spent in the People's Republic of China between 2006 and 2011. During that time, I was mainly based in Shanghai, first as a language student, and later as an M.A. and then a PhD researcher. My stay also included one year working in Beijing for a Sino-German development organization. In addition, I visited the countryside of most of China's rice-growing provinces, as well as the major coastal cities and common destinations for rural migrants, for instance, in the Pearl River Delta. My repeated casual conversations with migrant and non-migrant farmers and the observations I made, both in the cities and in the countryside, have provided valuable additional insights.

During everyday life in cities such as Shanghai or Beijing, rural-urban migrants are encountered virtually everywhere, as street vendors, cleaners, rubbish clearers, security guards or construction workers, as well as employees in shops, restaurants, canteens, massage parlours, hairdressers, and hospitals. Being a foreigner and speaking Chinese generally made it easy to strike up conversations with people in cities and villages, as they were eager to satisfy their curiosity about Western people. In this way, I also had the chance to practise more in-depth participant observation of migrants' everyday working lives, for example in cleaning, street vending and restaurant work, which are typical occupational fields for Chinese internal migrants, but also academic and office work.

Some of these contacts, with whom I established deeper and more regular relationships during my initial research into the urban side of migration (Kaufmann 2011; 2016), later became the main interlocutors for this ethnographic inquiry into rice farming and migration. Among these, are, notably, two extended families that I mentioned in the Introduction. One is the Wu family from Anhui, who I met in Shanghai and accompanied home in 2008; the other is Yuemei's family from Green Water, Hunan. My account from Hunan is strongly influenced by the perspectives of Yuemei's family, their relatives, neighbours, and friends. I first met Yuemei as a colleague in Beijing in 2009-2010, where we shared not only a desk, but many aspects of everyday life. Gradually, we also became close friends. When Yuemei heard about my plans to engage in a research project focusing on rice farming and migration, she immediately offered to take me home to stay with her parents. Shortly thereafter Yuemei and I met in Changsha, the provincial capital, and she took me to her rural home. Yuemei turned out to be a highly dedicated research assistant. She also helped me to acquire some rare written sources, and assisted me in collecting additional data during subsequent visits. Apart from sleeping and eating with Yuemei's family, I took part in a whole range of everyday life activities and agricultural tasks, from watering the fields to milling and eating the harvested rice. I also had the opportunity to pay overnight visits to members of their extended family in various nearby villages, and to participate in special occasions such as engagement, wedding, and funeral ceremonies, as well as the Chinese New Year. We began our research journey to Hunan just before the New Year celebrations. This period around the New Year was unique, as it offered the rare opportunity to meet 'complete' households, being the time when migrants return home to congregate with their left-behind family members.

Having arrived in Green Water Village before the wave of incoming migrant workers, I was able to witness the amazing differences in local population structure that ensued. It was enlightening to experience how the de-populated villages filled up step by step with more and more returning villagers, to share the waiting and anxieties during a time when smartphones did not yet exist and landline phone communication was expensive for many rural inhabitants. I also participated in the emotional reunions between family members who had missed each other and had not met for months or even years, especially those of migrant mothers and their left-behind children. It was remarkable to see how, in the wake of the celebrations, people changed their appearances by dressing themselves in new clothes and fashions that were perceived as urban and modern, especially the returnees.

Ethnographic field research made it possible to closely observe and participate in such situations. It provided in-depth insights into the ways farmers between farming and migration did things, and how they made sense out of this. Moreover, I also accessed the rural-urban community of practice through a range of written qualitative and historical sources. Some of these provide insights into farmers' knowledge and its transformation, while others help to understand the historical transformation as well as the official constraints that contemporary Chinese farmers face. The former include two anthologies of Chinese oral vernacular literature, the *minjian wenxue* (CZ 1988; XT 1988). These stem from a state-supported mass movement of oral literature collection in the 1980s, described in Chapter 3. My interest in this medium of knowledge representation was sparked by a proverb I saw painted on a farmer's house in Green Water Village (see Figure 4). The definite ethnographic value of oral literature has previously been discussed and demonstrated by scholars such as Chard (1990) and Flitsch (1994; 2004), who have suggested that the 1980s' mass attempt to collect such heritage does, indeed, deserve fresh attention. To access this particular medium, I drew on several established methods of folk literature analysis. Inspired by Ruth Finnegan (1992), these included stylistical analysis, textual analysis, the construction of typologies, and contextual analysis. I touched upon several of these to provide an exemplary kind of material way to frame questions of rice knowledge transmission, the transformation, and negotiation of knowledge. As most of the proverbs were not studied in interaction, however, there are clearly limitations in my scrutiny, from the lack of their performative aspect.

With regard to understanding the official perspective of the state and the related structural constraints that farmers face, I drew on local gazetteers (*difangzhi*), complemented by yearbooks and agricultural reports.[21] The

21 For some problems with Chinese agricultural statistics see OECD (2005, 51-52). While I have no way to judge the accuracy of statistics gleaned from these sources, my triangulation of available national, provincial, county and township statistics with the qualitative data obtained from my

gazeteers were county, township, and industrial gazetteers from my field site (ACIGCC 1993; ACGCC 1996; 2011; Wu 2010). Taken together, they covered the period between the 1840s and the early 2000s, with the main focus on PRC history. Similar PRC gazetteers have been evaluated in detail by various China scholars discussing their official nature and constraints, as well as their value for studying local history and contemporary China (Thøgersen and Clausen 1992; Vermeer 1992; Looney 2008). As sources for this book, the local gazetteers provided important insights into the official perspective of the state, which relates to the structural constraints that farmers face. Moreover, they not only provided ample overview data at the county and township levels that helped to contextualize the setting in which the paddy field predicament emerged, but also local historic details of periods that lack contemporary witnesses and where sources are difficult to obtain. Importantly, since agriculture is a key topic in Chinese gazetteers and fits well into the overarching narrative of development and progress constructed by their editors, rice farming practices are thoroughly described in great detail. Throughout this book, however, I aim to deconstruct the narrative of almost-linear progress and social and technological development that pervades the agricultural sections of the gazetteers, as well as much of our common-sense understanding of technology more generally.[22]

Structure of the book

This book aims to show the value of adopting a socio-technical perspective to understand migration processes, through the example of rice farming and migration in China. The book sets out from analyzing the important policy and knowledge transformations since the 1950s that have given rise to the particular situation that farmers currently face, before describing farmers' contemporary responses to these transformations.

This chapter has introduced the basic predicament being faced by rice farmers in post-reform China, i.e. the conflicting pressures to both migrate into cities and yet preserve their family resources in the countryside. It posits that paddy fields play a crucial role in shaping farmers' migration strategies.

interlocutors, local gazetteers and proverb collections shows that the data are fundamentally consistent. Besides, the overall trends, e.g. regarding the spread of agricultural mechanization, are so obvious that minor mistakes would not alter them.

22 For valuable critiques of this common-sense notion, see Pfaffenberger (1992) and Edgerton (2007) in general, as well as Bray (1994) and Sigaut (1994) on farming technology in particular.

More generally, it proposes that socio-technical resources and related skills are key factors in understanding migration flows and the characteristics of migrant-home relations. Furthermore, this introduction has proposed a socio-technical approach to investigating this paddy field predicament and explained how this approach contributes to existing literature at the intersection of the literature on migration, agriculture, and skilled practice. Finally, it has introduced the main field site of Green Water, a rice-farming village in southern China, and briefly discussed my study's data and sources.

Chapter 1 describes the political setting since the 1950s in which the paddy field predicament has emerged. It shows that the Chinese state has been a major driver of the current situation through its rural policies, which provide both constraints and opportunities with regard to possible household strategies at the nexus of farming and migration. In unfolding this argument, special attention is paid to the widespread adoption of modern farming technologies that have set free agricultural labour. These policy-based transformations in agricultural technology are further placed into the context of de-collectivization and marketization, the abolition of the collective welfare system, the new urban economy, and loosened migration restrictions – all of which have pushed farmers to migrate and enhanced their precarity, which in turn makes them want to protect their fields as a safety net.

Chapters 2 to 5 constitute the qualitative-ethnographic body of the book. In order to better understand the problems farmers face, and the options they can call on to deal with their situation, Chapter 2 considers how paddy field knowledge is transmitted and how this has changed over recent decades. The chapter shows that there has been a complex reconfiguration of the repertoire of rice knowledge. On the one hand, this has created challenges for the future preservation of the paddy fields, such as deskilling in the young migrant generation. On the other hand, it has provided farmers with an extended repertoire of knowledge they can use to handle their paddy field predicament.

Chapter 3 describes one specific verbal medium of paddy field knowledge transmission, farming proverbs, discussing the role these proverbs play in the context of the paddy field-migration predicament. The chapter asserts that these agricultural maxims not only provide additional evidence for the transformations described in Chapter 3. It also explains that, first, the strength of these sayings lies precisely in their flexibility, which has made them a platform for knowledge negotiation between farmers and the state; and, second, that these proverbs have the potential to serve as a back-up resource for retaining paddy field knowledge.

Based on my ethnographic field research, Chapters 4 and 5 both analyse the socio-technical strategies that rice farmers use to manage their farmland. Chapter 4 focuses specifically on the strategic decisions made in farming technology. It does so mainly through the example of one left-behind woman, Mrs. Luo, and her choice of harvesting technologies. This sheds light on the diverse factors behind decision-making. It argues against a linear perspective of technological development, showing why it makes sense for farmers to simultaneously draw on a repertoire of old and new technologies, rather than simply opting for mechanization in order to compensate for the migrated labour. This also provides additional insights into the complex relationship between farming technology and migration, the causality of which has been much debated.

Examining several cases of both migrant and left-behind household members, Chapter 5 provides a rare, comprehensive overview of twelve land-use and land-arrangement strategies. These include social strategies such as leaving behind close family members to take care of the paddy fields, as in the case of Mrs. Luo. They also comprise more technical options, as in the case of Granny Li, who has switched from cultivating rice to growing a particular type of cash crop that is easier to manage, in view of her household situation and available labour and skills. Furthermore, it includes a brief analysis of the response of the Chinese central and local state to each of the twelve strategies. Overall, it demonstrates how farmers draw on a wide repertoire of available resources to handle their complex situation. Shedding new light on the logics behind land-use decisions, it shows that, in taking seemingly technical farming decisions, farmers are in fact pursuing various long-term and short-term projects that best match their fluctuating current and anticipated future household situation.

The Conclusion discusses four general advantages of investigating migration settings from a socio-technical skill perspective. First, it provides an understanding of a particular form of peasant agency that is commonly overlooked, because it is rooted in often-tacit everyday material practices. Second, focusing on skill allows us to better understand the reasons behind farmers' decision-making. Third, a skill perspective provides new insights into technology and Chinese modernity. Finally, the chapter argues that taking such a skill perspective contributes to understanding migration beyond the common dichotomies such as between migrant people and things, or migrants and left-behind family members. It concludes that even those who move to the cities remain part of their village communities of practice, sustaining relationships with their families and friends through visits and interactions. Moreover, they maintain their ties to the land through the ongoing management of their paddy fields – whether hands-on in person or at a distance using other household farming strategies.

The Conclusion is followed by an Appendix which comprises: (I.) A Glossary including Chinese characters, (II.) a list of the names and dates of the solar terms that structure farming activities throughout the agricultural year, (III.) the 'Song of the 24 Solar Terms', which is used to memorize this calendrical structure and, (IV.) annotated examples of about 150 local rice farming proverbs and encoded knowledge, to provide a clearer illustration of the points made in Chapter 3.

References

Abranches, Maria. 2013. 'When People Stay and Things Make Their Way: Airports, Mobilities and Materialities of a Transnational Landscape'. *Mobilities* 8 (4): 506-527.

ACBS, Anren County Bureau of Statistics. 2020. '安仁县2019 年国民经济和社会发展统计公报 Anren xian 2019 nian guomin jingji he shehui fazhan tongji gonggao (Anren County 2019 Statistical Communiqué of the Development of the People's Economy and Society)'. 13 March 2020. Accessed 3 July 2020. http://www.anrenzf.gov.cn/15/51015/51019/content_3126960.html.

ACGCC, 安仁县志编纂委员会 Anren Xianzhi Bianji Weiyuanhui (Anren County Gazetteer Compilation Committee). 1996. 安仁县志 *Anren Xianzhi* (*Anren County Gazetteer*). Beijing: Zhongguo Shehui Publishing.

—. 2011. 安仁县志 *Anren Xianzhi, 1989-2003* (*Anren County Gazetteer, 1989-2003*). Changsha: Hunan Renmin Publishing.

ACIGCC, 安仁县工业志编纂委员会 Anren Xian Gongyezhi Bianji Weiyuanhui (Anren County Industrial Gazetteer Compilation Committee). 1993. 安仁县工业志 *Anren Xian Gongye Zhi* (*Anren County Industrial Gazetteer*). N.p.: Huangshan Publishing.

ACWF, All China Women's Federation. 2013. '我国农村留守儿童、城乡流动儿童状况研究报告 Woguo nongcun liushou ertong, chengxiang liudong ertong zhuangkuang yanjiu baogao (Research Report on the Situation of Left-behind Children and Migrant Children in the Chinese Countryside)'. *Zhongguo Funü Xinwen*. May 10, 2013. Accessed 9 December 2017. http://acwf.people.com.cn/n/2013/0510/c99013-21437965.html.

Ahearn, Laura M. 2001. *Invitations to Love: Literacy, Love Letters, and Social Change in Nepal*. Ann Arbor: University of Michigan Press.

Andreas, Joel, and Shaohua Zhan. 2016. 'Hukou and Land: Market Reform and Rural Displacement in China'. *The Journal of Peasant Studies* 43 (4): 798-827.

Appadurai, Arjun. 1986a. 'Introduction: Commodities and the Politics of Value'. In *The Social Life of Things: Commodities in Cultural Perspective*, edited by Arjun Appadurai, 3-63. Cambridge: Cambridge University Press.

—. 1986b. *The Social Life of Things: Commodities in Cultural Perspective*. Cambridge: Cambridge University Press.

—. 1999. *Modernity at Large: Cultural Dimensions of Globalization*. Minneapolis: University of Minnesota Press.

Barlett, Peggy F. 1980. 'Adaptive Strategies in Peasant Agricultural Production'. *Annual Review of Anthropology* 9 (1): 545-573.

Barth, Fredrik. 2002a. 'An Anthropology of Knowledge'. *Current Anthropology* 43 (1): 1-18.

—. 2002b. 'Toward a Richer Description and Analysis of Cultural Phenomena'. In *Anthropology Beyond Culture*, edited by Richard Fox and Barbara King, 23-36. Oxford: Berg.

Basu, Paul, and Simon Coleman. 2008. 'Introduction: Migrant Worlds, Material Cultures'. *Mobilities* 3 (3): 313-330.

Boserup, Ester. 1965. *The Conditions of Agricultural Growth: The Economics of Agrarian Change under Population Pressure*. London: Allen & Unwin.

Bourdieu, Pierre. 1977. *Outline of a Theory of Practice*. Cambridge: Cambridge University Press.

Bray, Francesca. 1984. *Agriculture*. In *Science and Civilisation in China*, Vol. 6, Part 2, edited by Joseph Needham. Cambridge: Cambridge University Press.

—. 1994. *The Rice Economies: Technology and Development in Asian Societies*. Reprint. Berkeley: University of California Press.

—. 2004. 'Rice, Technology, and History, the Case of China'. *Education About Asia* 9 (3): 14-20.

Bray, Francesca, Peter A. Coclanis, Edda L. Fields-Black, and Dagmar Schäfer, eds. 2015. *Rice: Global Networks and New Histories*. New York: Cambridge University Press.

Brettell, Caroline B. 2008. 'Theorizing Migration in Anthropology: The Social Construction of Networks, Identities, Communities, and Globalscapes'. In *Migration Theory: Talking Across Disciplines*, edited by Caroline B. Brettell and James F. Hollifield, 113-160. New York: Routledge.

Burrell, Kathy. 2011. 'Going Steerage on Ryanair: Cultures of Air Travel for Migration from Poland to the UK'. *Journal of Transport Geography* 19 (5): 1023-1030.

Castles, Stephen, Heine de Haas, and Mark J. Miller. 2014. *The Age of Migration: International Movements in the Modern World*. 5th ed. New York: Guilford.

Chard, Robert L. 1990. 'Folktales of the God of the Stove'. *Hanxue Yanjiu* (*Chinese Studies*) 8 (1): 149-182.

Chen, Chuanbo, and C. Cindy Fan. 2016. 'China's Hukou Puzzle: Why Don't Rural Migrants Want Urban Hukou?' *The China Review* 16 (3): 9-39.

Chen, Chen, and C. Cindy Fan. 2018. 'Rural-Urban Circularity in China: Analysis of Longitudinal Surveys in Anhui, 1980-2009'. *Geoforum* 93 (July): 97-104.

Chen, Nancy N. 2001. 'Health, Wealth, and the Good Life'. In *China Urban: Ethnographies of Contemporary Culture*, edited by Nancy N. Chen, Constance D. Clark, Suzanne Z. Gottschang, and Lyn Jeffery, 165-182. Durham: Duke University Press.

Chen, Ruishan, Chao Ye, Yunlong Cai, Xiaoshi Xing, and Qiong Chen. 2014. 'The Impact of Rural Out-Migration on Land Use Transition in China: Past, Present and Trend'. *Land Use Policy* 40: 101-110.

Cheng, Tiejun, and Mark Selden. 1994. 'The Origins and Social Consequences of China's Hukou System'. *The China Quarterly* 139: 644-668.

Chu, Julie Y. 2010. *Cosmologies of Credit: Transnational Mobility and the Politics of Destination in China*. Durham: Duke University Press.

Cohen, Jeffrey H., and Ibrahim Sirkeci. 2011. *Cultures of Migration: The Global Nature of Contemporary Mobility*. Austin: University of Texas Press.

Coy, Michael W., ed. 1989. *Apprenticeship: From Theory to Method and Back Again*. Albany: SUNY Press.

CZ, 郴州地区民间文学集成编委会 Chenzhou Diqu Minjian Wenxue Jicheng Bianweihui (Editorial Board of the Comprehensive Collection of the Folk Literature of Chenzhou District), ed. 1988. 中国民间谚语集成湖南卷: 郴州地区分卷 *Zhongguo minjian yanyu jicheng Hunan juan: Chenzhou diqu fen juan* (*Hunan Volume of the Comprehensive Collection of Chinese Folk Proverbs: Section Volume of Chenzhou District*). Chenzhou: N.p.

Day, Alexander F. 2013. *The Peasant in Postsocialist China: History, Politics, and Capitalism*. Cambridge: Cambridge University Press.

Douglass, Mike. 2006. 'Global Householding in Pacific Asia'. *International Development Planning Review* 28 (4): 421-446.

Edgerton, David. 2007. *The Shock of the Old: Technology and Global History since 1900*. Oxford: Oxford University Press.

Eyferth, Jacob. 2009. *Eating Rice from Bamboo Roots: The Social History of a Community of Handicraft Papermakers in Rural Sichuan, 1920-2000*. Harvard: Harvard University Press.

Fan, C. Cindy. 2008. *China on the Move: Migration, the State, and the Household*. London: Routledge.

—. 2016. 'Household-Splitting of Rural Migrants in Beijing, China'. *Trialog* 1/2 (116/117): 19-24.

Fan, C. Cindy, and Tianjiao Li. 2019. 'Familization of Rural-Urban Migration in China: Evidence from the 2011 and 2015 National Floating Population Surveys'. *Area Development and Policy* 4 (2): 134-156.

Fan, C. Cindy, Mingjie Sun, and Siqi Zheng. 2011. 'Migration and Split Households: A Comparison of Sole, Couple, and Family Migrants in Beijing, China'. *Environment and Planning* 43: 2164-2185.

Fan, C. Cindy, and Wenfei Winnie Wang. 2008. 'The Household as Security: Strategies of Rural-Urban Migrants in China'. In *Migration and Social Protection in China*, edited by Ingrid Nielsen and Russel Smyth, 205-243. New Jersey: World Scientific Publishing.

Farquhar, Judith. 2006. 'Food, Eating, and the Good Life'. In *Handbook of Material Culture*, edited by Christopher Tilley, Webb Keane, Susanne Küchler, Mike Rowlands, and Patricia Spyer, 145-160. Los Angeles: Sage Publications.

Feuchtwang, Stephan. 2002. *An Anthropological Analysis of Chinese Geomancy*. 2nd ed. Bangkok: White Lotus.

Finnegan, Ruth. 1992. *Oral Traditions and the Verbal Arts: A Guide to Research Practices*. London: Routledge.

Flitsch, Mareile. 1994. *Der Ginsengkomplex in den Han-chinesischen Erzähltraditionen des Jiliner Changbai-Gebietes*. Frankfurt am Main: Peter Lang.

—. 2004. *Der Kang: Eine Studie zur materiellen Alltagskultur bäuerlicher Gehöfte in der Manjurei*. Wiesbaden: Harrassowitz.

—. 2008. 'Knowledge, Embodiment, Skill and Risk: Anthropological Perspectives on Women's Everyday Technologies in Rural North China'. *East Asian Science, Technology and Society* 2 (2): 265-288.

Frykman, Maja Povrzanović. 2009. 'Material Aspects of Transnational Social Fields: An Introduction'. *Dve Domovini / Two Homelands* 29: 105-114.

Gaibazzi, Paolo. 2015. *Bush Bound: Young Men and Rural Permanence in Migrant West Africa*. New York: Berghahn.

Geertz, Clifford. 1963. *Agricultural Involution: The Processes of Ecological Change in Indonesia*. Berkeley: University of California Press.

Gell, Alfred. 1998. *Art and Agency: An Anthropological Theory*. Oxford: Clarendon.

van Gijn, Annelou, John Whittaker, and Patricia C. Anderson, eds. 2014. *Exploring and Explaining Diversity in Agricultural Technology*. Oxford: Oxbow.

Glick Schiller, Nina, Linda Basch, and Cristina Szanton Blanc, eds. 1992. *Towards a Transnational Perspective on Migration: Race, Class, Ethnicity, and Nationalism Reconsidered*. New York: New York Academy of Sciences.

Grasseni, Christina. 2009. *Developing Skill, Developing Vision: Practices of Locality at the Foot of the Alps*. Oxford: Berghahn.

Gray, Clark L. 2009. 'Rural Out-Migration and Smallholder Agriculture in the Southern Ecuadorian Andes'. *Population and Environment* 30 (4-5): 193-217.

Hanks, Lucien M. 1972. *Rice and Man: Agricultural Ecology in Southeast Asia*. Chicago: Aldine.

Hannerz, Ulf. 1996. *Transnational Connections: Cultures, People, Places*. London: Routledge.

Haudricourt, André-Georges, and Mariel Jean-Brunhes Delamarre. 1955. *L'homme et la charrue à travers le monde*. Paris: Galimard.

He, Congzhi, and Jingzhong Ye. 2014. 'Lonely Sunsets: Impacts of Rural-Urban Migration on the Left-behind Elderly in Rural China'. *Population, Space and Place* 20 (4): 352-369.

Hickey, Maureen. 2016. 'Modernisation, Migration, and Mobilisation: Relinking Internal and International Migrations in the "Migration and Development Nexus"'. *Population, Space and Place* 22 (7): 681-692.

Ho, Peter. 2001. 'Who Owns China's Land? Policies, Property Rights and Deliberate Institutional Ambiguity'. *The China Quarterly* 166: 394-421.

Hoang, Lan Anh, and Brenda Yeoh, eds. 2015. *Transnational Labour Migration, Remittances, and the Changing Family in Asia*. Basingstoke: Palgrave Macmillan.

HPBS, Hunan Provincial Bureau of Statistics. 2012b. 'Statistical Communiqué of Hunan Province on the 2011 National Economic and Social Development'. Accessed 19 August 2015. http://hntj.gov.cn/sjfb/tjnj/12tjnj/html/gbe.htm.

—. 2014. *Hunan Statistical Yearbook 2014*. Beijing: China Statistics Press.

—. 2019. *Hunan Statistical Yearbook 2019*. Beijing: China Statistics Press.

Huang, Martin W. 1995. *Literati and Self-Re-Presentation: Autobiographical Sensibility in the Eighteenth-Century Chinese Novel*. Stanford: Stanford University Press.

Hull, James R. 2007. 'Migration, Remittances, and Monetization of Farm Labor in Subsistence Sending Areas'. *Asian and Pacific Migration Journal* 16 (4): 451-484.

Hunan Government. 2015. 'About Hunan'. Accessed 31 August 2015. http://www.enghunan.gov.cn/AboutHunan/HunanFacts/.

Ingold, Tim. 2000. *The Perception of the Environment: Essays on Livelihood, Dwelling and Skill*. London: Routledge.

—. 2006. 'Walking the Plank: Meditations on a Process of Skill'. In *Defining Technological Literacy: Towards an Epistemological Framework*, edited by John R. Dakers, 65-80. New York: Palgrave Macmillan.

Jacka, Tamara. 2012. 'Migration, Householding and the Well-Being of Left-behind Women in Rural Ningxia'. *China Journal* 67 (2012): 1-22.

—. 2014. 'Left-behind and Vulnerable? Conceptualising Development and Older Women's Agency in Rural China'. *Asian Studies Review* 38 (2): 186-204.

Kaufmann, Lena. 2011. *Mala Tang – Alltagsstrategien ländlicher Migranten in Shanghai*. Wiesbaden: Harrassowitz.

—. 2016. 'Everyday Strategies of Rural Migrants: Assembling Skills in Mala Tang Production'. *Journal of Chinese Dietary Culture* 12 (2): 63-112.

Kelly, Philip F., ed. 2013. *Migration, Agrarian Transition, and Rural Change in Southeast Asia*. London: Routledge.

Kleinhenz, Volker, Wilfried H. Schnitzler, and David J. Midmore. 1996. 'Diversification and Transformation of Asian Paddy Rice Fields to Upland Vegetable Production'. *Plant Research and Development* 43: 81-94.

Latour, Bruno. 1988. *The Pasteurization of France*. Cambridge, MA: Harvard University Press.

—. 1999. *Pandora's Hope: Essays on the Reality of Science Studies*. Cambridge, MA: Harvard University Press.

Lave, Jean, and Etienne Wenger. 1991. *Situated Learning: Legitimate Peripheral Participation*. Cambridge: Cambridge University Press.

Lawson, Victoria A. 1998. 'Hierarchical Households and Gendered Migration in Latin America: Feminist Extensions to Migration Research'. *Progress in Human Geography* 22: 39-53.

Lemonnier, Pierre, ed. 1993. *Technological Choices: Transformation in Material Cultures Since the Neolithic*. London: Routledge.

Leroi-Gourhan, André. 1964. *Le geste et la parole I: Techniques et langage*. Paris: Albin Michel.

Linares, Olga F. 2003. 'Going to the City ... and Coming Back? Turnaround Migration Among the Jola of Senegal'. *Africa* 73 (1): 113-132.

Looney, Kristen E. 2008. 'Village Gazetteers: A New Source in the China Field'. *The China Journal* 60: 135-147.

Lu, Ming, and Yiran Xia. 2016. 'Migration in the People's Republic of China'. ADBI Working Paper No. 593. Tokyo: Asian Development Bank Institute, 1-35.

Malthus, Thomas. 1798. *An Essay on the Principle of Population*. Reprinted in 1986. Harmondsworth: Penguin Classics.

Manivong, Vongpaphane, Rob Cramb, and Jonathan Newby. 2014. 'Rice and Remittances: Crop Intensification Versus Labour Migration in Southern Laos'. *Human Ecology* 42 (3): 367-379.

Marchand, Trevor H.J., ed. 2010. 'Making Knowledge'. *Special Issue, Journal of the Royal Anthropological Institute* 16 (May).

McKay, Deirdre. 2005. 'Reading Remittance Landscapes: Female Migration and Agricultural Transition in the Philippines'. *Danish Journal of Geography* 105 (1): 89-99.

Miller, Daniel. 1987. *Material Culture and Mass Consumption*. Oxford: Basil Blackwell.

Müller, Daniel, and Thomas Sikor. 2006. 'Effects of Postsocialist Reforms on Land Cover and Land Use in South-Eastern Albania'. *Applied Geography* 26: 175-191.

Murphy, Rachel. 2002. *How Migrant Labor Is Changing Rural China*. Cambridge: Cambridge University Press.

—. 2006. 'Citizenship Education in Rural China: The Dispositional and Technical Training of Cadres and Farmers'. In *Chinese Citizenship: Views from the Margins*, edited by Vanessa L. Fong and Rachel Murphy, 9-26. London: Routledge.

Naughton, Barry. 2007. *The Chinese Economy: Transitions and Growth*. Cambridge, MA: MIT Press.

NBSC, National Bureau of Statistics of China. 2019. *China Statistical Yearbook 2019*. Beijing: China Statistics Press.

Nyíri, Pál. 2010. *Mobility and Cultural Authority in Contemporary China*. Seattle: University of Washington Press.

Oakes, Tim, and Louisa Schein. 2005. *Translocal China: Linkages, Identities and the Reimagining of Space*. London: Taylor & Francis.

OECD. 2005. *OECD Review of Agricultural Policies: China*. Paris: OECD Publishing.

Ong, Aihwa. 1999. *Flexible Citizenship: The Cultural Logics of Globalization*. Durham: Duke University Press.

Ortner, Sherry B. 2006. *Anthropology and Social Theory: Culture, Power, and the Acting Subject*. Durham: Duke University Press.

Pfaffenberger, Bryan. 1992. 'Social Anthropology of Technology'. *Annual Review of Anthropology* 21: 491-516.

Pieke, Frank N., Pál Nyíri, Mette Thunø, and Antonella Ceccagno. 2004. *Transnational Chinese: Fujianese Migrants in Europe*. Stanford: Stanford University Press.

van der Ploeg, Jan Douwe, and Jingzhong Ye, eds. 2016. *China's Peasant Agriculture and Rural Society: Changing Paradigms of Farming*. London: Routledge.

Postill, John. 2010. 'Introduction: Theorising Media and Practice'. In *Theorising Media and Practice*, edited by Birgit Bräuchler and John Postill, 1-34. New York: Berghahn.

Qin, Hua, and Tim F. Liao. 2016. 'Labor Out-Migration and Agricultural Change in Rural China: A Systematic Review and Meta-Analysis'. *Journal of Rural Studies* 47: 533-541.

Resurreccion, Bernadette P., and Ha Thi Van Khanh. 2007. 'Able to Come and Go: Reproducing Gender in Female Rural-Urban Migration in the Red River Delta'. *Population, Space and Place* 13 (3): 211-224.

Rice, Tom. 2010. 'Learning to Listen: Auscultation and the Transmission of Auditory Knowledge'. *Journal of the Royal Anthropological Institute* 16: S41-S61.

Richards, Paul. 1985. *Indigenous Agricultural Revolution: Ecology and Food Production in West Africa*. London: Hutchinson.

Rigg, Jonathan. 2019. *More than Rural: Textures of Thailand's Agrarian Transformation*. Honolulu: University of Hawai'i Press.

Salazar, Noel B. 2017. 'Key Figures of Mobility: An Introduction'. *Social Anthropology* 25 (1): 5-12.

Santos, Gonçalo. 2017. 'Multiple Mothering and Labor Migration in Rural South China'. In *Transforming Patriarchy: Chinese Families in the Twenty-First Century*, edited by Gonçalo Santos and Stevan Harrell, 91-110. Seattle: University of Washington Press.

Santos, Gonçalo, and Stevan Harrell, eds. 2017. *Transforming Patriarchy: Chinese Families in the Twenty-First Century*. Seattle: University of Washington Press.

Schippers, Thomas K. 2014a. 'Agricultural Practices: Change and Stability'. In *Exploring and Explaining Diversity in Agricultural Technology*, edited by Annelou van Gijn, John C. Whittaker, and Patricia C. Anderson, 339-341. Oxford: Oxbow.

—. 2014b. 'Seasonal Variations in Crop Tending and Folk Knowledge in Southern France'. In *Exploring and Explaining Diversity in Agricultural Technology*, edited by Annelou van Gijn, John C. Whittaker, and Patricia C. Anderson, 79-80. Oxford: Oxbow.

—. 2014c. 'Skills as Identity Markers'. In *Exploring and Explaining Diversity in Agricultural Technology*, edited by Annelou van Gijn, John C. Whittaker, and Patricia C. Anderson, 276-277. Oxford: Oxbow.

—. 2014d. 'The Contemporary Use of Iberian Threshing Sledges: Some Ethnographic Observations about an Obsolete Choice'. In *Exploring and Explaining Diversity in Agricultural Technology*, edited by Annelou van Gijn, John Whittaker, and Patricia C. Anderson, 152-154. Oxford: Oxbow.

Schmalzer, Sigrid. 2016. *Red Revolution, Green Revolution: Scientific Farming in Socialist China*. Chicago: University of Chicago Press.

Schneider, Mindi. 2015. 'What, Then, Is a Chinese Peasant? Nongmin Discourses and Agroindustrialization in Contemporary China'. *Agriculture and Human Values* 32 (2): 331-346.

Scoones, Ian. 2009. 'Livelihoods Perspectives and Rural Development'. *The Journal of Peasant Studies* 36 (1): 171-196.

Scott, James C. 1977. *The Moral Economy of the Peasant: Rebellion and Subsistence in Southeast Asia*. New Haven: Yale University Press.

Shakespeare, Tom. 2006. *Disability Rights and Wrongs*. London: Routledge.

Sheller, Mimi, and John Urry. 2006. 'The New Mobilities' Paradigm'. *Environment and Planning* 38: 207-226.

Sigaut, François. 1994. 'Technology'. In *Companion Encyclopedia of Anthropology*, edited by Tim Ingold, 420-459. London: Routledge.

Stark, Oded. 1991. *The Migration of Labor*. Oxford: Blackwell.

Stark, Oded, and David E. Bloom. 1985. 'The New Economics of Labour Migration'. *American Economic Review* 75: 173-178.

Stone, Glenn Davis. 2001. 'Theory of the Square Chicken: Advances in Agricultural Intensification Theory'. *Asia Pacific Viewpoint* 42 (2-3): 163-180.

—. 2007. 'Agricultural Deskilling and the Spread of Genetically Modified Cotton in Warangal'. *Current Anthropology* 48 (1): 67-103.

Thøgersen, Stig, and Søren Clausen. 1992. 'New Reflections in the Mirror: Local Chinese Gazetteers (Difangzhi) in the 1980s'. *The Australian Journal of Chinese Affairs* 27: 161-184.

Tilley, Christopher. 2006a. 'Introduction'. In *Handbook of Material Culture*, edited by Christopher Tilley, Webb Keane, Susanne Küchler, Michael Rowlands, and Patricia Spyer, 2-8. Los Angeles: Sage Publications.

—. 2006b. 'Objectification'. In *Handbook of Material Culture*, edited by Christopher Tilley, Webb Keane, Susanne Küchler, Michael Rowlands, and Patricia Spyer, 60-73. Los Angeles: Sage Publications.

Tolia-Kelly, Divya P. 2004. 'Materializing Post-Colonial Geographies: Examining the Textural Landscapes of Migration in the South Asian Home'. *Geoforum* 35 (6): 675-688.

Toyota, Mika, Brenda S.A. Yeoh, and Liem Nguyen. 2007. 'Bringing the "Left behind" Back into View in Asia: A Framework for Understanding the "migration-Left behind Nexus." *Population, Space and Place* 13 (3): 157-161.

Vermeer, Eduard B. 1992. 'New County Histories: A Research Note on Their Compilation and Value'. *Modern China* 18 (4): 438-467.

Vertovec, Steven. 2009. *Transnationalism*. London: Routledge.

Wang, Cangbai. 2016. 'Introduction: The "Material Turn" in Migration Studies'. *Modern Languages Open* 26: 1-11.

Wang, Nora. 2003. 'Hunan'. In *Das grosse China-Lexikon*, edited by Brunhild Staiger, Stefan Friedrich, and Hans-Wilm Schütte, 319-320. Darmstadt: Wissenschaftliche Buchgesellschaft.

Wang, Qianxi, and Xiaoling Zhang. 2017. 'Three Rights Separation: China's Proposed Rural Land Rights Reform and Four Types of Local Trials'. *Land Use Policy* 63: 111-121.

Wang, Zhiling, and Lu Chen. 2019. 'Destination Choices of Chinese Rural-Urban Migrant Workers: Jobs, Amenities, and Local Spillovers'. *Journal of Regional Science* 59 (3): 586-609.

Wenger, Etienne. 1998. *Communities of Practice: Learning, Meaning, and Identity*. Cambridge: Cambridge University Press.

Whyte, Susan Reynolds. 2009. 'Epilogue: Insecurity, Contingency and Uncertainty'. In *Situating Uncertainty in Contemporary Africa*, edited by Liv Haram and Bawa Yamba, 213-216. Uppsala: Nordic Africa Institute.

Wolf, Diane Lauren. 1992. *Factory Daughters: Gender, Household Dynamics, and Rural Industrialization in Java*. Berkeley: University of California Press.

Wolf, Eric. 1966. *Peasants*. Englewood Cliffs: Prentice Hall.

Wu, Chunyuan 伍春元, ed. 2010. 益相史话 *Yixiang shihua* (*Historical Narrative of Yixiang*). Beijing: Jiuzhou Press.

Wu, Yi. 2016. *Negotiating Rural Land Ownership in Southwest China: State, Village, Family*. Honolulu: University of Hawai'i Press.

Xiang, Biao. 2007. 'How Far Are the Left-behind Left behind? A Preliminary Study in Rural China'. *Population, Space and Place* 13 (3): 179-191.

—. 2016. 'Beyond Methodological Nationalism and Epistemological Behaviouralism: Drawing Illustrations from Migrations within and from China'. *Population, Space and Place* 22 (7): 669-680.

Xie, Yong, and Quanbao Jiang. 2016. 'Land Arrangements for Rural-Urban Migrant Workers in China: Findings from Jiangsu Province'. *Land Use Policy* 50: 262-267.

XT, 湘潭市民间文学集成编委会 Xiangtan Shi Minjian Wenxue Jicheng Bianweihui (Editorial Board of the Comprehensive Collection of the Folk Literature of Xiangtan City), ed. 1988. 中国谚语集成湖南卷: 湘潭市分卷 *Zhongguo yanyu jicheng Hunan juan: Xiangtan shi fen juan* (*Hunan Volume of the Comprehensive Collection of Chinese Proverbs: Section Volume of Xiangtan City*). Xiangtan: N.p.

Xu, Dingde, Shili Guo, Fangting Xie, Shaoquan Liu, and Sha Cao. 2017. 'The Impact of Rural Laborer Migration and Household Structure on Household Land Use Arrangements in Mountainous Areas of Sichuan Province, China'. *Habitat International* 70: 72-80.

Ye, Jingzhong. 2018. 'Stayers in China's "Hollowed-out" Villages: A Counter Narrative on Massive Rural-Urban Migration'. *Population, Space and Place* 24: 1-10.

—. 叶敬忠. 2019. '农村留守人口研究: 基本立场、认识误区与理论转向 Nongcun liushou renkou yanjiu: Jiben lichang, renshi wuqu yu lilun zhuanxiang (Research on the Rural Left-behind Population: Basic Positions, Mistakes, and Theoretical Turn)'. 人口研究 *Population Research* 43 (2): 21-31.

Zavoretti, Roberta. 2017. *Rural Origins, City Lives: Class and Place in Contemporary China*. Seattle: University of Washington Press.

1 How the Predicament Arose

Abstract

This chapter describes how the paddy field-migration predicament has emerged. It argues that the Chinese state has been a major driver of the current situation through its rural policies, which provide both constraints and opportunities with regard to possible household strategies at the nexus of farming and migration. Special attention is paid to the widespread adoption of post-Green Revolution farming technologies that have set free agricultural labour. These transformations are placed into the context of de-collectivization and marketization, the abolition of the collective welfare system, the new urban economy, and loosened migration restrictions – all of which have pushed peasant farmers to migrate and enhanced their precarity, which in turn makes them want to protect their fields as a safety net.

Keywords: China, rural state policy, paddy field predicament, Green Revolution farming technologies, rural-urban migration, land use decisions

When I tried to revisit Mr. Wu's street restaurant in Shanghai in spring 2010, it was gone. Not just the restaurant – the whole block was about to disappear. The buildings were in ruins, without doors and windows. A long plastic tarpaulin covered what had once been the entrance to the restaurant. The place was deserted. The inhabitants of the block had moved on. The only person around was a street cleaner with his wheelbarrow who had stopped for a cigarette break, contemplating the empty scene.

This transformation was brought about by a beautification campaign that Shanghai had undergone. It occurred in the advent of a major world fair called 'Expo', which the city hosted in 2010 under the slogan 'better city, better life'. In the course of this campaign, the shabby block where Mr. Wu and numerous other migrant workers had lived and worked were found not to fit the image of a better city. Even though Mr. Wu had known several months in advance that the buildings might be torn down, he had

Kaufmann, Lena, *Rural-Urban Migration and Agro-Technological Change in Post-Reform China*. Taylor & Francis Group 2021
DOI: 10.5117/9789463729734_CH01

not been informed about the exact time so, when it happened, he and his family did not have anywhere else to live. Mr. Wu relied on the support of his extended family and fellow villagers in the city to get him through this difficult time and establish a new business. When that enterprise failed, however, his wife Li Cuiping returned home for several years, taking up rice farming again. She did not rejoin him in the city until her husband finally managed to set up a new business that could support them both.

Around Chinese New Year 2016, I had a video conversation with Mrs. Luo, who had been left behind in Green Water Village. While she was bottle-feeding her little granddaughter and putting her to bed, she told me that her husband Zhou Wenlu was there too. Instead of returning just for the festival, as he usually did, he had been back for the whole year. As stated in my Introduction, Zhou Wenlu had worked as a construction worker in various provinces for many years to finance his three children's education. But now he was home. When I inquired about this further, I learned that he had been unable to continue his migrant work due to an illness. Without any access to affordable health services in the city, he was compelled to rely on his home resources until he recovered and could migrate again.

The experiences of Mr. Wu and Zhou Wenlu, both working in the city in such a precarious situation that Mr. Wu's wife and Zhou Wenlu himself had been forced to return home when confronted with adversity, are closely related to a specific policy context that has been implemented since the People's Republic of China (PRC) was established in 1949, and especially in the reform period after 1978. This context comprises: the promotion of modern agriculture, embedded in establishing a socialist market, abolishing the collective welfare system, the new urban economy, and loosened migration restrictions. I argue that the central and local state has played a crucial role in shaping the conditions of rural households. Implementing these policies has significantly contributed to the current situation of predicament, providing the frame of institutional constraints or 'structures' (Ortner 2006) that define the scope of farmers' strategic actions. While this chapter only focuses on the aspect of structural constraints, mainly describing the political side and general content of the policies, I will argue in the following chapters that these policies have simultaneously provided farmers with certain possibilities, e.g. technological options, to cope with the situation.

My starting point is Anren County, where Zhou Wenlu's and Mrs. Luo's home village, Green Water is located, and the state-led introduction of modern agricultural technologies that have enabled and compelled farmers to migrate in the first place.[1] In this regard I suggest that agricultural policy and, in particular, the promotion of modern agriculture is in fact a major, though largely ignored factor influencing rural-urban migration decisions, and one that deserves closer attention.

The details of agricultural policy provided in this chapter need to be understood within the framework of the wider socio-political and economic developments and the overall political climate of the PRC. For most of China's history, most of its people have worked as farmers. Transforming rural society was therefore seen as an important component of the journey towards a socialist revolution. Hence, the practical issue of fighting rural poverty and producing sufficient crops to feed the population has been the basis of every political development action.[2] Considering that China needs to feed around one fifth of the global population with less than 10 percent of the world's arable land (FAO 2020), the push towards agricultural modernization has been a continuous political issue in the nation's efforts towards grain security, self-sufficiency and sovereignty. It is therefore no coincidence that agricultural policy is closely connected to a series of key moments in the PRC's political and economic development (Watson 2001, 57).

However, even though the debates around agriculture have been a crucial concern in PRC politics, this does not mean that the countryside has necessarily always been a priority for policy makers. On the contrary, agricultural production was seen as the basis for fostering the nation's industrial development. Prioritizing industrialization, the PRC has followed a dual rural-urban development scheme. Thus, urban industrial development has ultimately grown at the expense of agriculture (ibid.; Day 2013, 2).

Generally speaking, there have been two main phases of PRC agricultural policy: collectivization from the 1950s onwards, followed by de-collectivization after 1978. First, in the early 1950s, farmers went through land reform, whereby poor and landless farmers were allocated land expropriated from rich farmers and landowners. At the same time, the first collectivization campaigns were implemented and agrarian trade was gradually monopolized by the state (Watson 2001, 57-58; Aubert 2003, 424-425). In addition, population movements were tightly restricted through the enforcement of the *hukou* system of household registration (Cheng and Selden 1994).

1 Aware that 'modern' is a highly value-laden term, I use it here descriptively to denote (post-) Green Revolution technology. At the same time, in the context described, the term carries the sense of an aspired scientific modernity, which has been an integral part of the political push for modernization.

2 For details see Song (1998, 155); Watson (2001, 57); Aubert (2003, 424); Santos (2011, 488); and Day (2013, 2).

Collectivization culminated in the establishment of people's communes in 1958. In the same year, my field site Longshi Township was renamed the Longshi People's Commune, and Green Water Village became part of this commune (Wu 2010, 5). Organizationally, a commune was constituted of several villages and was subdivided into production brigades, each of which normally comprised one natural village. The brigades were further subdivided into production teams of about seven to eight households. At the brigade level, households were allocated collectively-owned land, labour, implements, draught animals, and reward in the form of work points. Green Water villagers stored the fruits of their collective labour in a communal granary. They also ate together. A side building of Zhou Wenlu's house was transformed into a canteen. How this was organized is still documented on a large, now-fading table on the wall inside the building, which he was using as a shed in 2011. The overarching communes carried out production decisions as defined in the government's plans and quotas. These envisaged a unified village economy, especially with regard to the production of rice. The communes were multifunctional, effectively combining local government, economic management, education, welfare services, and public security (Potter and Potter 1990; Watson 2001).

Large-scale collectivization took place within the framework of the Great Leap Forward, a utopian development scheme that aimed to rapidly transform China into an industrialized nation, but was implemented at a high human cost (Perkins 1991, 478). Ultimately, the Great Leap Forward led to a serious agrarian crisis and disastrous famine from 1958 to 1961. As a result, collectivization was subsequently continued on a smaller scale. Moreover, offering a way out of this political and humanitarian crisis, the Chinese Green Revolution (1964-1967) gained momentum (Stavis 1974). This occurred parallel to the Great Proletarian Cultural Revolution (1966-1967), a major political movement launched by Chairman Mao Zedong, which came to an end on his death in 1976.

This foreshadowed the consecutive reform period, i.e. the second major phase of agricultural policy. Launched at the end of 1978 under the new leadership of Deng Xiaoping, these reforms produced a major and rapid transformation. In less than five years, between 1979 and 1983, China went from a collectivized agricultural system that was controlled by state-set quotas and slogans to one of individual household agriculture, operated mainly through indirect market mechanisms. The speed of change towards a market-oriented system was much faster in the agricultural than the industrial sector (Perkins 1991, 537), suggesting that it must have felt more dramatic for farmers than for people in the cities. Since the early 1980s, agricultural development has gained a new emphasis in central policy documents such as China's 13th Five Year Plan (2016-2020) or the government's annual policy statement, 'No. 1 Document'. It is clear that, particularly since the 2000s, solving the problems of the three *nong* – agriculture, countryside and peasants – has become top priority for Chinese rural policy.[3]

The local situation in my field site in Anren County, Hunan Province, is closely linked to the national policy framework, and Anren farmers have generally followed the national transitions (ACGCC 1996, 285-288; Wu 2010, 5). Hunan, the home of Mao Zedong, was initially somewhat slower to implement the reforms of the 1980s than the rest of the country (Tregear and Falkenheim 2015). This contributed to the province lagging behind in terms of economic prosperity, and has made it a major sending area of rural migrants. In addition, Hunan's substantial reliance on rice production suggests that the impact of the new rice farming technologies and techniques has been particularly significant for farmers in Hunan. Moreover, Hunan is the national centre of hybrid rice development and, along with the Philippines, the global centre. Farmers in Hunan have therefore adopted this new technology more rapidly and enthusiastically than farmers elsewhere.

Modern agriculture in Anren County

The first policy that has affected the current situation of Chinese farmers is the vigorous government promotion of modern agriculture. Its labour-saving technologies have set millions of farmers free from the land and opened up their opportunities to migrate. This has reinforced the push for migration and, more generally, the paddy field predicament. At the same time, it has also offered farmers more options for dealing with the situation (see Chapters 4 and 5).

Most of the agricultural practices described in the following chapters originate from the Chinese Green Revolution of the 1960s and 1970s. This Green Revolution was not just a political answer to the Great Famine (Stavis 1974, 98), but was part of a larger development scheme, which envisioned – in Mao Zedong's words – that 'man must conquer nature' (*ren ding sheng tian*) (Shapiro 2001). In line with this objective, everything traditional (*chuantong*) was rejected as being backward (*luohou*) and superstitious (*mixin*). The aim was to create new, secular and rational agricultural production systems that

3 For details see MoA (2012; 2016); Day (2013, 3); OECD (2013, 122); Xinhua News Agency (2015); Ye (2015); Central Government of the People's Republic of China (2016).

would increase and stabilize China's food resources (Santos 2011, 487-488). Technologically speaking, the Green Revolution meant a rapid popularization of the first generation of modern agricultural technologies. These new and China-developed technologies included mechanized irrigation, farm chemicals, mechanization and rural electrification, along with newly bred high-yielding varieties of crops (Stavis 1974, 22-60).

As the China historian Sigrid Schmalzer (2016, 7) notes, there has been a lot of creative hybridity in implementing this shift in practice, for both practical and political reasons. In practice, this means that customary manual intensification techniques and new technologies such as mechanization coexisted as a means of leveraging production in the Mao era. Schmalzer therefore reminds us that it is, in fact, more appropriate to speak about 'a patchwork of methodologies, [in which] the patches themselves cannot easily be characterized as "modern" or "traditional"' (ibid., 13). For instance, some methods were well-established in China, but new to certain localities; while, in other cases, the innovation was simply an increase of scale or different methods of applying the technologies (ibid.).

On a wider scale, the Chinese Green Revolution should be seen as part of broader technological modernization efforts in China. This kind of creative hybridity was not confined to the realm of agriculture, nor just to technologies of Chinese origin. In fact, there was considerable technology transfer between various foreign (e.g. Russian, Japanese, British, German) technologies and China, which then inspired and led to the development of Chinese versions during the Great Leap Forward.[4] Many of the related agricultural technologies, e.g. in the field of crop science, and the general layout of the Chinese agricultural extension system, originated from American-Chinese cooperation in the early twentieth century (Stross 1986; Schmalzer 2016, 32), and others may be traced back to Soviet assistance in the 1950s, for instance, in the field of conducting soil surveys. In addition, some senior Chinese scientists who had been trained abroad prior to 1949 also played a role in appropriating and developing new farming technologies (Stavis 1974, 81-87).

Against this wider background of modernization, the overall diffusion of Green Revolution technologies took place at breakneck speed, despite local variations. When these technologies were introduced in the early and mid-1960s, only about 20 percent of China's cultivated land received a full input of these technologies (ibid., 22). However, due to state promotion, Green Revolution farming technologies had largely replaced conventional

4 See, e.g., Cortada (2012) for information technology, or TU Berlin (2017) for steam and ordnance technologies.

ones within two decades. Hence, the agricultural extension services and scientists who were sent to the countryside played a crucial role in this change (ACGCC 1996; Schmalzer 2016).

The vigorous state advocacy of modern agriculture was not just a feature of the Great Leap Forward, but has been sustained at the national and local levels. Since 1994, the resources mobilized by central government to promote agricultural science and technology have increased significantly. Since 2000, it has directly promoted the usage of modern farming technology through a range of subsidies (see p. 92). In Anren County, between 1989 and 2003 alone there were 296 local government projects to promote technology (ACGCC 2011, 293). Such projects in Anren included aspects like new cultivation patterns, new plant varieties, improved irrigation and drainage, farm chemicals, and mechanization.

Changing cultivation patterns

The first policy-induced transition in the field of modern agriculture was changing cultivation patterns. Cultivation patterns deserve special attention, because of their direct correlation to labour size and input. In line with the national objective of attaining grain self-sufficiency and stability, the quantitative aspect of rice production has been of continuing interest for Chinese policy makers (Li, Xin, and Yuan 2009, 15). It was thought that changing cultivation patterns would be an efficient way to achieve this, so it has received substantive political attention – although this has not always been through simple top-down decision making, and was far from a linear process.

As the *Anren County Gazetteer* (on which the accounts of modern agriculture promotion below are mostly based) narrates, farmers in Anren have changed their dominant rice farming practices from single-cropping in 1949 to double-cropping and, eventually, multi-cropping today. This has resulted in evident production gains, encouraged by the County People's Government. Implementing this policy has entailed various experiments with different cultivation systems, including farmers' own methods (ACGCC 1996, 289-290).

For a long time, single-cropping, i.e. planting a single crop of rice a year, used to be the dominant practice in Hunan. After the new Han settlers reclaimed land in Hunan around 350 B.C.E. (Wang 2003), farmers only planted one rice crop. But by the time land was fully reclaimed in the eighteenth century, along with the introduction of early rice seeds, a new cropping pattern was established. This pattern entailed the cultivation of rice, followed by a so-called 'miscellaneous grain crop' (*zaliang*) (Perdue 1987, 38, 114). In Anren County this crop consisted of stubble soy beans or sweet potatoes. This cropping pattern continued to dominate in Anren County until 1949 (ACGCC 1996, 298).

Although since the Song Dynasty (960-1280) there was a general trend in China towards double-cropping, i.e. two rice harvests a year, farmers in Hunan Province were hesitant to adopt this practice. Hunan was a turbulent frontier area that was marked by war and rebellions towards the end of the Song and Ming dynasties, in the thirteenth and seventeenth centuries respectively. Therefore, although they were probably familiar with the practice, some farmers in Hunan began double-cropping only in the nineteenth century, when migrants from neighbouring provinces such as Jiangxi moved in, leading to a higher population density (Rawski 1972, 120-142).[5] However, farmers in southern, mountainous Hunan generally resisted local officials' efforts to introduce double-cropping. This refusal has been attributed to climate, topography, market considerations, and especially labour constraints (see Rawski 1972, 138; Perdue 1987, 132).[6]

Double-cropping proliferated more rapidly after collectivization, with encouragement from agricultural production cooperatives and, by 1958, was practiced everywhere in the county except for a few mountain areas (ACGCC 1996, 289-290). Moreover, double-cropping was encouraged by growing population pressures in the twentieth century (Perdue 1987, 114, 131-132). However, as the *Anren County Gazetteer* concedes, production between 1956 and 1965 was comparatively low and unreliable due to a lack of experience, natural disasters, bad varieties, the Great Leap Forward policies' failure and because double-cropping had spread too quickly (ACGCC 1996, 290). As Grandpa Zhou, the father of construction worker Zhou Wenbao who was mentioned in the Introduction, remembers: 'That time was very bitter (*ku*). We had to eat grass roots, there was so little to eat. There were no pigs or chickens either' (personal interview, 28 January 2011).

Subsequently, the local government introduced various cropping patterns that alternated double-cropped rice with a non-rice crop. The rice-rice-rape seed cultivation system, introduced in 1976, emerged as the most successful of these and local governments continued to promote and expand this

5 For details on these early migrant movents to – as well as from – Hunan, see Ho (1959).

6 Up to the 1960s, there was also the occasional practice of a customary local double-cropping technique called *yahe* (Wu 2010, 2, 7; ACGCC 1996, 290, 298), in which early and late rice were planted simultaneously in the same field, not one after another, as in true double-cropping. Some interview partners still remembered the technique from Hunan and Jiangxi.

right up until recent times (ACGCC 1996, 290, 298; 2011, 293; HPBS 2012a; 2013; 2014). With regard to this pattern, the farmer couple Zhou Wenlu and Mrs. Luo confirmed the relationship between available labour and intensity of rice cropping. Moreover, they attributed double-cropping to machinery replacing manual labour, which has sped up the cultivation process: 'Today there are ploughing machines, that's quicker!' (personal interview, 27 January 2011). Meanwhile, according to both the gazetteer and the Green Water villagers, farmers have begun trying out their own cropping patterns, experimenting with cultivation systems that combine rice with cash crops such as tobacco, peanuts, water melons or lilies (ACGCC 1996, 298). This is linked to the introduction of household farming, increased freedom in production decisions, and local government initatives.

Modern rice varieties

The development and dissemination of modern rice varieties has been a second central concern in the official promotion of modern agriculture. The aim is to produce higher yields under conditions of decreasing arable land (Li, Xin, and Yuan 2009, 18). In fact, over the last decades, cereal production in the PRC has increased greatly (FAO 2019). By 1988 Anren County's rice yield was already 4.4 times higher than in 1949. This means that, in line with population growth, the average yield per person almost doubled, from 562 *jin*[7]*/mu* (about 4215 kg/ha) to 1156 *jin/mu* (about 8670 kg/ha) (ACGCC 1996, 291, 297). In comparison, in 2011 Green Water villagers reported that the hybrid rice they cultivated yielded 1500 *jin/mu* (about 11.25 mt/ha), while previously, conventional rice produced only meagre yields of 300 to 400 *jin/mu* (about 2250 to 3000 kg/ha), and the 'varieties were not good'. My interview partners, some of whom had experienced the Great Famine in the early 1960s, obviously appreciated the higher yields. In particular, they valued being able to achieve higher yields with less physical input, which is one consequence of the new seed varieties.

Up to 1949, farmers in Anren County used to breed and select their own varieties. Sticky and glutinous rice strains were important local rice crops until the beginning of the twentieth century. It is stated that 30 rice varieties were recorded in Anren during the Qing Dynasty (1644-1911), including sticky and glutinous varieties. To breed and select these, farmers used techniques such as ear selection, grain selection, and the 'single harvest, single sow' (*dan shou dan zhong*) method, and the seeds had to be exchanged regularly to

7 One *jin* corresponds to 500 grams.

prevent their rapid degeneration. One way to transmit knowledge about this was via proverbs: e.g. 'it is better to change seeds than to increase fertilizer' (*zeng fei buru huan zhong*); and 'when preparing the field you need to put in skill, you need to change [seeds] twice in three years' (*zuo tian mao qiao, san nian liang tiao*) (ACGCC 1996, 296; see also Chapter 3).

Although seed diversity has decreased in the long run (Schmalzer 2016, 11), there was initially an enormous number of new varieties because of their local government promotion. For example, before 1949 local government efforts had already led to a move to non-glutinous *indica* rice as the dominant crop, which produced higher yields. After 1949, however, central and local governments pushed these new varieties more forcefully. In just under four decades from 1950 to 1988, more than 700 types of agricultural products had been selected, bred, and introduced in the County of Anren. These included 252 varieties of rice and 365 cash crops (ACGCC 1996, 290, 297). This is related to three main phases of PRC rice variety development; a fourth phase is currently in progress.

First, in the 1950s, emphasis was placed on the selection and promotion of the best local varieties, leading to a switch from single-cropping to double-cropping. In Anren County, the initial strategy was to import good seeds from other counties and provinces, with newly-established seed stations becoming responsible for introducing, identifying and popularizing fine varieties (ibid., 296, 309). The second phase encouraged the farming of improved varieties (*liangzhong*), then the third phase encouraged the use of hybrid rice (*zajiao shuidao*).[8] Improved varieties and hybrid varieties – two distinct technologies – were the most influential.

Improved varieties (also known as high-yielding varieties, HYV) rely on conventional breeding methods. Therefore, farmers and officials often simply call them 'conventional rice' (*changgui shuidao*), in comparison to hybrid rice. Farmers can perform this conventional breeding successfully, as long as they have good seeds to breed from. In addition to farmers, important advances in breeding improved varieties have come from scientists, who have produced improved short-stalked varieties. These enable mechanic processing, are resistant to 'lodging' (i.e. falling over), and suitable for close planting. Their other characteristics include requiring high volumes of water and being particularly responsive to nitrogen fertilizer. Moreover, these improved varieties have shorter ripening times, enabling double rice cropping as well as multi-cropping with other crops (Stavis 1974, 278).

8 The recent development of genetically modified rice (Shen 2010; MoA 2015) can be regarded as a fourth phase. However, this 'golden rice' is not yet being distributed commercially.

The extensive distribution of Green Revolution-improved short-stalked varieties took place in the early 1960s. In 1961, the Anren County Agricultural Science Research Institute supplied the first generation of such varieties to local farmers. By 1966, the institute had introduced 12 varieties of *indica* and four varieties of *japonica* rice with short stems (ACGCC 1996, 296-297). Three of these are also mentioned by Stavis. In contrast to the gazetteer's focus on statistics, Stavis informs us about a range of practical problems related to adopting these varieties, such as their taste, ripening time, and disease resistance (Stavis 1974, 27). Nevertheless, they were vigorously promoted throughout the county and, according to the *Anren County Gazetteer*, readily welcomed by the farmers (ACGCC 1996, 297). Generally, on the national level there seems to have been both cases of farmers' resistance and acceptance of such varieties and other Green Revolution technologies (see Schmalzer 2016, 151; Oxfeld 2017, 41).

In contrast, it is more complex to breed hybrid rice. Hybrid rice research in Hunan Province began in 1964 under Yuan Longping (born in 1930), nicknamed the 'Father of Hybrid Rice'.[9] His efforts led to the commercial distribution of hybrid rice seeds in 1976 (Lin 1991, 355-356). Simply put, the production of hybrid rice entails crossbreeding two genetically distinct parent lines. Because rice is a self-pollinating crop, this is not easy, because a line usually pollinates itself instead of propagating with another. While there is also the possibility of sterilizing each individual rice plant by hand, a practice that was already well-known in China, this does not allow for the mass production of seeds (Schmalzer 2016, 75). It is therefore usually up to scientists to use rare male sterile plants to achieve crossbreeding. When replanted, though, the second generation of these hybrids only gives a low yield, meaning that the seeds cannot be reused, and complicating production even further.

The production of hybrid seeds in China is linked to a massive central government agricultural campaign in 1975. Under this framework, local technical staff and farmers – including those from Anren – were sent to the warmer provinces of Guangxi and Hainan, where climatic conditions enable more rapid rice production. Nationwide, in the winter of 1975 more than 30,000 people from rice-growing provinces were sent to tropical Hainan Island for training and seed production. This was followed by zealous information and propagation campaigns, as well as the assignment

9 See Schmalzer (2016, 73-99) for a discussion of the narrative surrounding the 'intellectual peasant' Yuan, and the publicity that he received in the reform period, contributing 'to uphold the rightness of Deng Xiaoping's new course for the Chinese political economy' (ibid., 87).

of high-ranking central and local officials to supervise the extension and commercialization of hybrid rice (Li, Xin, and Yuan 2009, 18; Schmalzer 2016, 137).

Moreover, the participating institutions in Anren County established their own local breeding grounds, including the County Improved Seed Breeding Space, the County Department of Agriculture, and the County Research Institute for Agricultural Science. Every year the latter supplied the whole county with more than 50,000 kilograms of improved conventional rice seeds and 35,000 kilograms of hybrid rice, winning numerous national, provincial, and local awards (ACGCC 1996, 297, 309).

Initially, this local production faced some setbacks, however. Large amounts of local hybrid seeds were wasted, because 'the planning got out of control and supply exceeded demand on the seed market; in addition, the area was big and distribution wide, the technical guidance was not able to keep up with the technology, and some of the seeds lacked purity' (ibid., 297).[10] Therefore, in the late 1970s and early 1980s, the county government decided to professionalize and standardize hybrid seed production, establishing the County Seed Company in 1979. While China's county seed companies have gradually been privatized from the mid-1990s onwards (Ho, Zhao, and Xue 2009, 357-358), at the end of the 1970s the company was staffed with cadres, technicians, and workers. Meanwhile farmers in townships and villages were forbidden from excessive independent breeding. Moreover, the seed production period was gradually moved, leading to a county-promoted adjustment to two-season hybrid rice (ACGCC 1996, 297, 309; 2011, 284-285).

The establishment of the County Seed Company and its new role needs to be considered in connection with the Chinese three-tier seed system and four-level research extension network. Through this binary framework, hybrid seeds could be produced on a commercial level and simultaneously spread rapidly. In the three-tier seed system, provincial seed companies specialized in parental line purification; prefectural seed companies were responsible for A line multiplication, and county seed companies undertook F1 hybrid seed production.[11] The four-level extension network included county, commune, brigade, and production teams, which all played a role in efficiently and speedily evaluating, selecting, and adopting hybrid rice.

10 In Chinese: '计划失控，种子市场供过于求，加上面积大，分布广，技术指导跟不上，部分种子纯度差' *jihua shikong, zhongzi shichang gongguoyuqiu, jiashang mianji da, fenbu guang, jishu zhidao genbushang, bufen zhongzi chundu cha* (ACGCC 1996, 297).

11 See Schmalzer (2016, 76) for related breeding diagrams and explanations.

Every commune had one or more agricultural technicians to give appropriate training to farmers (Li, Xin, and Yuan 2009, 19). In this four-level system, in which Hunan's Huarong County served as an important model, policies and technologies could effectively flow in a top-down manner. At the same time, however, it was embedded in a paradigm of mass science, which implied that policies and technologies could also be developed at the grassroots level and extended upwards (Schmalzer 2016, 44). Hence, new field management practices were spread rapidly, including that of 'Tons-Rice-Grain-Production', which the gazetteer states that Anren County was particularly successful at (Li, Xin, and Yuan 2009, 7; ACGCC 1996, 291, 297).

Overall, whilst introducing the short-stalked improved varieties has doubtless been revolutionary and marked the beginning of modern Chinese farming, the innovation of breeding hybrid rice can be seen as even more revolutionary. As shown in Chapter 2, the farming of both improved and hybrid rice varieties has had far-reaching socio-technical consequences. Like the other technologies described below, the impact of these two varieties was so far-reaching because of the speed and scope of their take-up. The new rice varieties have vigorously spread at all levels, leading to sharp rises in their adoption rates. Although the improved varieties were only introduced in the early and mid-1960s, by 1982 they were sown on 92 percent of the national rice land (MoA 1989, 348). Since hybrid rice was commercially introduced in 1976, however, the acreage of improved varieties has shrunken. While they continue being cultivated and bred by both farmers and scientists, since 1991, more than half of China's rice acreage has been used for hybrid rice. Due to the locally-based research facility, Hunan was particularly quick to adopt the new hybrid seeds, which accounted for two thirds of Hunan's total rice acreage by the early 2000s. In fact, Hunan has adopted more hybrid rice, more rapidly than anywhere else in China (Lin 1991, 357, 363; Li, Xin, and Yuan 2009, 2-3).

Improved irrigation

A third component of the official promotion of modern agriculture is improved irrigation and drainage. At the national level, between 1952 and 2007, the proportion of China's cultivated areas being irrigated increased from 18 percent to approximately 50 percent (see Huang and Rozelle 2009, 101). This was also due to the increasing mechanization and electrification of irrigation (Stavis 1974, 23; ACGCC 2011, 301).

Irrigation in Anren County has been affected by these transitions as well. According to the county gazetteer, there were many low-yield and bad quality fields before 1949. These fields lacked modern large-scale irrigation, drainage facilities and machinery. Earlier efforts made during the Northern Song and in Republican China were not as far-reaching as the collective efforts undertaken after 1949 (ACGCC 1996, 295, 346-348). At that time, improved irrigation and drainage, particularly of low-yield fields, became one of the county government's main priorities. It was considered important for stabilizing and increasing yields in at least two ways. Firstly, irrigation – along with fertilizer application – was seen as part of agricultural soil improvement. Modern soil science therefore contributed to improved irrigation. Moreover, the new rice varieties are not drought resistant, so require conditions of careful watering; their high yields depend directly on specific, detailed irrigation. Therefore, the success of the new varieties was built on improved irrigation (Chang 2000).

Irrigation and drainage works in Anren after 1949 consisted of three components: firstly, constructing large-scale irrigation and drainage facilities; secondly, conducting soil surveys; and, thirdly, introducing new irrigation and drainage machinery. Initially, Anren's county government only focused on the first component, building numerous reservoirs, ditches, dykes, and canals. The apex of these construction efforts occurred under the Great Leap Forward framework and in the subsequent collective era. Every year from 1959 on, the County Department of Agriculture arranged for people to participate in the 'dig three ditches, drain three waters' movement, which aimed to improve low-yield fields (ACGCC 1996, 295-296, 346).[12] As Grandpa Zhou recalls: 'My mother [then in her mid-thirties] dug out the reservoir [close to Green Water Village]. They went in groups of two or three. The reservoir has several levels. They dug it out with their bare hands!' (personal interview, 28 January 2011). Up until 1976, participants in Anren dug more than five million cubic metres of soil, working on an area of more than one million *mu*. They improved substantial amounts of low-yield fields and built terraced fields. As a result, more than 95 percent of Anren's arable land was irrigated by 1988, earning the county national and provincial recognition (ACGCC 1996, 295-296, 346).

In the early 1980s, during de-collectivization, the government's focus shifted to the second component of the irrigation works, launching agricultural

12 In Chinese: '挖三沟（围山沟、灌溉沟、排水沟），排三水（黄沙水、冷浸水、地下水）' *wa san gou* (*weishan gou, guangai gou, paishui gou*), *pai san shui* (*huangsha shui, lengjin shui, dixia shui*) (ACGCC 1996, 296). The three ditches refer to enclosed mountain ditches, irrigation ditches, and drainage ditches; the three waters refer to yellow sand water, cold flood water, and groundwater.

zoning works and a soil survey between 1982 and 1984. Consequently, paddy soil – one of the ten listed soil types in the county – was classified into five subtypes, 26 soil categories and 87 soil kinds (ibid., 296).

The third component of irrigation and drainage works entailed the successive propagation of new irrigation machinery, along with hydropower facilities. The new machinery included mechanical water pumps in the 1950s, and electric water turbine pumps and sprinkler machinery in the 1960s and 1970s. In the 1980s, the irrigation and drainage machinery was re-organized and turned into a water conservancy conveyance system (ibid., 307).

The improvement of irrigation and drainage facilities has continued in recent years. Hunan Province had more reservoirs than any other Chinese province by 2018 (NBSC 2019a, sec. 12-7). Chenzhou City alone has a total of 1,084 water reservoirs, and 86,719 agricultural drainage and irrigation machines (HPBS 2019, secs. 21-20/21). On average, about 14 rural households share one of these machines – a sharp increase from the small-scale irrigation there at the beginning of the twentieth century.

Farm chemicals

Farm chemicals (*nongyao*) are the fourth component of the new rice farming technologies that have spread rapidly in the PRC. These mainly include chemical fertilizers and pesticides (insecticides), along with some herbicides. Due to government promotion, the use of chemical fertilizers and pesticides has risen from almost zero before the 1950s to environmentally alarming amounts in the following decades (MoA 1989, 344, 346; Smil, 2004). For example, there was a 60-fold increase in the national use of chemical fertilizers between 1960 and 2005 (Greenpeace China 2010). Similarly, at the provincial level, the consumption of all farm chemicals combined in Hunan increased by a factor of 50 between 1957 and 2009 (HPBS 1984, 139; NBSC 2010, sec. 3-15). The province now ranks third nationally in pesticide use (Statista 2018). In the first decade of the new millennium, farmers in Hunan used on average 28.7 kg of pesticides per hectare (Liang 2010, 150). This has contributed to the contamination of three quarters of the rice fields in Hunan, which has alarmed the central government (Patton 2014). Therefore, in 2015 the government announced the Zero-Growth Action Plan for Chemical Fertilizers and Pesticides. According to the Ministry of Agriculture and Rural Affairs (MARA, which superseded the Ministry of Agriculture, MoA, in 2018), this target was reached ahead of time in 2017 (OECD 2019, 188). Nevertheless, China still uses more chemical fertilizers, herbicides and pesticides than any other country in the world, i.e. more than

30 percent of global fertilizers and pesticides on less than 10 percent of the world's agricultural land (Wu et al. 2018). In the 2010s, farm chemicals had become an integral part of the household strategies of my interlocutors from Hunan and Anhui. In Anren County, the usage of fertilizers, pesticides, and herbicides evolved as follows.

Fertilizers

Before the systematic production and usage of chemical fertilizers in the 1960s, China's farmers had almost exclusively used conventional fertilizers (see Stavis 1974, 40; Bray 1984, 289-297). According to the *Anren County Gazetteer*, in the Qing Dynasty (1644-1911) and Republican eras (1912-1949), farmers used manual techniques to enhance their soil quality. These included treading green plants into the soil, mixing it with other soils, applying lime, night soil (sewerage), barnyard manure, ash, cake fertilizer, bone meal, and gypsum (ACGCC 1996, 295, 299).[13]

In 1946 Republican China, Anren farmers came into contact with artificial fertilizers for the first time. As a relief measure, the province allocated 44 tons of sulphuric acid to the county. However, farmers were not familiar with it and were not told how to apply it, so their crops grew too high and lodged – a highly undesirable condition. This resulted in what was known as 'sulphuric acid fear' (*liusuan pa*), deterring farmers from trying chemical fertilizers again for about a decade (ibid., 299).

Since 1949, there have been continuous experiments and changes in types of fertilizers, both organic and synthetic. Inspired by Mao Zedong, there was a national policy of 'relying mainly on farmers' fertilizers [i.e. organic fertilizer], and secondarily on chemical fertilizers' (*yi nongjiafei weizhu, yi huafei weifu*) (Schmalzer 2016, 116). At the same time, China expanded its chemical fertilizer industry, signing contracts with US, Dutch and Japanese firms in the early 1970s to establish ten ammonia factories (Stavis 1974, 44). The Anren County Supply and Marketing Cooperative played the leading role in distributing both organic and inorganic fertilizers to local farmers. Meanwhile, cooperative farmers' experiments took place throughout the collective period (ACGCC 1996, 299, 440). Despite this mixed approach to different fertilizers, the total amount of chemical fertilizers consumed has increased significantly over the years. Chemical fertilizer application in Anren County rose in connection to four particular occurrences.

13 Cake fertilizer (*bingfei*) refers to the cake-like organic residues from oil or soybean milk production.

First, following the first successful official experiments of the early 1950s, farmers were instructed in line with the slogan 'setting up an example with a model, popularizing steadily' (*dianxing shifan, wenbu tuiguang*). In practice, this meant forming working groups who brought fertilizer to the fields. There, they explained how to use it, citing their own experiences as examples. Favourable policies such as loans, or selling fertilizer on credit supported these efforts and, according to the county gazetteer, farmers increasingly welcomed sulphuric acid fertilizer because of the evident gain in yields it produced (see ibid., 299, 440-441).

Second, the usage of, and requirement for, fertilizer increased abruptly when the semi-dwarf rice varieties were developed in the 1960s. Anren established its own nitrogenous fertilizer factory in 1971-1972 to meet higher demands for these, as well as importing various other types of fertilizer (ibid., 299, 369, 441).

Third, chemical fertilizer consumption soared again after 1975, with the gradual spread of hybrid rice, and the resultant change in cropping patterns. Because hybrid rice was newly intercropped with rape, the cultivation and, hence, availability of green manure[14] decreased. According to the county gazetteer, this meant that Anren farmers increasingly welcomed synthetic fertilizers, eagerly mixing and matching different types in their enthusiasm for the new technology (ACGCC 1996, 299).

Fourth, fertilizer consumption in Anren rose again in the first half of the 1980s, when de-collectivization encouraged farmers to invest more into farming. In addition, the double-cropped hybrid rice and new cash crops required higher levels of fertilizer. Moreover, farmers began to use chemical fertilizer in the 1980s not just as base fertilizer, but also for deep and sprinkled applications (ibid.).

As a consequence, fertilizer application reached its peak in 1989, mirroring the provincial situation (HPBS 2018, sec. 12-6). This forced Anren County to allocate and apply amounts of chemical fertilizer to specific districts and soil types and strictly regulate its marketing and supply. Therefore, the use of chemical fertilizer decreased radically in the following years, although this decline has been steadily reversing since 2012 (Hexun 2020). Hunan now ranks tenth nationally in terms of fertilizer application (NBSC 2019a, 12-5).

In Green Water Village in the 2010s, villagers creatively made use of both conventional organic and chemical fertilizers. Organic fertilizing

14 Green manure refers to nitrogen-fixing plants which are grown specifically to fertilize soil for the subsequent crop (in this case, rice). These plants were referred to as early as the sixth century, in the major agricultural treatise *Qi Min Yao Shu* (see Bray 1984, 293).

substances included the remaining dry stubble of the harvested rice plants, burned or rotten rice straw, burned rice chaff, and animal excrement. The latter came from chickens, ducks, pigs, and preferably oxen, as well as from fish inside the fields. Moreover, human excrement was used for producing the so-called 'night soil'. It was taken out of toilet houses and applied in a highly laborious process (see Santos 2011, 494-495). As the local proverbs make clear (see Appendix IV, A.7), farmers in the area near and around Green Water Village had a rich knowledge of producing night soil. The resulting manure is called 'pit fertilizer' (*dangfei*). Here, substances such as grass, leaves, weed, mud, household waste, and human excrement were mixed with water and soaked in a pit, where they fermented. This fertilizer was mainly used as a base fertilizer in rice fields. At the same time, Green Water farmers also made ample use of chemical fertilizers. They perceived these as more modern and applied them especially on rice grown for sale.

Pesticides

There were virtually no chemical pesticides in Anren before 1949, and harmful insects were countered with manual techniques and natural products. Throughout the PRC, there has been a lot of experimentation with various techniques and technologies for pest management (see ACGCC 1996, 300). State support was given to both biological and chemical products, because of a shortage of chemical insecticides and worries about resistance and toxicity (Schmalzer 2016, 12-13). Despite this, there has still been a sharp increase in pesticide consumption generally, related to changes in cropping patterns and varieties, as well as the large-scale official promotion and coordination of pesticides.

In the early days of the PRC, there were 57 types of pests and diseases which regularly affected wet rice, but this has increased significantly, due to the transformation of cultivation techniques and technologies. Moreover, the dominant types of pests and diseases, as well as weeds, have changed continuously. For example, the increase of double-cropping in Anren since 1955 has fostered certain pests, such as the yellow rice borer, while the promotion of short-stalked varieties and small-pocket close planting in the 1970s has led to a rise of relative humidity in the fields. This has resulted in an upsurge in rice-leaf rollers and plant hoppers, as well as more seasonal febrile rice diseases (ACGCC 1996, 300; see also Labrada 2003).

Despite such challenges, plant protection became more controlled when several official institutions became engaged in the subject. First, following the Great Famine, plant protection grew more coordinated when a Plant

Protection Station was set up in 1963 by the Department of Agriculture (ACGCC 1996, 300). Establishing such a plant protection station, along with seeds, technology, soil, and fertilizer stations, all stemmed from a central government initiative which aimed to disseminate hybrid rice technology at different administrative levels, including the county level (see Li, Xin, and Yuan 2009, 19).

Second, with the advent of de-collectivization in 1982, the Plant Protection Company was established by the County Department of Agriculture, to manage the supply of pesticides and agricultural machinery. In the same year, the government also began restricting the use of hypertoxic and polluting pesticides, promoting less harmful pesticides. During the 1980s, the Plant Protection Company publicly announced set times when farmers should apply specific chemicals to counteract the diseases it observed. It did this through meetings, wall newspapers, radio and television broadcasts, which the gazetteer asserts successfully reduced certain pests and diseases (ACGCC 1996, 308, 310, 441).

Finally, in the early 1980s, each township established a Plant Hospital which included a salesroom and a training centre. Villages set up technician posts and organized model households. In line with the national extension system mentioned above, a four-level plant protection system was established in Anren County, which comprised 248 newly-trained plant protection members (ibid., 300, 310).

The promotion of chemical pest management obviously showed success in terms of reducing pests in the short run. In the 2010s, Green Water villagers commonly controlled harmful insects by using insecticides that were applied with the help of an atomizer carried on their back. Farmers sprayed insecticides through a tube onto their plants whilst walking through the field. According to Grandpa Zhou, today 'every family and every household' owns this implement (personal interview, 28 January 2011).

Herbicides

Traditional rice weeding methods in China consisted of a combination of water management and hand weeding, as well as particular cultivation techniques (Labrada 2003). In the early 1980s manual weeding – with the help of hands, feet, and tools – used to be the most common weeding practice (Bray 1984, 314). Young migrant woman Yuemei and her father's construction worker colleague Zhou Wenbao reported that, in their childhoods in the 1960s-1970s and 1980s respectively, children had to 'collect pig weed' (*da zhucao*) with their bare hands. In 2011, this practice had largely been supplemented with the application of herbicides, although manually pulling out those weeds that resisted herbicides nevertheless remained important. In addition, treading weeds into the field by feet was also common (personal interviews, January 2011). This is a proven technique that was already being used as early as in Song Dynasty China (960-1279). Here, weeds are trodden into the mud of the rice field, which gains nutrients from the rotting weeds. Moreover, manual weeding is performed with hoes, harrows and other tools (see Chang 2000, 141; Bray 1984, 314-318). Their importance was downplayed by the villagers, however, and in one case migrant worker Xiao Chen could not even remember the name of a weeding tool – a non-trivial fact, in view of agricultural deskilling.

Herbicides have been introduced and promoted somewhat later than other farm chemicals. They gained popularity since official field trials and field demonstrations were carried out in China's main rice-growing regions during the 1970s (see Zhang 2003). Accordingly, herbicide consumption in China has risen from 1067 tons in 1970 to more than 1 million tons in 2015 (Gianessi and Williams 2011; Huang, Wang, and Xiao 2017, 615). Around the beginning of the new millennium, herbicides were being applied to almost three quarters of China's rice acres, much more than for other crops (Zhang 2003). There are currently no consumption figures available for Anren County. It is worth noting, however, that in its section on weeds, the new *Anren County Gazetteer* only lists chemical products, and does not mention any manual techniques (ACGCC 2011, 295). My interlocutors from both Hunan and Anhui said that the most common technology used in 2011 was herbicides.

There is a close link between the use of herbicides and other farm chemicals and migration. According to migrant worker Xiao Chen, spraying herbicides (*Nongda*, i.e. 'Roundup' from the biotech company Monsanto, now part of Beyer) today releases farmers from the task of manual weeding (personal interview, 9 April 2011). In fact, since the mid-2000s, the use of herbicides in China has risen sharply and migrant work has been identified as one major driving factor for this (Huang, Wang, and Xiao 2017). Moreover, as different technologies in the agricultural system affect each other, the use of herbicides in particular has fostered a shift from the rice cultivation technique of transplanting to direct seeding (Zhang 2003, 198; Labrada 2003). The latter is another labour-saving technique linked to migration (see Chapter 5). In view of environmental protection, these findings clearly imply that if related policies are to be successful, policy makers need to take a much broader perspective on the issue and include areas such as migration, rather than merely focusing on the reduction of farm chemicals alone.

Mechanization

The fifth and last component of the suite of new rice farming technologies is mechanization. Mechanization received much stronger state commitment than farm chemicals, where organic methods continued to receive strong state approval. This was, on the one hand, because of the environmental costs of farm chemicals. On the other hand, Mao Zedong considered mechanization to be related much more directly to his objective of large-scale agricultural collectivization (Stavis 1978, 170; Schmalzer 2016, 116). Therefore, although the mechanization of Chinese wet rice farming occurred later and less comprehensively compared to other crops, farm machinery has been vigorously distributed in the PRC, at the national, county, and township levels (ACGCC 1996, 306; Wu 2010, 244; Eisenman 2018, 255-256). Just like the rest of China, there was virtually no mechanized agriculture in Anren County prior to 1949, in the sense of electricity or diesel-powered machinery (Stavis 1974; ACIGCC 1993, 108). The *Anren County Gazetteer* (1996, 306, 441) lists the most common farm implements in Qing (1644-1911) and Republican times (1912-1949) as:

- tilling: ploughs (*li*), harrows (*ba/pa*), six types of hoes (*chu*)[15];
- transplanting: hands;
- harvesting: sickles (*liandao*);
- threshing: wooden barrels (*bantong*);
- winnowing: winnowing machine (*fengche*), bamboo sieves (*shaizi* or *zhushai*);
- processing: axes (*fu*), knives (*dao*), saws (*ju*), planers/diggers (*bao/pao*), rice hullers (*long*), treadle-operated tilt hammers for hulling rice (*dui*), rollers (*nian*), grinders (*mo*);
- transportation: square-bottomed bamboo baskets (*luo*), winnowing baskets (*ji*), shoulder poles (*biandan*), wheelbarrows (*dulun che*).

Since the beginning of the PRC, the county has continuously tried to update tools and introduce new ones. Thus, it has focused on the whole gamut of farm operations, from irrigation and drainage to cultivation, processing, and transportation. The individual production and supply of farm implements has changed into a collective system of supply and marketing, run by the Anren County Supply and Marketing Department (ACGCC 1996, 306-307, 441; ACIGCC 1993, 108).

15 These include the round mouth hoe (*yuankou chu*), board hoe (*banchu*), one line hoe (*yizi chu*), two teeth hoe (*erchi chu*), three teeth hoe (*sanchi chu*), and four teeth hoe (*sichi chu*) (ACGCC 1996, 306).

In 1954-1955, the department organized the large-scale supply of 150,000 wood, bamboo and iron tools, corresponding to an average of 3.1 tools per household. It provided farmers with an increasing range of items, from medium and small agricultural tools to fertilizers, pesticides, oxen for ploughing, seeds, water wheels, semi-mechanic farm implements, processing machines, and tools. In 1956, the department introduced the first semi-mechanic threshing machines (*dadaoji*), improved ploughs (*gailiang li*), paddle instruments (*huaxingqi*), transplanting machines (*chayangji*), tilling machines (*pugun*), intertilling machines (*zhonggengqi*), and grain cutting machines (*geheqi*). Ten years later, in 1966, it provided 1000 threshing machines and 340,000 medium and small agricultural implements. This was accompanied by diverse local experiments to modernize farm implements in the mid-1950s (ACGCC 1996, 306, 440-441).

The pinnacle of agricultural tool reform is linked to the formation of people's communes in 1958. At that time, the county set up a Farm Tool Reform Steering Group which worked under the slogan 'the whole Party starts action, nationwide mobilization, and comprehensive reform of old-style farm implements'. The Steering Group proposed an ambitious reform scheme that focused simultaneously and equally on the 'assembly and repair, additional purchase, changing use, new construction, introduction, promotion, supervision, and maintenance' of farming tools. In addition, Anren followed the principle to 'give priority to the soil by combining native with foreign methods; give priority to changing the old by combining changing the old with creation and innovation' up until 1959. On this basis, a total of 42,500 farm implements were brought in from elsewhere or manufactured locally (ibid., 306).[16]

The success of these new farm implements was variable, however. Even though Hunan Province used more than 13,000 mechanical transplanters in 1970 (Stavis 1974, 50-51), this was not the case in Anren. According to the *Anren County Gazetteer*, some of the products, such as a rice seedling transplanting machine, were unsuitable for local conditions because findings from a survey were missing and experiences of its use were not shared in field demonstrations (ACGCC 1996, 306). According to some Green Water villagers, this is because the area is not flat and even, and the fields are

16 In Chinese: '全党动手，全民动员，全面改革旧式农具' *quan dang dongshou, quanmin dongyuan, quanmian gaige jiushi nongju*; '拼修、添置、改用、新造并举，引进、推广、管理、维修并重', *pin xiu, tianzhi, gaiyong, xinzao bingju, yijin tuiguang, guanli, weixiu bingzhong*; and '以土为主，土洋结合；改旧为主，改旧与创新结合' *yi tu weizhu, tu yang jiehe; gai jiu weizhu, gai jiu yu chuangxin jiehe* (ACGCC 1996, 306).

small, which renders mechanical transplantation difficult. In contrast, an atomizer (*penwuqi*) which was carried on your back to spray farm chemicals, and human-powered threshing machines (*dadaoji*) were the most popular items in the 1960s and 1970s. By 1975, people in the county owned 10,300 atomizers. Moreover, there were 4817 '737 type' human-powered threshing machines, which made up 23 percent of the threshing tools used (ACGCC 1996, 306). In 1973, 80 percent of all threshing in Hunan Province was done mechanically or semi-mechanically (Stavis 1974, 50).

With de-collectivization, machine ownership has moved to the household level. In the 1980s, this led to a steady decrease in supply by the Supply and Marketing Department, especially with regard to the big, formerly collectively-owned equipment. Nevertheless, some implements, such as small atomizers and threshing machines suitable for small-scale applications, have flourished (see below) (ACGCC 1996, 306, 441).

The new millennium has seen a continued rise in the number of mechanical farm implements, along with growing electric power consumption and diesel engines. In 2003, Anren County had 18,260 farming machines, equating to an average 82 machines in every administrative village, or one machine shared between ten people working in the primary sector (ACGCC 2011, 49, 300).[17] At the same time, provincial statistics show that electric power consumption per hectare of cultivated land more than doubled in Hunan Province between 2000 and 2012, from 1135.00 to 2659.26 Kw.h (HPBS 2014, sec. 1-9). Today, Hunan has the fifth highest amount of agricultural machinery and diesel engines of any province in China (NBSC 2018, sec. 12-4).

Anren County also increasingly engaged in locally producing industrial farm machinery, with a particularly sharp rise in farm implement production during the 1960s and 1970s. A basic precondition for this was the arrival of electricity. While private households on the village level only gained electricity in the early 1980s, Anren County received electricity in 1950, when the People's Liberation Army established a rice mill there. This was followed by the founding of an agricultural tool processing plant in 1952, the Anren County Farm Implements Factory in 1957, and the state-owned but locally administered Anren County Agricultural Machinery Factory in 1962 (Wu 2010, 245-246; ACGCC 1996, 306, 474).

While the implements were initially manufactured in a simple manner, and most of them were unusable, the Machinery Factory was more successful.

17 Here and in the following statistics, the 'primary sector' refers to agriculture, forestry, animal husbandry, fishing, and irrigation (ACGCC 2011, 49).

It became the main producer of hundreds of machines until it was closed down and moved to Hengyang City at the beginning of the reform period. Its final task was to produce 70 '3/5-type' walking tractors, with the help of state investment of 150,000 Yuan (about 21,200 USD). Overall, from 1965 to 1988, Anren County produced 4552 rice threshing machines, 2616 tilling machines (of a type called *pugun*) and 481 tilling machines (the *ji gunchuan* type), 333 mini water turbines (*weixing shuilunji*), 62,730,000 fittings for various kinds of agricultural machinery, and 103 walking tractors (*shoufu tuolaji*) (ACIGCC 1993, 108-109; ACGCC 1996, 368). The latter are used to till paddy fields (see Figure 6). They have two handles and the farmer (usually male) walks behind the machine. They do not have much in common with actual tractors, and villagers also call them a 'power plough' (*dianli*), or 'ploughing machine' (*gengtianji* or *litianji*).

The spectrum of different machines generally reflects the different steps of the rice cultivation cycle (see Chapter 2). Each of them has their own biography. Machine tillage has increased since the 1980s, despite initial fluctuations when privatization rendered large tractors unsuitable for small-scale farming. There were some other additional setbacks, because only a few of the machines were fit for local soil conditions, so they got stuck in the mud. Nevertheless, in the long run, the trend towards mechanical tillage has continued, with the area of Hunan Province ploughed by tractor more than doubling between 1988 and 2013, and an average of five Anren farming households sharing one tractor by 2003 (see ACGCC 1996, 301, 307; 2011, 301; HPBS 2014, sec. 19-28).

When I visited Green Water Village in 2011, farmers mostly tilled their fields with the help of mechanical 'power ploughs', and only few farmers relied on customary ploughs and animal power.[18] In the words of village woman Zhou Meijuan: 'Previously everybody used hoes, today there are power ploughs' (personal interview, 25 January 2011). Villager Zhou Wenxiang recalled that power ploughs had become common in Green Water since 1993 (personal interview, 21 January 2011). This was more than two decades later than the initial production efforts around this technology in the county seat during the Cultural Revolution.

Around Green Water, the multifunctional power plough has not only replaced wooden animal ploughs, but also the need for harrows and human-powered rollers (*tuopen*) that are used to smooth and flatten the surface of the soil. Zhou Meijuan, who practises seasonal migration, explained that she had stopped using her roller in 2009, when she began renting a power plough.

18 For details on customary ploughing techniqes see Bray (1984, 138-195).

As explained in Chapter 4, technology choice is complex and influenced by several factors, but migration certainly plays into it.

Regarding the machines that support pest and disease management, with de-collectivization, large farm chemical application machines – as well as other large farming machines – lost importance, because they could not fit into the smaller allocated fields or due to the high cost of running them for individual households. However, by 2003 there were still 61 big, powered atomizers in the county. Longshi Township was one of the five townships in which they were concentrated. The area they have been applied in corresponds to almost one quarter of the total ploughed area in the county (ACGCC 1996, 306-307, 441; 2011, 284, 301). If small individual atomizers are included in this figure, the proportion is much higher.

With regard to threshing, according to the county gazetteer, human-powered threshing machines already counted as an essential tool in every household by 1982, and demand for them grew rapidly. Eventually, these human-powered machines almost completely replaced wooden threshing barrels (ACGCC 1996, 306). Meanwhile, threshing machines have continued to thrive throughout the new millennium and, in 2012, Hunan Province had one of the highest number of motorized threshing machines in China (NBSC 2015).

In 2011, some villagers continued to store large wooden threshing tubs about one and a half to two metres wide in their homes.[19] Two women from Hunan and Anhui provinces both remembered the use of such threshing technology in their childhoods, in the 1950s and 1970s respectively. During the harvest, they were carried out into the fields and the cut bushels of rice were then forcefully hit against the rim of the tub. Granny Xu, in her early sixties, recalled this sonorously and with accompanying gestures: 'flinging down the paddy, bushel by bushel, *beng beng*' (personal interviews with Granny Xu, 27 January 2011, and Xiao Chen, 9 April 2011).[20] When I visited the village, most households owned foot-operated threshing machines that are operated by two people pressing on a pedal. According to villager Zhou Wenxiang, however, this kind of threshing machine is no longer used much, since its work is now done by a combine harvester (personal interview, 21 January 2011). Nevertheless, I found one in almost every household I visited,

19 For an overview of non-mechanical threshing practices, see Bray (1984, 345-358). Most threshing machines in East Asia originate from an eighteenth-century Japanese model, which was probably introduced to China in the early twentieth century through the Japanese possessions in Taiwan and Manchuria (ibid., 361-362).

20 See also Bray (1984, 349).

and the rice stubble on fields that had been cut with a sickle indicated that this type of machine was still commonly being used (see Figure 5). Some farmers also equipped their machines with diesel motors.

Processing machines such as various types of mills spread gradually. By the end of the 1960s, every natural village in Anren County had rice-milling machines, which gradually replaced rice hullers and hammers. These machines were initially powered by coal, then later by diesel or electricity (ACGCC 1996, 308-309, 474). Farmers in Green Water in 2011 preserved conventional processing implements such as stone mortars, mills, grinders, and hammers, but had stopped using them. For example, 20-year-old Lanying stated that the people from her grandfather, Grandpa Zhou's generation, had husked rice with such implements, while Mrs. Luo claimed that she had never the eaten brown, unpolished rice that results from this technology (personal interviews, 31 January 2011 and 1 February 2011). According to Mrs. Luo's brother-in-law Zhou Wenxiang, people had only used these manual husking implements up the 1950s. He added that he had never seen one being used (personal interview, 21 January 2011). Instead, he and others used electric rice mills, even though very few households could afford to buy one, so had to rent the milling service. The mills perform two tasks at once – husking and polishing the rice. Some machines have an additional pulveriser installed in them. Similar to the farmers from Anhui, however, the Green Water farmers continued to use manually-operated wooden winnowing machines that are used to clean the milled grains.

The means of transporting agricultural goods has also gradually shifted towards the use of powered vehicles, as the de-collectivization of the 1980s led to increasing numbers of them. There was a rapid emergence of specialized households engaged in the transportation business, and many individual car purchases (ACGCC 2011, 307). When I visited Green Water Village in 2011, however, there was only one paved road. Besides, hardly anybody could afford to buy a car, so farmers continued to use non-motorized transporting equipment for short and medium distances.

In summary, the county government of Anren has actively and successfully promoted modern farming technology for almost every single cultivation stage. The only exception is transplanting where, as explained above, the uneven landscape and small field sizes were unsuitable for machine harvesting. In 2017 I was informed that transplanting machines have, in fact, made their way into Green Water Village due to the pressure from increased migration to seek new solutions, along with the migrant incomes that have made these machines more affordable. Yet while migration compels farmers to find new technological solutions, at the same time, the technologies described above have also considerably added to farmers' migration pressure because of their labour-saving capacities. This happened at the precise moment when migration became a larger issue under the major paradigm shifts of de-collectivization and marketization; abolishing the collective welfare system; and the new urban economy and loosened migration restrictions, as turning points in modern Chinese agriculture.

De-collectivization and marketization

Modern agriculture in Anren County has developed, first, in a setting of de-collectivization and the establishment of a socialist market economy. In 1984, Anren County's 21 people's communes were administratively transformed into one town (*zhen*) and 20 townships (*xiang*), and the former production brigades replaced by villagers' committees (*cunmin weiyuanhui*) (ACGCC 1996, 7).[21] The related policies have further contributed to the emergence of the paddy field predicament, influencing both the legal situation of farmland allocation, and production and occupational decisions. They have increased farmers' freedom, but also their uncertainty vis-à-vis grain production.

The most important implication of de-collectivization with regard to the paddy field situation concerns land use rights. Between the 1950s and early 1980s, land was mainly farmed collectively, so there was little scope for individual farming.[22] Farmers had to meet fixed production quotas, while all farming decisions were taken by the production team leaders, who pursued and implemented quotas prescribed by central government (Song 1998, 155). It can be inferred from the *Anren County Gazetteer* (1996) that, during the collective era, the production emphasis was on maximizing the intensification of rice farming as far as possible, at the expense of other crops and a diversified rural economy.

As a result of de-collectivization, today farmland is contracted to individual farmer households on a per capita basis. Since the beginning of the reform period, the central government has implemented various tenure experiments and introduced a series of policies targeting specific aspects of land use and land ownership. Nevertheless, the interpretation and

21 Currently there are six towns and 15 townships.

22 Rural households also had private plots (*ziliu di*), except during the Cultural Revolution (1967-1977), when the practice was denounced as capitalist and forbidden. Use rights were provided on a per capita basis to grow crops and raise farm animals for family consumption. The amount was negligible, however, constituting only 5 to 7 percent of the production team's entire farmland (Li 1995, 597-598).

implementation of these policies differs starkly across regions.[23] Generally speaking, de-collectivization has given households the right to occupy, use, and profit from their tenured land (OECD 2013, 126-127). The 2002 Law on Rural Land Contracts extended farmers' land use rights to a period of at least 30 years (NPC 2002). Further legal support for farmers is provided by the Land Administration Law, amended for the second time in 2004; the 2007 Property Law; and the 2009 Mediation and Arbitration of Rural Land Contract Disputes Law (NPC 2004; 2007; 2009). Farmland has therefore become a key asset for rural households, and farmers are now able to take land-use and production decisions increasingly autonomously. It has also become a space where farmers can strategically bring the new scope of agricultural technology into play.

Despite these new opportunities, though, farmers still continue to face certain constraints. First, they do not own the land they farm and live on, as the land is de facto owned by the village collective.[24] Individual farmer households lease land from the collective based on written contracts, although each exact leasing arrangement may differ even among neighbouring villages (OECD 2005, 38-39). Farmers are not allowed to sell the leased land, use it as collateral for construction and industrial purposes, or for business development. They can merely transfer their use rights to other farmers within the contracted period (NPC 2002). While these regulations are being revised (Xinhua 2017; Xinhua News Network 2019), the current regulations continue to have implications for migrating farmers. For example, farmers still lack the options of selling their land, mortgaging it to a bank, or transferring management rights in their absence while retaining the contract right if they wish to.

Moreover, tenure insecurity often persists. In practice, the leasing contracts do not always carry much meaning. Since the 1990s, there have been numerous cases of unofficial land redistribution or local governments' reclamation of farmland. This occurs for industrialization and urbanization purposes, infrastructure projects, and increasing revenues – often in the

23 For some examples see Kong and Unger (2013); Long (2014); Wu (2016); Brandt et al. (2017); and Wang and Zhang (2017).

24 As Ho (2001) points out, the law does not make clear who represents the 'collective' and who really owns the land. The term is intentionally left vague for fear that conflicts may arise from consolidating ownership with one specific administrative level. In the early 1960s, land ownership was attributed to the production team, while in the reform period it has become vested in higher administrative levels. Nowadays, the 'collective' may refer to the township, administrative village, natural village, or village group, with leaders from these levels assuming the role of landowners. Generally, many levels of authority have a say when it comes to farmland (see OECD 2005, 38-40).

name of development and under considerable protest.[25] For many farmers, including those in Hunan, the loss of land has caused a significant social security problem (Wang 2014). This insecure tenure situation has had a twofold effect. On the one hand, it has affected farmers' willingness to invest in the maintenance of their fields. In Green Water Village, this happened especially in the early reform period, when reallocations were more common. On the other hand, and more importantly, in Green Water and other places farmland is regarded as a crucial component of social insurance. The tenure insecurity affects farmers' willingness to abandon agriculture in the long term, in the fear that local governments may follow a 'use-it-or-lose-it' rule (Brandt et al. 2017, 1035) – a topic I take up again in Chapter 5.

Second, allocating fields during de-collectivization entailed creating several fragmented small, scattered plots of farmland. After some initial readjustments, in 2010 the per capita average in Longshi Township was 0.69 *mu* (less than 0.05 ha) of wet land, i.e. paddy fields, and 0.11 *mu* (less than 0.01 ha) of dry land (Wu 2010, 278). Green Water villagers prefer wet land because of its better quality and suitability for rice farming. According to them, the per capita amount was close to the amount actually cultivated in 2011. The average cultivated area per agricultural labourer has generally been increasing in line with the rural exodus (HPBS 2019, sec. 1-8). The land area allocated in Green Water was much lower than the national average of 0.6 ha, a figure that includes the large dry crop farms in northern China (see Huang and Rozelle 2009, 106). It was also slightly lower than the provincial average of 0.9 *mu* (0.06 ha). These small sizes affect farming options, for instance limiting the possibility of using big machines, and, thus, farmers' potential responses to the paddy field predicament.

Generally, due to population growth and other factors, the amount of available farmland has been decreasing. For example, at the end of the fourteenth century, when Hunan Province was still sparsely settled, the average per capita cultivated land area was 5.1 *mu* (Perdue 1987, 46). This was about five times as much as the per capita cultivated amount in Green Water Village in 2011. Therefore, in the 1980s, the non-agricultural use of arable land was officially restricted in Anren County. Nevertheless, production output has grown in the reform period, more than doubling between 1979 and the mid-1990s alone. According to the county gazetteer, this was due to factors including new farming technologies, their intensified usage, and increased economic benefits (ACGCC 1996, 289).

25 See, e.g., Sargeson (2012); Kong and Unger (2013, 9); Wu (2016, 152), as well as more recently Chen (2020), and Heger (2020).

Despite these constraints, therefore, a key outcome of de-collectivization has been giving farmers the right, and responsibility, to maintain their land resources. This change went hand in hand with a second major change: marketization, which influences production decisions as well as migration pressures on the paddy field question. The economic transitions of the reform period have led to a dramatic change in occupations among China's rural population. Whereas in 1952, 84 percent of people were employed in the primary sector, this dropped to 60 percent in 1980 and 35 percent in 2013 (UNDP 2014, 96; OECD 2013, 119). Likewise, the share of the primary industry (including agriculture) in the Gross Domestic Product declined from 27.7 percent in 1978 to 7.2 percent in 2018 (NBSC 2019a, sec. 1-3). The millions of farmers set free by the new economic system's incentives and productivity could only partly be absorbed by the newly-created Township and Village Enterprises (TVEs). Therefore, many farmers have turned their backs on agriculture and migrated.

The new economic situation has also created widening social disparities and severe rural-urban differences. For example, like the income discrepancies mentioned by Zhou Wenbao in my introduction, where an urban income averaged around 19,595 Yuan (almost 2800 USD) compared to a rural income of around 6723 Yuan (about 950 USD), the 2013 per capita annual net incomes in Shanghai were almost three times as high as those in rural Anren (NBSC 2015; HPBS 2014, sec. 20-25). While incomes and living standards in China in general, and Anren in particular, have continued to rise over the long term, the urban-rural gap remains. In 2018 the per capita disposable income of urban households (including all urban areas, not just high-end cities) was still more than twice as high as rural households, i.e. 39,250 Yuan (about 5530 USD) compared to 14,617 Yuan (about 2060 USD) (NBSC 2019b, 6-16; ACBS 2020). This is another contributory factor prompting farmers to migrate to the cities.

Moreover, marketization was followed by sharp fluctuations in grain prices and related policies, with implications for farmers' repertoires around rice cultivation. In 1979, the central government raised grain prices substantively and took the first steps towards establishing a free grain market. Subsequently, in the early 1980s, grain output and sales grew rapidly – to the point where there was such a surplus of grain that farmers had trouble selling their harvests. In response, the state's grain monopoly was abolished in 1985, and the collective era's mandatory unified grain procurement system was replaced by a contractual dual-track grain pricing system. In this new system, state procurement quotas were lowered and state procurement prices further raised. Individual households now gave a set amount of their grain harvest to the state in taxes and sales. In addition, they had the possibility to sell surplus grain at (usually higher) negotiated or market prices. However, the particular price setting of the new contractual procurement system actually discouraged farmers from grain production. This led to a situation of grain scarcity, as farmers felt encouraged to plant more economically-rewarding crops (Oi 1989, 155-163, 175).

In order to be able to control the resulting huge increase in grain prices, in 1995 the government installed a 'governor responsibility system', which assumed increased state control over purchasing and marketing grain at the provincial level, and installed various incentives for grain production (Tian and Zhou 2018, 11-13). Although this strategy was intended to produce higher rural incomes, the result was, again, a grain surplus, which did not benefit farmers. Therefore, in 1999 the Chinese government abandoned their domination over grain, as well as price subsidies. At the same time, with growing fiscal decentralization, local cadres looking for alternative revenues encouraged or even compelled local farmers to plant new cash crops. Meanwhile, China's increasing integration into global markets augmented farmers' vulnerability to market fluctuations and cheaper imports, leading to a sharp decrease in their incomes. These vulnerabilities remained, despite once again achieving higher grain prices through a strategic reduction in grain acreage in 2002-2003 (see Murphy 2006, 10-12, 19). In 2004, the government reformed and liberalized the grain market. It abandoned its direct role in the grain market in favour of an indirect one, limited to buying and selling reserves to maintain food security and stabilize prices (Gale 2013, 3). Taken together, these ups and downs in grain prices brought considerable insecurities to famers, which caused them to diversify their household strategies and look for alternative ways of earning an income, such as turning to other products, seeking employment in TVEs, or migrating.

On top of these grain price fluctuations, farmers also suffered from excessive rural taxation, which had further implications for migration. While there were explicit official government fees, price scissors continued to be the major taxation instrument throughout the 1980s and early 1990s. This was inherited from the collective era as a way to squeeze the agricultural sector in favour of promoting the country's industrial sector. Since local governments controlled the transaction channels and prices of both agricultural inputs and outputs, these price scissors meant that they could collect taxes tacitly by raising the prices of agricultural inputs such as fertilizers, pesticides, water and electricity, while depressing the prices of agricultural outputs, including grain (Lin, Tao, and Liu 2007, 4-5). Although throughout the reform period the central government repeatedly raised the grain prices paid to farmers, this could not fully compensate for the rising input prices, and more generally, the total production costs for agriculture (Gao and Fennell 2018, 72-73). In addition, especially from the mid-1990s onwards, township governments and village community organizations short of funding often over-procured grain by adding a margin above the state quota. Moreover, local officials charged villagers hefty informal fees for all types of services (Lin, Tao, and Liu 2007, 6-8). For example, in 1984, a farmer in Hunan had to pay twelve different fees just to operate a tractor, corresponding to almost 30 percent of their annual income, in addition to spending 41 percent of their income on costs for fuel and maintenance (see Oi 1989, 209-210).

Such fees, as well as the nature of rural taxes (levied mostly on arable land) meant that farmwork had a particularly high tax burden, compared to other sources of income. While some of these taxes were paid in kind, there was nevertheless an increasing need to acquire cash to cope with the situation. This need was augmented by policies spurring on farmers' consumption of consumer goods (Oi 1989, 159), all of which rendered migration especially attractive. There were two additional benefits to migration. On the one hand, migrant remittances were not subject to taxation (Lin, Tao, and Liu 2007, 11-12). On the other hand, local cadres faced more difficulties in collecting taxes from villagers who had migrated, while it also became easier for migrants to resist paying levies, as they were less likely to encounter the tax collectors (Takeuchi 2014, 107).

To counteract these vulnerabilities and inequalities experienced by farmers, the Chinese government has recently taken further measures to improve the situation of the rural population. These include abolishing direct taxes on farmers and staple crops between 2002 and 2006, making direct payments to farmers since 2004, raising rice prices, and investing in 'building the new socialist countryside' since 2006. Additional policies in the last two decades have been specified in the Strategic Plan for Rural Revitalization 2018-2022 (Xinhua News Network 2018), as well as every annual 'No. 1 Document', which include further direct payments to grain producers. These are paid according to unit of land, in the framework of the 'agricultural support and protection subsidy', which also includes subsidies for purchasing agricultural inputs, including fertilizers, diesel fuel, and pesticides, as well as an improved seed variety subsidy. Moreover, there are subsidies for agricultural machinery, land consolidation, and agricultural infrastructure such as irrigation construction.

The amounts of the subsidies, as well as the ways in which they are calculated and actually received by farmers, differ according to specific commodities and localities.[26] For instance, Green Water villagers told me that they received a rather symbolic subsidy of 20 Yuan (less than 3 USD) in 2011 for every *mu* of paddy land they cultivated. However, this would merely correspond to about two percent of what Green Water villagers reported as the household income they earned from rice farming at that time. They certainly did not consider the payments sufficient, especially considering the high costs of inputs such as fertilizer, farm machinery, fuel, and seeds, in addition to the labour of family members who could earn much more in off-farm jobs. While payments on sown areas have constantly been increasing in recent years, in general Market Price Support is the most important support mechanism for farmers. This is achieved through both domestic policies, such as the minimum purchase price for rice, and various import and trade policies.[27] Nevertheless, for the villagers I interviewed, the incentives for farming – especially grain – remained low, and many saw it as vital to have part of the household working away from the farm, to augment income from farming.

Abolition of the collective welfare system

In addition to all the factors described above, the promotion and wide-scale adoption of modern farming technologies took effect against a background of the abolition of the collective rural welfare system. This has created an insecure, if not precarious situation for the rural population. In collective times, welfare was state planned and was delivered in rural China through the communes and production brigades (see Dixon 1981). As elsewhere, in Anren, the so-called 'barefoot doctors' and collective medical stations provided virtually free treatment to the villagers (ACGCC 1996, 604).

In the reform period of the 1980s, however, the state retreated from the welfare system, divesting responsibility for it to impoverished local governments. Health care became subject to marketization, so it has become costly and unaffordable for many rural Chinese people, especially when in the city, as in Zou Wenlu's case cited at the beginning of this chapter.[28]

26 For details see Gale (2013, 8-13); Huang, Wang, and Rozelle (2013, 126-127).

27 See OECD (2019, 177, 184). The OECD defines Market Price Support as 'an indicator of the annual monetary value of gross transfers from consumers and taxpayers to agricultural producers arising from policy measures creating a gap between domestic producer prices and reference prices of a specific agricultural commodity measured at the farm-gate level' (OECD 2003).

28 For details on this transformation see Duckett and Carrillo (2011, 1-6); Lora-Wainwright (2011, 107-109); UNDP (2014, 90); ACGCC (1996, 604).

Aware of such situations, the central government has introduced new rural welfare schemes such as the New Rural Cooperative Medical Scheme. Nevertheless, even though 97.1 percent of Anren's inhabitants had health insurance in 2011 (HPBS 2012a), regional differences in insurance cover persist (Duckett and Carrillo 2011, 9). These disparities are even more evident with regard to the minimum livelihood guarantee (*dibao*), a scheme which aims to reduce the number of people living in poverty. While only urban Chinese people were eligible at first, it has recently been extended to those in the countryside. In 2011, only 6.05 percent of people in rural Anren were covered by this scheme. Moreover, with a monthly subsidy of 68 Yuan (less than 10 USD), rural beneficiaries in Anren have been given less than half what urban citizens in Anren County have received (HPBS 2012b). These kinds of differences are more stark between rural areas and big cities outside Anren County.

In this situation of welfare gaps, it has become ever more necessary for people to rely on their own families for support, but this has been impacted by the strict implementation of birth control policies since 1978. Even though in October 2015, the Chinese government officially announced the end of its one-child policy, allowing two children per couple from 2016 onwards, the outcomes of the policy remain tangible. As a result, household sizes have shrunk to an average of 3.73 people, compared to the 5.8 people that constituted an average household in Anren in 1816. This means that the working population now has to support more old people and fewer children than before (ACGCC 1996, 97-101; NBSC 2019a, sec. 22-1). Together with the problems posed by an ageing society (see Cuhls et al. 2016), this has posed significant challenges to customary family care arrangements and fostered new insecurities (Madsen 1991, 674; Goh 2013, 4). In this context, substantial out-migration puts additional strain onto the situation.

The new urban economy and increased migration

Finally, the large-scale promotion and adoption of modern agriculture has occurred in a policy setting where rural and urban China have become more and more integrated economically through the establishment of a new urban economy and increased domestic migration, through loosened restrictions on population movement. Together, these policies play into the paddy field predicament by pushing farmers to migrate on the one hand, and making resource preservation at home mandatory in a context of migrant precarity on the other.

Economic development and liberalization since the 1980s, along with growing urbanization and industrialization, has created an urgent demand for cheap labour to work in city sectors such as manufacturing or construction. Meanwhile, millions of rural farmers have been set free from the land by the growing efficiencies of the household responsibility system, by mechanization, and by an overall expansion of the rural population. In Anren, more than half the labour force had been released from full-time agricultural labour by the mid-1990s. These people have turned either to local enterprises in an increasingly urbanized countryside, or to migrant work elsewhere (ACGCC 1996, 284-289), whilst still retaining their ties to the land.

With the loosening of migration policies, rural citizens have gradually been allowed – and to some degree encouraged – to move to the cities. The year 1984 marked the beginning of tremendous population movements when, for the first time since the 1950s, rural labourers were allowed to take up temporary work in the cities (State Council 1984). In 2014, the number of China's 'floating population', i.e. people commonly called migrants, reached a peak.[29] The number rose to 253 million and then declined slightly afterwards (NBSC 2019a, sec. 2-3). Consequently, since 2011, for the first time in its history, less than 50 percent of China's population live in rural areas, while most people who migrate for work move to areas outside their home province (OECD 2013, 119; Wang and Chen 2019). Thus, by the early 2000s, at least one member of every rural household was working away from their farm (Huang and Rozelle 2009, 104). Accordingly, during my research period in 2010 and 2011, Chinese farmers derived an average of 50 to 70 percent of their annual income from off-farm sources including migrant work and full-time or part-time non-agricultural activities.[30]

Even though migration generates income, it also has some pitfalls, as exemplified by the two cases of Mr. Wu and Zhou Wenlu at the start of this chapter. Despite the ongoing gradual abolition of the *hukou* system, rural Chinese people face numerous inequalities while working in cities, since the differentiation between rural and urban registration means that they are only granted temporary residence rights (Chan 2019). This implies that rural Chinese workers in many metropolitan areas today continue to suffer from a lack of access to the substantively better urban welfare system.

29 Here, 'floating population' refers to people who have left their place of registered residence for more than six months, except for intra-urban migrants.

30 Estimations vary, see Huang, Wang, and Qiu (2012, 17, 35); Zhou and Liu (2012); and Chen et al. (2014).

Moreover, schooling migrant children remains a problem, usually inducing rural children to attend schools in their hometowns, while their parents migrate for work (Ye 2018, 3). At the same time rural migrants, particularly older female migrants, experience various forms of inequality and job insecurity in the urban labour market (see Cooke 2011, 263). This leaves migrants in a highly precarious situation, making resource preservation at home all the more critical.

The insecurities of migrant work became especially apparent during the global financial crisis of 2008-2009, which principally affected migrant workers, especially those working in manufacturing and construction.[31] By the end of 2008, thousands of Chinese factories producing export products and relying primarily on migrant labour had closed. In early 2009, at least 23 million migrant workers, or more than 15 percent of all migrant workers in China, were estimated to have lost their jobs and gone home. To cope with the situation, the Chinese government encouraged the laid-off rural workers to return home in a campaign called 'back to the village to construct the new countryside'. Without alternatives, many of the returnees thus first turned to agriculture, although some struggled with this because they either lacked farming skills or had lost access to their land during their absence (Chan 2010, 665-668; Kong, Meng, and Zhang 2010, 234, 253). Nevertheless, for most of the returned unemployed workers, their home resources were crucial for coping with the crisis, and their land entitlements provided an important safety net. These land resources will possibly regain importance in view of the economic effects of the Covid-19 pandemic at the time of writing (Zhou 2020).

Taken together, the government's rural policies have been a major driver of the situation that many rural Chinese people have found themselves in, as well as more generally of the transformation of the countryside. They provide the institutional context in which the paddy field predicament has emerged, and the overarching structure within which the actions of individual farmers described in Chapters 4 and 5 take place. On the one hand, central and local

31 When I completed the first part of my fieldwork in Shanghai in April 2008, the topic did not come up prominently in my interviews. On the one hand, it was still too early, given that the crisis only hit China seriously after the collapse of the US investment bank Lehman Brothers in September 2008. On the other hand, the migrant workers I interviewed at that time were mainly working in the service sector, which was less affected by the crisis, although some of them reported reductions in their incomes. Moreover, Shanghai got off more lightly than other export-oriented eastern cities (Kong, Meng, and Zhang 2010, 247). When I began researching in Green Water Village in early 2011, those villagers who had lost their factory or construction jobs had already found new migrant jobs, while some had stayed home for other reasons, such as giving birth. In the meantime, they had relied on their home resources to get by.

government promotion of modern agriculture has liberated farmers to take on work away from their land. The resulting pressure to migrate has been further intensified by the policies of de-collectivization and marketization, and the new urban economy's demand for cheap labour, along with relaxed controls of population movement. On the other hand, the abolition of the collective welfare system and the precarious position of migrants in the cities have increased the importance of sustaining migrants' land assets in the countryside. As farmer Li Xiangshen succinctly summarized this situation: 'In the worst situation, we can at least return to our land and make a livelihood from the soil' (Wu 2016, 123).

References

ACBS, Anren County Bureau of Statistics. 2020. '安仁县2019 年国民经济和社会发展统计公报 Anren xian 2019 nian guomin jingji he shehui fazhan tongji gonggao (Anren County 2019 Statistical Communiqué of the Development of the People's Economy and Society)'. 13 March 2020. Accessed 3 July 2020. http://www.anrenzf.gov.cn/15/51015/51019/content_3126960.html.

ACGCC, 安仁县志编纂委员会 Anren Xianzhi Bianji Weiyuanhui (Anren County Gazetteer Compilation Committee). 1996. 安仁县志 *Anren Xianzhi* (*Anren County Gazetteer*). Beijing: Zhongguo Shehui Publishing.

—. 2011. 安仁县志 *Anren Xianzhi, 1989-2003* (*Anren County Gazetteer, 1989-2003*). Changsha: Hunan Renmin Publishing.

ACIGCC, 安仁县工业志编纂委员会 Anren Xian Gongyezhi Bianji Weiyuanhui (Anren County Industrial Gazetteer Compilation Committee). 1993. 安仁县工业志 *Anren Xian Gongye Zhi* (*Anren County Industrial Gazetteer*). Hefei: Huangshan Publishing.

Aubert, Claude. 2003. 'Landwirtschaftspolitik'. In *Das grosse China-Lexikon*, edited by Brunhild Staiger, Stefan Friedrich, and Hans-Wilm Schütte, 424-427. Darmstadt: Wissenschaftliche Buchgesellschaft.

Brandt, Loren, Susan H. Whiting, Linxiu Zhang, and Tonglong Zhang. 2017. 'Changing Property-Rights Regimes: A Study of Rural Land Tenure in China'. *The China Quarterly* 232: 1026-1049.

Bray, Francesca. 1984. *Agriculture*. In *Science and Civilisation in China*, Vol. 6, Part 2, edited by Joseph Needham. Cambridge: Cambridge University Press.

Central Government of the People's Republic of China. 2016. 'The 13th Five-Year Plan for Economic and Social Development of the People's Republic of China (2016-2020)'. Accessed 9 January 2020. https://en.ndrc.gov.cn/newsrelease_8232/201612/P020191101481868235378.pdf.

Chan, Kam Wing. 2010. 'The Global Financial Crisis and Migrant Workers in China: "There Is No Future as a Labourer; Returning to the Village Has No Meaning"'. *International Journal of Urban and Regional Research* 34 (3): 659-677.

—. 2019. 'China's Hukou System at 60: Continuity and Reform'. In *Handbook on Urban Development in China*, edited by Ray Yep, June Wang, and Thomas Johnson, 59-79. Cheltenham: Edward Elgar Publishing.

Chang, Te-Tzu. 2000. 'Rice'. In *The Cambridge World History of Food*, edited by Kenneth F. Kiple and Kriemhild Coneè Ornelas, 132-149. Cambridge: Cambridge University Press.

Chen, An. 2020. 'The Impact of Land Requisition on Peasant Life in China'. *Modern China* 46 (1): 79-110.

Cheng, Tiejun, and Mark Selden. 1994. 'The Origins and Social Consequences of China's Hukou System'. *The China Quarterly* 139: 644-668.

Cooke, Fang Lee. 2011. 'Labour Market Disparities and Inequalities'. In *China's Changing Workplace: Dynamism, Diversity and Disparity*, edited by Peter Sheldon, Sunghoon Kim, Yiqiong Li, and Malcolm Warner, 259-276. New York: Routledge.

Cortada, James W. 2012. *The Digital Flood: The Diffusion of Information Technology across the U.S., Europe and Asia*. Oxford: Oxford University Press.

Cuhls, Kerstin, Cheng Fan, Bruno Gransche, Erduana Shala, and Benjamin Teufel. 2016. *Zusammenfassender Endbericht schrumpfende Gesellschaften im Vergleich*. Karlsruhe: Fraunhofer ISI.

Day, Alexander F. 2013. *The Peasant in Postsocialist China: History, Politics, and Capitalism*. Cambridge: Cambridge University Press.

Dixon, John. 1981. *The Chinese Welfare System, 1949-1979*. New York: Praeger.

Duckett, Jane, and Beatriz Carrillo. 2011. 'China's Changing Welfare Mix: Introducing the Local Perspective'. In *China's Changing Welfare Mix: Local Perspectives*, edited by Beatriz Carrillo and Jane Duckett, 1-19. New York: Routledge.

Eisenman, Joshua. 2018. *Red China's Green Revolution: Technological Innovation, Institutional Change, and Economic Development Under the Commune*. New York: Columbia University Press.

FAO, Food and Agriculture Organization. 2019. 'Cereal Yield (Kg per Hectare): China'. The World Bank Data. 2019. Accessed 9 January 2020. https://data.worldbank.org/indicator/AG.YLD.CREL.KG?locations=CN&view=chart.

—. 2020. 'FAO in China: China at a Glance'. 2020. Accessed 9 January 2020. http://www.fao.org/china/fao-in-china/china-at-a-glance.

Gale, Fred. 2013. *Growth and Evolution in China's Agricultural Support Policies*. Washington DC: USDA Economic Research Service.

Gao, Yan, and Shailaja Fennell. 2018. *China's Rural-Urban Inequality in the Countryside*. Singapore: Springer.

Gianessi, Leonard, and Ashley Williams. 2011. 'China's Agricultural Success: The Role of Herbicides'. Washington DC: Crop Protection Research Institute.

Goh, Esther C. L. 2013. *China's One-Child Policy and Multiple Caregiving: Raising Little Suns in Xiamen*. London: Routledge.

Greenpeace China. 2010. 'The Real Cost of Nitrogen Fertilizer'. Accessed 31 July 2013. http://www.greenpeace.org/canada/Global/canada/report/2010/8/Summary-En.pdf.

Heger, Isabel. 2020. 'More Than "Peasants Without Land": Individualisation and Identity Formation of Landless Peasants in the Process of China's State-Led Rural Urbanisation'. *Journal of Current Chinese Affairs* 49 (1): 1-25.

Hexun. 2020. '农用化肥施用量 Nongyong huafei shiyong liang, 1984-2018 (Agricultural Chemical Fertilizer Consumption, 1984-2018)'. Accessed 13 July 2020. http://calendar.hexun.com/area/dqzb_430000_D0610000.shtml.

Ho, Peter, Jennifer H. Zhao, and Dayuan Xue. 2009. 'Access and Control of Agro-biotechnology: Bt Cotton, Ecological Change and Risk in China'. *The Journal of Peasant Studies* 36 (2): 345-364.

Ho, Ping-ti. 1959. *Studies on the Population of China, 1368-1853*. Cambridge, MA: Harvard University Press.

HPBS, Hunan Provincial Bureau of Statistics. 1984. *1982湖南省统计年鉴 Hunan sheng tongji nianjian* (*Hunan Province Statistical Yearbook 1982*). Changsha: Hunan Renmin Publishing.

—. 2012a. '郴州·安仁县2011 年国民经济和社会发展统计公报 Chenzhou, Anren xian 2011 nian guomin jingji he shehui fazhan tongji gongbao (Chenzhou, Anren County 2011 Statistical Communiqué of the Development of the People's Economy and Society)'. Accessed 9 March 2016. http://www.hntj.gov.cn/tjgb/xqgb/czgb/201204/t20120401_92218.htm.

—. 2013. '郴州·安仁县2012 年国民经济和社会发展统计公报 Chenzhou, Anren xian 2012 nian guomin jingji he shehui fazhan tongji gongbao (Chenzhou, Anren County 2012 Statistical Communiqué of the Development of the People's Economy and Society)'. Accessed 9 March 2016. http://www.hntj.gov.cn/tjgb/xqgb/czgb/201304/t20130417_99892.htm.

—. 2014. *Hunan Statistical Yearbook 2014*. Beijing: China Statistics Press.

—. 2018. *Hunan Statistical Yearbook 2018*. Beijing: China Statistics Press.

—. 2019. *Hunan Statistical Yearbook 2019*. Beijing: China Statistics Press.

Huang, Jikun, and Scott Rozelle. 2009. 'Agricultural Development and Nutrition: The Policies behind China's Success'. *Asian Journal of Agriculture and Development* 7 (1): 93-126.

Huang, Jikun, Shukun Wang, and Zhihua Xiao. 2017. 'Rising Herbicide Use and Its Driving Forces in China'. *The European Journal of Development Research* 29 (3): 614-627.

Huang, Jikun, Xiaobing Wang, and Huanguang Qiu. 2012. *Small-Scale Farmers in China in the Face of Modernisation and Globalisation.* London: International Institute for Environment and Development.

Huang, Jikun, Xiaobing Wang, and Scott Rozelle. 2013. 'The Subsidization of Farming Households in China's Agriculture'. *Food Policy* 41: 124-132.

Kong, Sherry Tao, Xin Meng, and Dandan Zhang. 2010. 'Impact of Economic Slowdown on Migrant Workers'. In *China's New Place in a World in Crisis*, edited by Ross Garnaut, Ligang Song, and Wing Thye Woo, 233-260. Canberra: ANU E Press.

Kong, Sherry Tao, and Jonathan Unger. 2013. 'Egalitarian Redistributions of Agricultural Land in China through Community Consensus: Findings from Two Surveys'. *The China Journal* 69: 1-19.

Labrada, Ricardo. 2003. 'The Need for Improved Weed Management in Rice'. In *Sustainable Rice Production for Food Security.* Rome: Food and Agriculture Organization of the United Nations.

Li, Jiming, Yeyun Xin, and Longping Yuan. 2009. 'Hybrid Rice Technology Development: Ensuring China's Food Security'. IFPRI Discussion Paper 00918, 2020 Vision Initiative.

Li, Kwok-sing. 1995. *A Glossary of Political Terms of the People's Republic of China.* Translated by Mary Lok. Hong Kong: The Chinese University Press.

Liang, Shumin. 2010. 'Agricultural Pollution Zoning and Comprehensive Control Measures'. In *The China Environment Yearbook, Volume 4: Tragedy and Hope – From the Sichuan Earthquake to the Olympics*, edited by Dongping Yang and Friends of Nature, 145-160. Leiden: Brill.

Lin, Justin Yifu. 1991. 'The Household Responsibility System Reform and the Adoption of Hybrid Rice in China'. *Journal of Development Economics* 36 (2): 353-372.

Lin, Justin Yifu, Ran Tao, and Mingxing Liu. 2007. 'Rural Taxation and Local Governance Reform in China's Economic Transition: Origins, Policy Responses and Remaining Challenges'. Stanford Center for International Development, Working Paper No. 317.

Long, Hualou. 2014. 'Land Use Policy in China: Introduction'. *Land Use Policy* 40 (September): 1-5.

Lora-Wainwright, Anna. 2011. '"If You Can Walk and Eat, You Don't Go to Hospital": The Quest for Healthcare in Rural Sichuan'. In *China's Changing Welfare Mix: Local Perspectives*, edited by Beatriz Carrillo and Jane Duckett, 104-125. New York: Routledge.

Madsen, Richard. 1991. 'The Countryside under Communism'. In *The Cambridge History of China*, Vol. 15, Part 2, edited by Roderick MacFarquhar and John K. Fairbank, 619-681. Cambridge: Cambridge University Press.

MoA, Ministry of Agriculture of the People's Republic of China. 1989. 中国农村经济统计大全 *Zhongguo nongcun jingji tongji daquan, 1949-1986* (*Comprehensive*

Collection of the Statistics of China's Rural Economy, 1949-1986). Beijing: Agricultural Press.

—. 2012. 'Excerpt of China's 12th Five-Year Plan: Agriculture Part'. Accessed 21 May 2015. http://english.agri.gov.cn/hottopics/five/201301/t20130115_9545.htm.

—. 2015. 'SCIO Briefing on Agricultural Modernization'. 3 February 2015. Accessed 6 June 2016. http://english.agri.gov.cn/hottopics/cpc/201502/t20150204_24960.htm.

—. 2016. 'China's No.1 Central Document Focuses on Agriculture for 13th Consecutive Year'. Accessed 28 January 2016. http://english.agri.gov.cn/news/dqnf/201601/t20160128_164966.htm.

Murphy, Rachel. 2006. 'Citizenship Education in Rural China: The Dispositional and Technical Training of Cadres and Farmers'. In *Chinese Citizenship: Views from the Margins*, edited by Vanessa L. Fong and Rachel Murphy, 9-26. London: Routledge.

NBSC, National Bureau of Statistics of China. 2015. 'National Data: Annual by Province'. Accessed 20 July 2015. http://data.stats.gov.cn/english/mapdata.htm?cn=E0103&zb=A0301.

—. 2018. *China Statistical Yearbook 2018*. Beijing: China Statistics Press.

—. 2019a. *China Statistical Yearbook 2019*. Beijing: China Statistics Press.

—. 2019b. 'National Data: Annual by Province'. Accessed 6 January 2020. http://data.stats.gov.cn/english/easyquery.htm?cn=E0103.

NBSC Rural Social and Economic Investigation Division. 2010. *2010*中国农村统计年鉴 *Zhongguo nongcun tongji nianjian* (*2010 Statistical Yearbook of the Chinese Countryside*). Beijing: China Statistics Press.

NPC, The National People's Congress of the People's Republic of China. 2002. 'Law of the People's Republic of China on Land Contract in Rural Areas'. Beijing: Database of Laws and Regulations.

—. 2004. 'Land Administration Law of the People's Republic of China'. Beijing: Database of Laws and Regulations.

—. 2007. 'Property Law of the People's Republic of China'. Beijing: Database of Laws and Regulations.

—. 2009. 'Law of the People's Republic of China on Mediation and Arbitration of Disputes over Rural Land Contract and Management'. Beijing: Database of Laws and Regulations.

OECD. 2003. 'OECD Glossary of Statistical Terms: Market Price Support (MPS)'. 7 March 2003. Accessed 6 May 2020. https://stats.oecd.org/glossary/detail.asp?ID=1600.

—. 2005. *OECD Review of Agricultural Policies: China*. Paris: OECD Publishing.

—. 2013. *Agricultural Policy Monitoring and Evaluation 2011: OECD Countries and Emerging Economies*. Paris: OECD Publishing.

—. 2019. *Agricultural Policy Monitoring and Evaluation 2019*. Paris: OECD Publishing.

Oi, Jean. 1989. *State and Peasant in Contemporary China: The Political Economy of Village Government*. Berkeley: University of California Press.

Ortner, Sherry B. 2006. *Anthropology and Social Theory: Culture, Power, and the Acting Subject*. Durham: Duke University Press.

Oxfeld, Ellen. 2017. *Bitter and Sweet: Food, Meaning, and Modernity in Rural China*. Oakland: University of California Press.

Patton, Dominique. 2014. 'China Looking to Curb Fertilizer, Pesticide Use'. *Reuters*. 5 December 2014. Accessed 6 December 2014. https://www.reuters.com/article/us-china-grain-fertilizers/china-looking-to-curb-fertilizer-pesticide-use-idUSKCN0JJ0SO20141205.

Perdue, Peter C. 1987. *Exhausting the Earth: State and Peasant in Hunan, 1500-1850*. Cambridge, MA: Harvard University Press.

Perkins, Dwight H. 1991. 'China's Economic Policy and Performance'. In *The Cambridge History of China*, Vol. 15, Part 2, edited by Roderick MacFarquhar and John K. Fairbank, 475-539. Cambridge: Cambridge University Press.

Potter, Sulamith Heins, and Jack M. Potter. 1990. *China's Peasants: The Anthropology of a Revolution*. Cambridge: Cambridge University Press.

Rawski, Evelyn Sakakida. 1972. *Agricultural Change and the Peasant Economy of South China*. Cambridge, MA: Harvard University Press.

Santos, Gonçalo. 2011. 'Rethinking the Green Revolution in South China: Technological Materialities and Human-Environment Relations'. *East Asian Science, Technology and Society* 5 (4): 479-504.

Sargeson, Sally. 2012. 'Villains, Victims and Aspiring Proprietors: Framing "Land-Losing Villagers" in China's Strategies of Accumulation'. *Journal of Contemporary China* 21 (77): 757-777.

Schmalzer, Sigrid. 2016. *Red Revolution, Green Revolution: Scientific Farming in Socialist China*. Chicago: University of Chicago Press.

Shapiro, Judith. 2001. *Mao's War against Nature: Politics and the Environment in Revolutionary China*. Cambridge: Cambridge University Press.

Shen, Xiaobai. 2010. 'Understanding the Evolution of Rice Technology in China: From Traditional Agriculture to GM Rice Today'. *The Journal of Development Studies* 46 (6): 1026-1046.

Smil, Václav. 2004. *China's Past, China's Future: Energy, Food, Environment*. New York: RoutledgeCurzon.

Song, Yiching. 1998. '"New" Seed in "Old" China: Impact of CIMMYT Collaborative Programme on Maize Breeding in South-Western China'. PhD diss., Wageningen University.

State Council. 1984. '国务院关于农民进入集镇落户问题的通知 Guowuyuan guanyu nongmin jinru jizhen luohu wenti de tongzhi (Notice of the State Council on the Issue of Peasants Entering Towns and Settling Down)'. Notice No. 141, 13 October 1984.

Statista. 2018. 'China: Pesticide Use by Region 2016'. Accessed 12 December 2019. https://www.statista.com/statistics/300970/china-pesticide-use-by-region/.

Stavis, Benedict. 1974. *Making Green Revolution: The Politics of Agricultural Development in China*. Ithaca, NY: Rural Development Committee, Cornell University.

—. 1978. *The Politics of Agricultural Mechanization in China*. Ithaca, NY: Cornell University Press.

Stross, Randal E. 1986. *The Stubborn Earth: American Agriculturalists on Chinese Soil, 1898-1937*. Berkeley: University of California Press.

Takeuchi, Hiroki. 2014. *Tax Reform in Rural China: Revenue, Resistance, and Authoritarian Rule*. New York: Cambridge University Press.

Tian, Wei-Ming, and Zhang-Yue Zhou. 2018. 'Developments in China's Grain Policies'. In *Grains in China: Foodgrain, Feedgrain and World Trade*, edited by Zhang-Yue Zhou and Wei-Ming Tian, 11-23. Oxon: Routledge.

Tregear, Thomas R., and Victor C. Falkenheim. 2015. 'Hunan'. In *Encyclopædia Britannica*. Accessed 20 July 2015. http://academic.eb.com/EBchecked/topic/276448/Hunan.

TU Berlin, Center for Cultural Studies on Science and Technology in China. 2017. 'DFG-Projekt "Making Technology Appropriate: Technology Transfer from Germany to China. The Case of Steam and Ordnance Technologies, 1860-1980"'. Accessed 7 February 2018. http://www.china.tu-berlin.de/menue/forschung/projekte/making_technology_appropriate/.

UNDP, United Nations Development Programme. 2014. *Human Development Report 2014 – Sustaining Human Progress: Reducing Vulnerability and Building Resilience*. New York: UNDP.

Wang, Dehua 王德华. 2014. '湖南省失地农民社会保障问题调查研究 Hunan sheng shidi nongmin shehui baozhang wenti diaocha yanjiu (A Study of the Social Security of Landless Peasants in Hunan)'. 怀化学院学报 *Journal of Huaihua University* 33 (2): 75-77.

Wang, Nora. 2003. 'Hunan'. In *Das grosse China-Lexikon*, edited by Brunhild Staiger, Stefan Friedrich, and Hans-Wilm Schütte, 319-320. Darmstadt: Wissenschaftliche Buchgesellschaft.

Wang, Zhiling, and Lu Chen. 2019. 'Destination Choices of Chinese Rural-Urban Migrant Workers: Jobs, Amenities, and Local Spillovers'. *Journal of Regional Science* 59 (3): 586-609.

Watson, Andrew. 2001. 'Agriculture (Policies, Organizations, Units)'. In *Dictionary of the Politics of the People's Republic of China*, edited by Colin Mackerras, Donald H. McMillen, and Andrew Watson, 57-61. London: Routledge.

Wu, Chunyuan 伍春元, ed. 2010. 益相史话 *Yixiang shihua* (*Historical Narrative of Yixiang*). Beijing: Jiuzhou Press.

Wu, Yi. 2016. *Negotiating Rural Land Ownership in Southwest China: State, Village, Family*. Honolulu: University of Hawai'i Press.

Wu, Yiyun, Xican Xi, Xin Tang, Deming Luo, Baojing Gu, Shu Kee Lam, Peter M. Vitousek, and Deli Chen. 2018. 'Policy Distortions, Farm Size, and the Overuse of Agricultural Chemicals in China'. *PNAS* 115 (27): 7010-7015.

Xinhua. 2017. 'China Considers Revising Rural Land Contract Law'. *China Daily*. 31 October 2017. Accessed 13 December 2017. https://www.chinadaily.com.cn/china/2017-10/31/content_33937185.htm.

Xinhua News Agency. 2015. 'China's Focus on Agricultural Reform for 12th Year in a Row'. 3 February 2015. Accessed 31 October 2017. http://english.agri.gov.cn/news/dqnf/201502/t20150203_24957.htm.

Xinhua News Network. 2018. '中共中央 国务院 印发"乡村振兴战略规划（2018—2022年）" Zhonggong zhongyang guowuyuan yinfa "Xiangcun zhenxing zhanlüe guihua (2018-2022 nian)" (The Central Committee of the Communist Party of China and the State Council Issue the "Strategic Plan for Rural Revitalization (2018-2022)"'. 26 September 2018. Accessed 7 August 2020. http://www.xinhuanet.com/politics/2018-09/26/c_1123487123.htm.

—. 2019. '全国人民代表大会常务委员会关于修改"中华人民共和国土地管理法"、"中华人民共和国城市房地产管理法"的决定 Quanguo renmin daibiao dahui changwu weiyuanhui guanyu xiugai "Zhonghua renmin gongheguo tudi guanli fa", "Zhonghua renmin gongheguo chengshi fangdichan guanli fa" de jueding (Decision of the Standing Committee of the National People's Congress on Amending the Land Management Law and the Law on Urban Real Estate Management of the People's Republic of China)'. 26 August 2019. Accessed 6 January 2020. http://www.xinhuanet.com/politics/2019-08/26/c_1124923935.htm.

Ye, Jingzhong. 2015. 'Land Transfer and the Pursuit of Agricultural Modernization in China'. *Journal of Agrarian Change* 15 (3): 314-337.

—. 2018. 'Stayers in China's "Hollowed-out" Villages: A Counter Narrative on Massive Rural-Urban Migration'. *Population, Space and Place* 24: 1-10.

Zhang, Ze Pu. 2003. 'Development of Chemical Weed Control and Integrated Weed Management in China'. *Weed Biology and Management* 3 (4): 197-203.

Zhou, Cissy. 2020. 'Coronavirus Has Hit China's Migrant Workers Harder than Sars and the Financial Crisis, but Worst yet to Come'. *South China Morning Post*, 25 May 2020. Accessed 29 May 2020. https://www.scmp.com/economy/china-economy/article/3085904/coronavirus-has-hit-chinas-migrant-workers-harder-sars-and.

Zhou, Xuesong, and Ying Liu. 2012. 'Peasants' Income Structure Evolution and Its Enlightenment'. *Chinese Agricultural Science Bulletin* 28 (14): 210-213.

2 Rice Knowledge Systems in Transition

Abstract

This chapter considers how paddy field knowledge is transmitted and how this has changed over recent decades in China, in order to better understand the problems that farmers face at the nexus of rice farming and rural-urban migration, and the options they can call on to deal with their situation. The chapter argues that there has been a complex reconfiguration of the repertoire of rice farming knowledge. On the one hand, this has created challenges for the future preservation of the paddy fields in the Chinese countryside, such as deskilling in the young migrant generation. On the other hand, it has provided peasants with an extended repertoire of knowledge they can use to handle their paddy field farming-migration predicament.

Keywords: China, rice farming, transformation of knowledge repertoires, agricultural deskilling, rural-urban migration, socio-technical system

I became acquainted with Grandpa Zhou while I was observing a tofu-making process in Green Water Village. Several villagers had gathered behind the house of a family who provided the use of an electrically-powered mill, as well as water, fire, and other tools, to help others transform their soy beans into tofu, in exchange for cash. Some people also took their rice there, which the mill owners turned into rice flour. While Grandpa Zhou smoked a cigarette and waited for his bean curd, we began talking. He was calm and friendly. His face was marked by the sun, the hardships of farming, and the famine of the 1960s. He was nearing his seventies and expressed worries about the young migrant population's lack of farming knowledge. Referring to the farmers' calendar,[1] which is the structural basis for all agricultural activities in Green Water Village and other parts of China, he stated:

1 See Appendix II and the section on the agricultural calendar below.

Kaufmann, Lena, *Rural-Urban Migration and Agro-Technological Change in Post-Reform China*. Taylor & Francis Group 2021
DOI: 10.5117/9789463729734_CH02

> Young people do not understand the 24 solar terms [i.e. reference points for farming, which divide the year into 24 two-week periods, two per month], because when they finish school they immediately [leave to] work [in the city]. (Personal interview, 28 January 2011.)

Caixia saw things differently from Grandpa Zhou. Born in 1995, she had been a left-behind twelve-year-old when I first met her in Shanghai in 2007. At that time, she was living all by herself in her parents' empty house back home, surviving on instant noodles and dinners at her father's sister's house. She spent her summer holidays in her parents' small street restaurant in Shanghai, crying bitterly when she had to leave them again. She put much of her energy into drawing and was listening to Korean pop music, dreaming of becoming a famous singer when she grew up.

Ten years later, she was in her early twenties and had just graduated as a nurse, being about to move to Shanghai to work in a hospital. Having acquired only limited rice farming experience from living with her grandmother when she was little, in her view:

> The people who stay home, only 50-60 year-olds, are all people whose families had no money to support their education, who have no culture (*wenhua*) and who don't have any experience of leaving home to make their own way in society, which is why they have to stay home and work in farming [...]. After all, they are not capable of doing anything else. (Text conversation, 5 September 2017.)

Emphasizing again that those old farmers lacked any culture, she added that it was a common perception at home that only those who were not capable stayed behind, while education was the only way to improve your life (video conversation, 5 September 2017).[2] Her words reminded me of similar discourse I have heard about farming and education, not just from young migrants, but also from urban Shanghai residents and on Chinese television. They are part of a wider pervasive discourse about 'uncivilized' and 'backward' Chinese farmers lacking human 'quality' or *suzhi* (Anagnost 1997, 76; Murphy 2004, 2; Schmalzer 2016, 107).

The experiences and opinions of Grandpa Zhou and the young migrant woman Caixia indicate a transformation of the Chinese rice knowledge system that has accompanied the policy-led modifications described

2 See also Croll and Ping (1997, 145).

in the preceding chapter. In this chapter, I argue that the profound transformation of the Chinese rural knowledge system has led to a complex reconfiguration of the repertoire of knowledge and skills that farmers in the in the twenty-first century use to manage their fields. This reconfiguration poses several challenges for the future preservation of rice fields. At the same time, however, it is precisely the mix of rapid technical transformation alongside the residues of what would be called 'old technologies' in China that has created a unique situation, offering farmers more strategic options to deal with their paddy field predicament. Such residues survive or stay valid due to the reluctance of the human body and cognition to quickly change working habits, and due to the particularities of 'skills in aging' (see Marchand 2014; VMZ 2017). In fact, farmers can now draw on an extended repertoire of knowledge, including that of pre-industrial, manual techniques, mechanization, chemicalization, and hybrid seeds.

I suggest that what is happening is a generational issue. As the examples of Grandpa Zhou and Caixia show, different generations view this issue from divergent perspectives. Moreover, their situations differ in practice. On the one hand, senior, left-behind farmers like Grandpa Zhou continue to perform, extend, test, ascertain, and refine their everyday technologies as well as their trained bodies through everyday practice. They have experience-saturated practical farming knowledge and skills. As scholars of skilled practice have shown, such embodied knowledge can only be acquired and embodied through many years of practice and body-sensual and cognitive experience; it cannot simply be learned in a short time frame (Sigaut 1994; Ingold 2000; Marchand 2010).

On the other hand, young migrants such as Caixia largely forget and, ultimately, lose farming knowledge, which is replaced in their consciousness by new knowledge gained for, and from, life in the city. This is a cohort of young people who are individually calibrated for contemporary life with potentials that only become visible when they are actually challenged to deal with their land resources. As more and more young people migrate early and long-term, this obviously poses challenges to the 'skill-producing group' (Sigaut 1994) that sustains agricultural production in China, especially since their children might eventually lack the necessary farming knowledge and skills to continue maintaining their paddy fields for themselves in the future. While in the following chapters I shift my focus to the older generation of skill-holders, in this chapter I take a closer look at the transformation of gradually acquired, embodied farming skills in the young village generation.

Transformation of agricultural knowledge transmission

To model the transformation of rice field knowledge, I draw on Fredrik Barth's (2002) model of knowledge transmission, which looks at the interplay of the three 'faces' of knowledge: corpus, medium, and social organization. The model is useful for understanding a crucial phase of the realignment of the Green Water rice knowledge system since the 1980s, providing a clearer picture of its challenges and opportunities.

To render the transformations more clearly, I discuss three distinct models of knowledge transmission, and accentuate their differences. As the political events in twentieth-century China have left a strong imprint on the rural system of knowledge transmission, my approach follows the standard periodization of modern Chinese history into: late imperial China and the Republic of China before 1949, the period of high socialism until the end of the 1970s, and the reform period since the 1980s. I call the three models respectively the pre-collectivization household system of knowledge transmission, the collective system of knowledge transmission, and the post-reform household system of knowledge transmission (see Table 1). My modelling focuses mainly on the latter two transition phases, since the paddy field predicament has emerged since then.

The pre-collectivization system (before 1949)

In what presents a very rough sketch, in the first half of the twentieth century Barth's three faces of knowledge may be identified as follows. A major part of the corpus consisted of the conventional knowledge that was developed and shared amongst farmers, much of which was embodied, contextualized knowledge. This comprised knowledge about manual weeding and harvesting techniques, as well as substantial seed selection and breeding knowledge. Notably, much of this knowledge was experience-saturated, preventive knowledge. Moreover, rural households were well integrated into the wider economic system in a diversified economy. Accordingly, the corpus consisted of sufficient knowledge about other economic activities that could be combined with rice farming and migration (Rawski 1972; Bray 1984; Skeldon 1996; Roberts 1997).

Rituals, the farmers' calendar and proverbs were all highly relevant media for representing this knowledge (XT 1988; CZ 1988; ACGCC 1996; Wu 2010). Moreover, the human body and proven local technologies were also important material media of knowledge. Apart from this, since both the state and academia extracted vernacular knowledge to standardize

Table 1 Simplified overview of the changing Chinese system of rice knowledge transmission

Rice knowledge system	Corpus	Media	Social organization
Pre-collectivization (before 1949)	embodied, contextualized conventional farmers' knowledge; manual techniques; seed and breeding knowledge; multifaceted knowledge	farmers' bodies; proven local technologies; customs, proverbs, farmers' calendar; agricultural treatises	households, agnatic groups, lineages, irrigation groups; gendered norms: women inside/men outside; local officials
Collective (1950s to early 1980s)	scientific knowledge; one-sided rice farming knowledge	Chinese 'Agricultural Technology Extension System', state-owned experimentation farms, model fields; newspapers, loudspeakers, slogans; Green Revolution technologies; body (collective)	collective, communes, brigades, production teams; scientists, local Department of Agriculture
Post-reform (mid-1980s to today)	conventional and scientific knowledge; knowledge about wider economy and migration	pre-industrial and industrial technologies; extension bureaus; body (individual); some customs and proverbs; TV, Internet	farmers' households and individuals vs. state, scientists, seed companies; feminization and greying of agriculture; agricultural deskilling of young migrants

into auditable scientific knowledge, there was a scholarly engagement with agricultural knowledge conservation and transmission, in the form of agricultural treatises. These were compiled by government officials, who were also trying to introduce and promote new methods of farming (Bray 1984; Perdue 1987; Eyferth 2009).

Among farmers, rice knowledge was transmitted among household members, within agnatic groups, lineages, irrigation and transplanting groups – in short, within specialized 'skill-producing groups' and 'communities of practice' (Sigaut 1994; Lave and Wenger 1991). Specific gendered norms prevailed, which influenced the division of labour and knowledge, asserting that men should ideally operate outdoors in agriculture, with women confined to the internal sphere of production. Accordingly, the occupation of the farmer and the bureaucrat characterized what was considered essentially male and men's knowledge. Women, in turn, were associated with textile production (Jacka 1997; Bray 2013).

The collective system (1950s to early 1980s)

Between the 1950s and early 1980s, Barth's three faces of knowledge were reflected in a collectivized, top-down system of rice knowledge transmission. Roughly speaking, this evolved into a knowledge system that was organized collectively, in which farmers, cadres, and scientists collaborated on the agenda of a new Chinese technological modernity (Schmalzer 2016). In this context, the government actively promoted scientific agricultural knowledge. While there had already been attempts to standardize agriculture in imperial China (see Perdue 1987; Bray 2008), in the collective system this occurred in line with unprecedented national endeavours to intensify and standardize production, to increase rice yields. Here, many of the media of transmission were strictly controlled. These ranged from state-owned experimental farms and model fields, to new farming schedules propagated through newspapers and loudspeakers, and collectively-owned industrial farm technologies introduced from the outside (see Chapter 1). Much of this happened in the framework of consolidating a Chinese Agricultural Technology Extension System, including scientists being sent to the countryside (Gao and Zhang 2010; Schmalzer 2016).

In this collective system, the corpus of knowledge surrounding the resource of paddy field up to the 1980s can be roughly characterized as follows. The corpus of rice knowledge was greatly transformed by the wide-scale introduction and promotion of hybrid seeds, farm chemicals, and machines from the mid-1960s onwards. Furthermore, with a strict household registration system in place that prevented out-migration, the corpus of knowledge in the collective system consisted solely of the knowledge needed to survive in the countryside. The focus on rice monoculture in Hunan Province encouraged one-sided rice farming knowledge, rather than multifaceted knowledge about the wider economy (see Chapter 1).

Farming books, newspapers, loudspeakers, slogans, radio broadcasts, films, and Green Revolution technologies were rapidly added as new media of knowledge transmission. At the same time, with the introduction of new rice varieties, the media of customs were condemned as superstitious, and the older cultivation schedules recorded in the farmers' calendar became less important (see Chapter 1; ACGCC 1996).

Concerning social organization, knowledge was organized in communes, brigades, and production teams. Everybody – male and female – was expected to labour in the fields (Schmalzer 2016, 120-121). Much rice cultivation knowledge was spread collectively, while some of the new farming knowledge was available mainly to scientists and local officials working in the local Department of Agriculture or in newly-created institutions such as plant hospitals and experimental farms (see Chapter 1).

The post-reform system (mid-1980s to today)

With the re-introduction of household farming and the allocation of individual land-use rights in the mid-1980s, the rural knowledge system once more underwent a remarkable shift. From the late 1980s onwards in rural China a knowledge system has prevailed which is rooted in households operating increasingly independently. This system is, nevertheless, still closely directed by the state, especially in terms of post-Green Revolution Chinese scientific farming technologies. In addition, this knowledge system is increasingly being guided by market demands in a gradually liberalizing rural market. Despite official regulations, farming households can take economic decisions more individually and move more freely. This implies that rural knowledge is again extending towards diversification and the outside world.

In the twenty-first century, the corpus of conventional farmers' applied and preventive knowledge mainly lies in the hands of senior villagers, the 'inheritor[s] of traditional farming knowledge' (He and Ye 2014, 364). Meanwhile, young villagers often lack the substantial agricultural knowledge that their parents and, especially, their grandparents' generation had. Instead, young people have expanded their corpus of knowledge towards the wider urban economy. Nevertheless, in Chinese village society today, some deeper buried, long-term knowledge still remains regarding how to uphold a system of rice farming under various conditions.

As for the media of knowledge transmission, both conventional and post-Green Revolution technologies are available nowadays. It is notable that a large share of control over the knowledge concerning the latter technologies lies in the hand of government agents and researchers. Moreover, knowledge is inscribed in the form of skill within individual farmers' bodies. As Caixia, the young migrant nurse explained, because the old farmers have engaged in farm work for such a long term, their bodies have adapted to the field environment (video conversation, 5 September 2017).

Despite the loss of certain ritual practices, such as the 'burning of seedbed paper' (see the sections on cropping patterns and rice rituals below), other ritual practices are being revived. Besides, proverbs continue to exist as a medium of potential knowledge conservation (see Chapter 3). Finally, extension services, agricultural demonstration sessions and school lessons, leaflets, notice boards, wall slogans, songs, newspapers, radio, TV, smartphones, and the Internet are additional media for knowledge distribution. State cadres use a range of these media for the technical education and guidance of farmers, carried out in the framework of a more general citizenship education (see Murphy 2006, 18-19).

Today, paddy field knowledge is organized and spread more individually, in the framework of a household economy with individual land-use rights and clear gendered patterns. In terms of gender, the social organization of knowledge has clearly been affected by the out-migration of male and/or young villagers. Up until the new millennium, migrants were mainly men, then women also joined in. However, women commonly return home for marriage and childbirth and migrate again when their child is old enough to live with their grandparents. Once the elderly grandparents themselves require old-age care, it is usually women, as customary care-givers, who take over this task at home.[3] This situation has contributed to what is commonly called the 'feminization of agriculture' in China, and to a concentration of everyday agricultural knowledge in female hands (Jacka 1997; Meng 2014; Kaufmann 2019). Nevertheless, scientists, government experts, and seed companies also control some of this knowledge.

With regard to preserving paddy fields as a resource, the transition from the tightly-controlled collective system of knowledge transmission to the more individual, yet technologically state-led and market-guided

3 For details on this gendered division of labour, see Bossen (1994; 2011); Gaetano and Jacka (2004); Jacka (2006); and Liu (2017).

post-reform system of knowledge transmission clearly brings a range of contradictions and challenges, especially around the role of the state and the distribution of knowledge. In both the collective and the post-reform systems, the state strongly influenced knowledge distribution through its policies. In the first, it did so directly through collectivization and its top-down approach to modernizing farming practices. In the second, its influences were somewhat more indirect. First, knowledge transmission was affected by the widespread promotion of post-Green Revolution farming technologies. Second, the introduction of household-based farming brought with it a change in the organizational units of knowledge transmission. Third, the loosening of strict migration restrictions impacted on the flow of experts and their fields of expertise. In the post-reform system, these factors have led to a different distribution of knowledge, control over this knowledge, and its transmission.

Furthermore, there has been a twofold transformation of knowledge distribution and scope. First, farming knowledge has changed from being held entirely by farmers, towards young farmers having increased wider economic and migrant knowledge at the expense of farming knowledge. Second, scientific knowledge has changed from being held by and distributed to farmers to now being owned by scientists and government experts.

Transformation of the repertoire of knowledge

Against this background of a more general transformation of the rural knowledge system, it is interesting to take a closer look at one specific case of transforming the repertoire of rice farming knowledge, as documented in Green Water Village and in the relevant literature on rice farming.

As the cases of Grandpa Zhou and young migrant nurse Caixia suggest, a generational difference seems to exist in relation to not only the distribution of practical farming skills, but also how this issue is perceived, and in terms of proposed solutions. While senior farmers like Grandpa Zhou are very concerned about it (see also Yuan and Niehof 2011; Meng 2014, 77), Caixia does not consider it problematic. She does acknowledge that some specific farming knowledge has been acquired and transmitted from generation to generation, however, in her view, this is no longer needed. Therefore, she does not believe the lost skills are a problem, because everybody can manage to farm somehow, even if that means asking other farmers, or even a company, to do the actual farm work, or switching to other crops, or using machines such as transplanters (video conversation, 5 September 2017).

In contrast, Grandpa Zhou believes in re-learning and improving the skills acquired earlier. When I asked what his migrant son – the construction worker Zhou Wenbao quoted in the Introduction – will do when he returns to the village lacking farming knowledge, he simply answered: 'He will learn again when he returns (*hui lai zai xuexi*)' (personal interview, 28 January 2011). It should be noted that Grandpa Zhou's son belongs to the middle generation, i.e. he is about the same age as Caixia's father. This generation is somewhere in-between the transition, in terms of their age and their skill repertoire. In 2011, Grandpa Zhou's son Wenbao had migrated only five years earlier, meaning that he had almost forty years of rice farming experience. This does not compare to Grandpa Zhou but is, nevertheless, much richer and longer than Caixia's farming experience, as she only spent her early childhood years in the village with her grandmother. Grandpa Zhou's son therefore has a solid foundation to build upon when he 'learns again'.

Caixia's account not only echoes the views of policy makers and agronomists, who see economies of scale and modern agro-technology as the solution, dependent on new types of scientific knowledge and skills, but it also indicates the discursive side of skill. It mirrors what Harry Braverman (1974) has described with regard to deskilling in the industrial workplace, which values educational skills over practical skills. Together, Caixia's and Grandpa Zhou's interpretations hint to the complex reconfiguration of the repertoire of rice field knowledge that has been taking place in recent decades. Much of this knowledge is reflected in specific agricultural practices (Schippers 2014).

Soil knowledge

In rice farming, knowledge about the soil is related, first, to assessing soil characteristics concerning its quality and improvement, and second, to physically working with and experiencing farming the soil. Much of the paddy soil knowledge ultimately concerns questions of the short-term and long-term usage of the paddy field resource in view of its ecological particularities (see Netting 1993, 50-51).

According to the Chinese knowledge system, knowledge about soil characteristics is, first of all, condensed in classificatory terms denoting different types of fields, an example of soil knowledge being represented through the medium of terminology (Barth 2002). As Gene Wilken's work shows, classificatory systems allow 'farmers to deal in a meaningful way with the otherwise unwieldy forces of nature [...,] describ[ing] propensities for change or manipulation, which in turn reflect levels of technology and management skills' (Wilken 1987, 4-5). There does not seem to have been a major transformation in the repertoire of classificatory knowledge in this regard, as it is generally difficult to alter a plot's quality thoroughly.

Instead, the differences lie in the objective of classification. While farmers were obliged to follow the government's target of immediate productivity under the collective system, in the pre-collective and post-reform household systems, farmers' short-term and long-term goals are more complex. In the collective system, the assessment of soil quality was fuelled by the national aim to reclaim additional farmland. It focused on enhancing the collectively cultivated and owned fields' characteristics in order to intensify the rice production that was intended to spur on China's economic development and feed the nation (ACGCC 1996, 295-296).

The post-reform household system does not challenge the basic conceptual distinction between 'low production fields' (*dichan tian*) and 'high production fields' (*gaochan tian*) (ibid.). Local officials used similar categories when land was allocated to villagers in the de-collectivization framework of the early 1980s. At that time, rice fields were classified into first, second, and third-class fields, and everybody was allocated a certain number of each. For example, Zhou Wenlu and his household, consisting of him, his wife, two daughters, and one son, have an allocated amount of 1.8 *mu* of first-class fields and 1.2 *mu* of third-class fields. During the allocation, his son was not taken into consideration, because he was born outside China's one-child policy, which was, in fact, officially a two-child policy in rural areas if the first child was a girl.

The official and local categories were virtually the same in 2011. Green Water villagers distinguished between fertile, literally 'fat fields' (*fei tian*) and unfertile or 'lean fields' (*shou tian*). Whereas fertile fields were suitable for growing rice (*zuo daogu*, literally 'doing paddy rice'), unfertile fields were cultivated as dry fields for growing vegetables and other crops (*zuo hantu*, literally 'doing dry earth'). As Zhou Wenlu explained, this classification into fertile and unfertile fields was related to each individual plot's capacity to retain water. Fertile fields had good water retaining capacities (personal interviews, 27 January 2011 and 2 February 2011).

Moreover, there are other aspects of knowing a paddy field, some of which relate to physically working the fields and concern their 'workability' (Netting 1993, 50). This term highlights a tacit, embodied type of soil knowledge, i.e. the skill needed to work the soil appropriately and, through this, ensure that the paddy field is preserved through continued cultivation. Although this knowledge is more difficult to document, it is still possible to provide some illustrations of its transformation.

One example concerns soil assessment. The *Anren County Gazetteer* (1996, 295) states that pre-1949 farmers used to pinch the soil with their hands and taste it, to ascertain whether it was sandy, sticky, acidic or alkali. Here, the tactile and gustatory senses were vital for assessing the soil with the aim of improving it. Since the introduction of farm chemicals from the 1950s onwards, which became widespread in the 1980s, this type of intimate soil knowledge has largely become obsolete. Chemical fertilizers are applied according to their instructions for use, rather than based on a farmer's personal assessment of particular soil types' requirements. This is one example of where new a medium of knowledge, farm chemical instruction manuals, has impinged on the corpus of embodied soil knowledge.

A second example relates to the action of ploughing. In Anren, as in other areas in rural China and Asia, the heavy task of ploughing has conventionally been a man's task,[4] so the knowledge resides in men's bodies. It requires a close bodily acquaintance with the right soil consistency, as confirmed by a proverb which states that, when ploughing, 'the mud [has to be] as soft as paste' (CZ 1988, 161), implying a feeling for the soil.[5] However, the large-scale out-migration of rural men in the 1980s and '90s, the resulting feminization of agriculture, and the wide-scale introduction and adoption of the medium of walking tractors instead of oxen-pulled ploughs means that the distribution and the corpus of this knowledge has altered substantially. According to Green Water migrant worker Yuemei, tractors have rendered ploughing less physically demanding, making it possible for women to carry out (email exchanges, 24 November 2016 and 13 March 2017).

Meng (2014) describes how a gender shift in Chinese rice farming may take place in practice – even though in her case it is still the man who performs the ploughing. Hence, women acquire farming knowledge mainly based on learning that is grounded in daily practice, as well as – to some extent – through learning from their parents and parents-in-law. She quotes a left-behind woman:

> He (the husband) does not know how to farm. For more than ten years, he has not done the managerial work in agricultural production. Only during the busy/harvest season, he came back to do the work I could not do, such as, driving tractors or doing heavy work. I know more things about farming than him. (Meng 2014, 77.)

4 For instance in Java (Schweizer 1989), Vietnam (Bergstedt 2016), and Anhui Province (personal interviews, 2011).

5 For the full proverb, see Chapter 3.

Moreover, the use of cattle used to be an integral part of ploughing and soil knowledge and there seems to be an important relation between cattle and children with regard to acquiring farming skills, as interlocutors from Anhui and Hunan provinces stated that they had regularly tended the oxen in their childhood. For example, Yongcai acquired fundamental skills from looking after his family's oxen, which prepared him for guiding and communicating with the animals during ploughing, a crucial task, without which proper paddy rice farming is not possible. More generally, this childhood task also prepared him for raising oxen himself, for caring for them, and for ensuring their continuity (see Smerdel 2014, 265).

For Yongcai's family, however, things ultimately turned out differently, as he became a university student and turned his back on farming. His family also stopped raising oxen, disrupting this particular way of acquiring and transmitting knowledge (personal interview, 23 January 2011). The low number of households raising cattle and the high number of electric ploughs today suggests that similar situations occurred also elsewhere in Green Water: when I stayed there, only one family had an ox.

The gradual replacement of conventional oxen ploughing technology with tractors and other ploughing machines, especially since the 1980s, has also affected the spiritual side of farming. According to the *Anren County Gazetteer*, this change – along with the introduction of collective farming after the 1950s – has progressively diminished the ritual of *qichun*, 'the beginning of spring'. This used to be performed with an ox on a chosen auspicious day after *lichun* (Beginning of Spring, see Appendix II) to ensure good weather and mark the beginning of the agricultural cycle. During the ritual, the ox was guided to plough a first few rounds. Afterwards, a red paper reading *wugu fengdeng* ('an abundant harvest of all food crops') that was fixed on a bamboo stick used to guide the ox was stuck into the paddy field (ACGCC 1996; Wu 2010). By engaging in certain techniques and technologies (guiding an oxen-pulled plough through the field), the ritual can be seen as a medium which represents precisely these techniques and technologies of ploughing knowledge.

These examples reveal a complex dynamic in the corpus of soil knowledge, which has become increasingly important since the 1980s. While they suggest that certain types of embodied soil knowledge have been lost in favour of scientific knowledge related to the use of farm chemicals and mechanic ploughing technology, new or adapted soil knowledge has also evolved. For instance, the combine harvester owner in Green Water, Hugen, told me that he would test if a field was too muddy to use the machine in by stepping on it and feeling the consistency with his bare feet. If he sank any deeper than his ankles, he knew that the machine would also sink in, so the field was unsuitable for machine harvesting (personal interview, 1 February 2011).

Water control

Knowledge about water and its control is critical in wet rice farming, to stabilize and raise yields and to make the most out of the paddy field resource. Knowledge is needed at the scale of an individual plot, as well as with regard to the overall irrigation structure that extends to several villages.

At the individual plot level, this knowledge involves having an intimate acquaintance with the field, in order to meticulously manage water. A central aspect here is controlling the water level, by opening and closing the adjustable inlets and outlets of the plot, according to the weather and cultivation cycle. The right water level is necessary for several reasons. In winter and spring, for example, as Mrs. Luo explained, blocking the inlets (*du kouzi*) allows the fields to drain properly, which stops them from silting up. Later on in the cultivation cycle, the right water level helps to deter and control weeds, to stabilize the roots of the rice plants, to prevent any vertical movement of the water that would counteract the soil leaching, and to reduce erosion by protecting the soil from high temperatures, wind and direct rainfall.[6] Generally, small fields require an even water level and a high degree of control over irrigation and drainage. This allows a degree of strategic retarding or spurring of the time when rice plants in individual fields ripen which, in turn, permits farmers to stagger their labour-intensive tasks (Netting 1993, 42-43). Moreover, it is possible to regulate the water's temperature by influencing its direction and speed (see Chang 2000, 141).

I assume that all wet rice farmers are aware of the centrality of scrupulous irrigation control, even though this was not made explicit in my interviews. The knowledge consists of a corpus of detailed technical knowledge, as well as embodied irrigation skills, for example, how to use a hoe to repair ridges or close inlets, as left-behind Mrs. Luo demonstrated to me. Here the body, tools and fields can be seen as media in which this corpus of irrigation knowledge is inscribed. On the basis of my available data, it is difficult to say how far this corpus of knowledge has been transformed. However, the similarity of the circumstances in Green Water Village and the village investigated by Yuan Juanwen in nearby Guizhou Province suggests that a

6 For technical details see Bray (1984, 318); Altieri (1987, 76); Netting (1993, 42); and Chang (2000, 141).

Green Water villager might well have made the following statement. Talking about the young, migrated generation, a senior farmer complained:

> I know how to prepare the paddy field well enough so that it contains more water. Every year, the first time you prepare the field and irrigate it is very important. If you do this well the first time, the field can contain good water the whole year round. But the younger people do not put an effort into learning agricultural technologies by practice. (Yuan and Niehof 2011, 420.)

At the level of large-scale irrigation structures, water control not only helps to counter droughts and floods, but also regulates the distribution of water within the village community and across several villages that form part of the wider irrigation structure. The social organization aspect of water control knowledge is more important for this, but I will not describe the complex social and technical details of water control here, as they have been discussed in depth elsewhere.[7]

Despite all the changes, the basic technological logic underlying the irrigation system in Green Water Village has remained the same. Mainly due to environmental factors, the fields around Green Water are irrigated through a system of terraced fields and gravity. This is different from the system of my Anhui interlocutors, who use water pumps instead. The Green Water system's structure is analogous to that of a leaf, with a main watercourse in the centre which branches spread out from. Apart from water from newer reservoirs, fields in Green Water continue to be irrigated with channelled water from natural sources such as springs, rivulets, and rain, as well as ponds. In contrast to rain water, the inflowing water from rivers and mountain springs fertilizes the plants, as it is nutrient-rich, containing salts, mould, fungi, bacteria, algae, and organic debris (Hanks 1972, 37).

Nevertheless, the irrigation facilities have undergone some transformation. The customary village ponds are multifunctional, being used for laundry, fish breeding, as a watering hole for the oxen, and to irrigate the rice fields. However, according to Grandpa Zhou's granddaughters Lanxiang and Lanying, the ponds' water was clearer in their childhood during the early 1990s, when it could also be used to wash in and to raise freshwater clams (personal interview, 28 January 2011). A more obvious transformation is the declining number of ponds, which require people to maintain them. Previously, the natural village of Green Water had three major ponds. Today,

7 See, among others, Chi (1936); Wittfogel (1957); Needham (1974); and Bray (1994, ch. 3).

these have survived only in the villagers' memories and on a geomantic map in the Zhou lineage book. In 2011, two of these ponds had already been filled in to build houses on, and the third one was about to disappear for the same purpose. Against the background of the current de-intensification of rice farming and the fact that fewer villagers are left in the village to carry out the maintenance work, this is an instance where out-migration has left an imprint on the material interface of the local irrigation structure.

Nonetheless, in contrast to other parts of Hunan (Li 2006, 221-223), the overall irrigation system in Green Water was still intact. This was also due to a nearby reservoir, that was managed by a government official. Farmers used its water in the summer and autumn for double-cropping. The presence of the government official suggests, however, that the distribution of irrigation knowledge may have been transferred out of the hands of farmers and into those of officials.

Cropping patterns

As Chapter 1 explained, cropping patterns have undergone considerable transformation ever since wet rice cultivation first began in Hunan. In Green Water Village's post-reform household system of 2011, the rice-rice-rape seed pattern, expounded by the local government in 1976, was the main pattern being used, and my younger and middle-aged interlocutors viewed it as the normal rice cultivation pattern. However, according to my interviewees, individual households have been converting this back to single-cropping in the context of a lack of labour due to migration. Here, the social organization of labour and knowledge obviously constitutes certain constraints with regard to intensification. Nevertheless, experimenting with various cropping systems over recent decades has also produced a rich repertoire of knowledge about the possibilities of de-intensification and intensification, which is relevant in a migration setting (see Chapter 5).

The transition to double-cropping has also affected the spiritual side of rice cultivation knowledge. According to the *Anren County Gazetteer*, rituals such as the 'burning of seedbed paper' (*shao yangtian zhi*), have disappeared with the introduction of double-cropping. During that ritual, the head of a household pulled out the first seedling, throwing it down on the ridge or inside the field. Afterwards, paper money and incense were burned at the side of the seedbed as an offering to the heavens, to ensure good seed growth and a plentiful harvest. In the past, sowing and, hence, the ritual, was performed around *guyu* (Grain Rain, see Appendix II). However, the new cropping pattern has shifted the ploughing season forward to an earlier date, which has made this ritual – which was tied to the specific *guyu* period in the farmers' calendar – redundant (ACGCC 1996, 298, 610; Wu 2010, 165-166). In this case, the post-Green Revolution cropping patterns and technologies appear to have disarranged the conventional representation of local knowledge as marked in the calendar, challenging the supremacy of the calendar as a fixed and established reference point for farming.

The agricultural calendar

The cultivation process is structured by the medium of the Chinese luni-solar agricultural calendar, simply called the farmers' almanac (*nongli*). I observed that Green Water villagers kept a printed copy of this at home and memorized important dates in related songs and proverbs.

The farmers' calendar has a long history. According to Chinese mythology, the Divine Farmer Shennong is said to have invented the calendar. He is praised for accomplishments including teaching humans how to use farm tools, dig wells, reclaim and irrigate land, and preserve seeds (Yang and An 2008, 70). With regard to historical evidence, the first written Chinese agricultural calendars date from the Zhou Dynasty (approximately 1045-256 B.C.E.) (Bray 1984, 53).

Today's calendar is based on Wang Zhen's well-known agricultural treatise *Nongshu*, from 1313. Wang Zhen's calendar is divided into the Heavenly Stems, the Earthly Branches, four seasons, twelve months, 24 solar terms, and 72 five-day periods. Moreover, it contains – in the condensed form of a circle diagram – information about 'each sequence of agricultural tasks and the natural phenomena which signal their necessity, stellar configurations, seasons, phenology, and the sequences of agricultural production' (Bray 1984, 53-54).[8] The calendar is issued nationally and adapted regionally to the individual climate zones. This implies that farmers have to memorize not only the calendar's terminology, but also how this links to their specific local characteristics.

At the everyday level, Green Water villagers use the 24 solar terms (*jieqi*, see Appendix II) as the main way to organize not just their agricultural, but also their everyday and ritual activities. In practice, the solar terms stand for seasonal climatic and weather changes. They occur twice a month, providing farmers with 24 reference points per year which designate specific farming and other tasks to be carried out at each point. Their calculation is based on

8 For a concrete explanation of the calendrical details based on translations of up-to-date calendar sheets, see the related blog by LaFleur (2020).

the Earth's orbit around the Sun (which are always 15° away from each other along the ecliptic). The 24 solar terms include age-old farmers' knowledge about the connections between agriculture, the changing seasons, and climatic conditions (Qi 1986, 139-140).

Importantly, Francesca Bray reminds us that the impression of a linear progression, whereby the observation of nature becomes more accurate over time, is inaccurate. On the contrary, people pay less attention to nature when new technologies enable them to become more independent from it. As a result, the references to plants or stars in early agricultural texts have increasingly been disappearing from modern calendars. The newer calendars focus instead simply on the dates of the lunar months and solar terms (Bray 1984, 52).

Bray describes an inverse development that has occurred in China over several centuries. I observed the more immediate outcome in Green Water Village, that some people – notably migrants – could not remember the solar terms. For example, Yuemei declared that she could not remember the solar terms, although she had memorized the 'Song of the 24 Solar Terms' (*Ershisi jieqi ge*) at school (personal interview, 28 January 2011; see Appendix III). While a printed calendar – or a smartphone calendar today – can indeed remind farmers about these dates, it cannot convey the same precise adaptation of the calendar for a specific micro-locality, regarding the range of cultivation decisions that have to be taken.

Sowing and transplanting

After preparing a field, the first step of rice cultivation in Green Water Village is sowing and transplanting. The whole process of growing the rice plant takes about four months, depending on the variety. Within this process, transplanting is a major agricultural peak period. It poses special challenges for the organization of labour and knowledge, which becomes a pressing concern in a context of emigrated labour. Transplanting knowledge used to be literally held in the hands of women who repetitively performed the minute task of pushing the rice plants' roots into the mud.

In accordance with the farmers' calendar and in a laborious process that is repeated for several days, depending on the number of plots and available helpers, Green Water farmers perform transplanting as follows (see Figure 3). In the second lunar month, they water and plough the paddy field – typically a male task. After the fourth solar term (the Vernal Equinox, around 20 or 21 March), when the weather is no longer too cold, the villagers soak a bag of rice seeds in lukewarm water. They immerse them for three days, until

Figure 3 Throwing the bundles of seedlings and transplanting the seedlings

Photograph taken by Zhou Yuemei

the seeds start to germinate. Once the sprouts are about one centimetre long, they are sown into a seedbed. After 25 to 30 days, the transplanting begins. The villagers pull the seedlings out of the seedbed by hand, wash their roots, and tie them into neat bundles. They evenly throw the bundles into rows in the watery field, open them, and plant three to four seedlings together – a women's task. For hours, they stand barefoot up their shins in the water, the mud squelching between their toes. To ensure that the seedlings are transplanted in even rows and strips, farmers from Green Water stretch a cord across the field and begin by planting a guideline – a task my interviewees from Anhui skipped, believing that that was 'way too much work'. Next, the women move backwards, their bodies almost continuously bent down towards the field in a repetitive movement where specific gendered bodies and meanings of place and work evolve (Bergstedt 2016, 135).

Transplanting is not only painstaking, but also requires specific knowledge and skills. Farmers need to know how to make the seeds germinate; they need to calculate how many healthy sprouts will grow from the seeds, and how many healthy seedlings will grow out of the sprouts. They need the know-how that allows them to calculate in advance how many seedlings are needed for a particular field. This is not always easy, because fields have different sizes, often with asymmetrical shapes. If not transplanted immediately, the dug-up seedlings would soon die; any miscalculation would lead to financial loss. Farmers also have to know when it is the right moment for transplanting, in view of the agricultural calendar and the weather. Moreover, knowledge about the correct planting distance and density, depth and water level, which all vary according to different rice varieties, is necessary. As Yuemei put it, the plants 'are not allowed to be too dense, but also not too scattered' (personal interview, 25 January 2011). Manual skills are needed to perform the movements with speed and dexterity. Pushing each plant straight and evenly into the ground without damaging it requires practice (Bergstedt 2016, 135). Last but not least, social skills are also required to organize labour effectively in this peak season activity.

At another level, farmers need to be aware of the advantages and disadvantages of this highly labour-intensive cultivation practice, and estimate whether it is worth the effort or not. This is particularly acute around out-migrated labour and the resultant loss of people with the necessary skills. According to Bray (1984; 2004) and Chang (2000), the major advantage of transplanting rice is high productivity. The preconditions for transplanting are that the seedbed and the paddy field must be well watered, ploughed, levelled, puddled, and fertilized. Planting and replanting the sprouts and seedlings twice (from a bag immersed in water into the nursery, and from the nursery into the wet field) allows a double selection process to choose the strongest and healthiest plants. In the short run this ensures a more productive harvest while, in the long run the practice also leads to breeding better quality varieties. Further, small amounts of fertilizer and water can be used efficiently; transplanting fosters more seed-bearing stalks and tillering capacity, leading to higher yields; because the plants spend about one month in the nursery, the wet field is available longer for cultivating other crops; because plants are transplanted in even rows, weeding is easier and, finally, plants also ripen evenly (Bray 1984, 288; 2004, 17; Chang 2000, 141). In sum, through this labour-intensive practice, fields may be cultivated intensively and productively, with efficient use of scarce land, water, and farm chemicals.

Due to the demands of organizing labour in the peak transplanting periods, this activity is an exemplary illustration of how the social organization of knowledge interrelates with the other aspects of knowledge. Depending on the circumstances, the technique may either be further intensified, as in the famous, although not undisputed Javanese case of 'agricultural involution'

(Geertz 1963).[9] Or, conversely, it might be completely abandoned, because of the emigration of skilled labour. Generally, it is possible to forgo manual transplanting, either through mechanization (transplanting machines) or by changing sowing techniques (switching to direct seeding). Around Green Water Village only the latter was relevant in 2011 and I discuss this in Chapter 5. Here it is sufficient to note that – while transplanting requires a large corpus of knowledge, especially skills – in the twenty-first century system of knowledge transmission there has been a partial transformation of this knowledge, related to opting for less complex techniques in the light of a transformed social organization of this knowledge.

Plant protection

Once the rice plants have been transplanted, they need to be protected and nurtured. Plant protection in Green Water Village and beyond includes four aspects. These may be summarized as follows, although in practice, they overlap in a complex interplay:

(1) Fertilizing: After the first week in the wet field, fertilizer (animal dung from fowls, pigs, and cows, human manure or chemicals) is applied for the first time. One month later it is applied for the second time. After each application of fertilizer, the field openings are closed to prevent the fertilizer from flowing out.

(2) Applying insecticides: After a fortnight in the wet field, the first round of insecticide is sprayed. This is repeated three times, each a fortnight apart. Several villagers claimed that three times had previously been sufficient, as opposed to four times today.

(3) Water control: After successful transplanting, just three to four centimetres of water in the field is sufficient. In the following weeks – during the rice plant's flowering and ripening stages – the water level has to be consistently five centimetres high. As soon as the rice plants have grown so densely that the gaps between the planted rows are no longer visible, the water in the field is drained and the field is dried in the sun for 30, or up to 40, days. During this time, the plants grow quickly and begin to produce seeds. Even after draining there is still enough humidity in the soil. If it is very dry, water is added.

(4) Weeding: Weeds obstruct the rice plants' growth by stealing their humidity, light and nutrients. The most common paddy field weeds are barnyard grass and various types of rushes and marsh plants, all

9 For a critical engagement with Geertz's concept see, e.g., Bray et al. (2015).

> of which are particularly stubborn and tenacious (Bray 1984, 299, 311, 314). Therefore, regular and effective weeding is crucial, not only for successful rice cultivation in the short term, but also to protect the paddy field as a resource in the long term.

Irrigation and fertilizing knowledge are particularly demanding, as reflected in the high number of proverbs about these processes (see Chapter 3). Weeding nevertheless provides an exemplary case of how plant protection knowledge has transformed. Weeding used to be a task for children, as Yuemei recalls from her own childhood in the 1980s and '90s:

> We were sent to the rice fields to help with weeding after school and during the summer holidays. We used to 'step into the fields' three times: first, before planting, we trod the rice stubbles [remaining from the previous year's harvested rice crop] into the field. This is mechanized today. After transplanting, we trod the weeds between the young rice plants into the mud. This also fertilized the field. Finally, just before the rice plants ripened, we broke down the barnyard grass by stamping on it. (Email exchange, 17 July 2013.)

This division of labour not only reflects a hierarchy of tasks in which simpler tasks are delegated to children, who are not yet full-grown experts. It also reveals a specific learning system which gradually familiarizes children with rice production, through sensory engagement with the paddy field and plants. There is a clear analogy between growing plants and growing people, which is also well-known in China (Jepson 2014, 160; Schmalzer 2016, 138).

Nevertheless, manual weeding also requires knowledge. This includes knowledge about various plants and their characteristics, about the negative effects of weeds on the growth of rice plants and for preserving the field's overall value, and the bodily skills needed to get rid of the weeds. By way of illustration, whereas local Green Water village children seem to naturally master this apparently easy task, the experiences of an urban couple from Shanghai gives a different picture.

The Zhao couple were sent down to Hunan's neighbouring province Jiangxi for ten years during the Cultural Revolution (1966-1976). They were part of the estimated twelve million Chinese urban young people who were sent to the countryside between 1968 and 1975 in a 'massive rustication movement' (Schmalzer 2016, 155). Along with two other classmates, Mr. and Mrs. Zhao had spent their youth living and working with the local farmers. They eventually married each other there, and had a son. Without any prior knowledge of rice farming, and working hard on a particularly beautiful rice field, the couple accidentally hoed away not just the weeds, but also the young rice plants, which caused them to wither the following day (personal interview, 13 February 2011).

With the wide-scale adoption of new ploughing technology and farm chemicals, manual weeding knowledge has undergone a transformation. Most obviously, the corpus of plant protection knowledge has expanded from predominantly manual techniques to the handling of chemical substances. In view of the social distribution of plant protection knowledge, this implies that there has been a shift towards scientific knowledge. However, farmers only have a partial knowledge of this. As a consequence, thousands of Chinese farmers suffer food poisoning and soil pollution every year because they do not know how to apply pesticides correctly (Sternfeld 2009, 2; Xin et al. 2009, 115). This is particularly worrying if we consider that weeding was a children's task, which trained them to grow up into farmers. Replacing their labour through herbicides also raises important questions about the further transmission of weeding knowledge and sustainability.

Rice varieties

Rice seeds are the foundation of successful rice cultivation. The seed varieties require a particularly rich knowledge, extending from preserving seeds, breeding varieties, and estimating their characteristics. It leads to strategically choosing the right varieties and combinations thereof for various reasons: whether socio-cultural, to be used in certain rituals or preferred foods; technical, in view of weather and ecological constraints, or harvesting technologies; or economic, as a result of increased sales options.

Rice seeds are a complex technology. Rice is part of the grass family (*Gramineae*) and belongs to the genus *Oryza*, which includes 20 wild and two cultivated species: African rice (*O. glaberrima*) and Asian rice (*O. sativa*). Asian rice is assumed to have originated in south China in the middle of the fifth millennium B.C.E., in the domestication process from a wild grass species (Sweeney and McCouch 2007; Gilbert 2015, 214). It is subdivided into two main varietal groups: long-grained *indica* (*xiandao*), which is well adapted to a tropical and subtropical climate, and round-grained *japonica* (*gengdao* or *jingdao*), which is better suited to temperate zones. Both are commonly cultivated in China and, in southern areas with two rice harvests a year, they are seasonally combined (Chang 2000, 138; Kolb 2003, 620-621; Bray 2004, 18).

The Green Water villagers' planting strategies are based on, first, the distinction between varieties of early rice (*zaodao*) and late rice (*wandao*). As there are early and late ripening types of rice in both varieties, these categories do not refer to *indica/xian* and *japonica/geng* rice alone. It is essential that early rice enables multi-cropping, whereas late rice produces higher yields (see Bray 1984, 490).

Second, there is a significant distinction between hybrid rice and 'conventional rice'. Notably, in 2011 hybrid rice was being marketed; only a small amount was kept for producing so-called 'rice tofu'. In Green Water, hybrid rice varieties have been common since the 1980s and are planted as early rice. Common hybrid rice in 2011 were, for example, varieties No. 388 and No. 520, or *Xianyou huajian*. These ripen in less than 130 days, so are suitable for double-cropping. In contrast, conventional rice such as the local *ziku mi* variety was kept for personal consumption, being appreciated for its softer consistency and flavour. Conventional varieties were planted as late rice, whereas they had been double-cropped in the past. The fact that Green Water farmers continued to grow conventional varieties despite the comparatively low yields (see Chapter 1) highlights the differing perspectives of markets and consumers which are taken into account in making cultivation decisions.

One major difference between hybrid and conventional varieties is that previously conventional seeds were preserved and bred by farmers. In contrast, hybrid seeds have to be purchased in seed shops. In 2011, the cost was about 50 Yuan (about 7 USD) per kilogram, with 1.5 kilograms needed to cultivate one *mu* (by using the technique of transplanting). For most farmers, this is expensive. As a reference, about one Yuan and a few Jiao (about 0.15 USD) may be earned from selling one *jin* (500 gr) of unhusked rice. One *jin* of unhusked rice, which corresponds to about half a *jin* of white rice, is also about the amount of rice that each villager consumes a day, in the form of three meals of rice with side dishes.

Apart from these key differentiations (early/late, hybrid/conventional), further broad categories in Green Water are 'middle rice' (*zhongdao*) and 'sticky rice' (*nuomi*). The former includes long-ripening varieties that need 150 days to mature. They have higher yields (up to 1200 *jin/mu* or 9000 kg/ha), but also need two additional applications of insecticides and can only be mono-cropped. Some of the latter is planted in both seasons, to produce certain culinary specialities.

With these and other factors in mind – which may, in fact, be seen as knowledge condensed into the medium of rice seed – Green Water villagers practiced the following cultivation schedule: early rice is sown in the second lunar month and harvested in the sixth month. It takes one week to harvest the fields with a combine harvester. Late rice is sown in the fifth month and harvested in the ninth month. Here too, it takes one week to complete the harvesting. Middle rice is also sown in the fifth month and harvested, somewhat later than the late rice, during the ninth month. However, harvesting with the combine harvester takes a fortnight, because the fields sown with middle rice are widely scattered and distant from each other.

Although the times differ slightly, generally, combining early and late rice suggests that the summer is the busiest time of the year, with simultaneous harvesting and planting. Working in high temperatures of up to 40 degrees Celsius renders this busy season even more exhausting, especially for senior farmers. Usually, several varieties are planted simultaneously, occasionally even in the same plot. This is an important strategy for risk reduction, which has been practiced by Chinese and other farmers for centuries (Bray 2004, 18; Chevalier, Marinova, and Peña-Chocarro 2014, 5).

The varieties mentioned are the result of continuous active selection and breeding processes that began several millennia ago, whereby preferred genetic traits have been selected to match environmental, climatic, consumption, and the requirements of other socio-technical environments.[10] The outcome of these selection processes is an enormous variety of rice.[11] It has been estimated that more than 100,000 Asian varieties have been cultivated over the last two centuries (including duplicates), before the arrival of science-bred cultivars from the 1950s onwards, (Chang 2000, 138). In China, this varietal composition was greatly increased with the adoption of the early-ripening and drought-resistant *Champa* rice, which was introduced in 1020 by the emperor from the area of today's Vietnam. This enabled double- and inter-cropping in southern China (Bray 1984, 491-495; Chang 2000, 139). Besides, in Hunan, it also served as an insurance policy against regular seasonal droughts and floods (Perdue 1987, 117, 121).

In the post-reform system of knowledge transmission, the corpus of rice variety knowledge has undergone a transformation. On the one hand, the older villagers, who grew up without hybrid varieties, are still familiar with conventional breeding practices. The continuity of planting small amounts of local rice is evidence of this. On the other hand, the introduction of new rice varieties that are suitable for modern cultivation methods has gone hand in hand with the increased use of mechanization and farm chemicals.

10 For details, see Chang (2000); Kovach, Sweeney, and McCouch (2007); and Zhang (2014).

11 However, when viewed over a longer time span, the number of plant varieties decreased when humans changed from hunting and gathering to farming, which made them focus on the domestication of certain plants only (see Zapata 2014, 16).

Moreover, the vast spread of hybrid rice from the 1980s onwards has also limited farmers' breeding ability to a certain degree. Because farmers cannot breed hybrid rice seeds themselves, cultivating this rice means relying on scientists and seed dealers for their supply. This constitutes a significant shift in the social organization of knowledge, independent from out-migration.

Harvesting and processing

Harvesting and processing knowledge has also undergone considerable transformation. I review harvesting and processing knowledge through the lens of the urban Zhao couple from Shanghai, who were sent down to rural Jiangxi Province in the 1970s. Being novices at farming, Mr. and Mrs. Zhao's description reveals the practical knowledge and skills which usually remain tacit among knowledgeable practitioners. According to my observations of farm tools and the *Anren County Gazetteer*, the harvesting and processing practices described were practically the same in Jiangxi, Hunan and Anhui provinces, although some technologies for other cultivation steps differed somewhat.

I visited the Zhao couple in the kitchen of their terraced house in Shanghai's Minhang district, where gated compounds are increasingly replacing Shanghai's rural areas. Mr. Zhao worked as an overseer on construction sites. He would soon follow his wife, a former garment saleswoman, into retirement. Mr. Zhao was smoking a cigarette, while their daughter-in-law was clearing remains of the lunch he had prepared from the table. The family had just returned from a trip to the village, enabled by the car their son had recently bought, and had been very impressed by their first visit there for thirty years. Speaking loudly and gesticulating, they described harvesting in the 1970s as follows:

> At first, we used a threshing tub. Later the tub was substituted with a foot-powered threshing machine. The machine was carried on your shoulders into the field, where it was operated by two people stamping. Meanwhile, the other harvesting helpers carried bushels of rice.

Mr. and Mrs. Zhao particularly remembered the physical hardship and exhaustion, which is not usually mentioned by people whose bodies are accustomed to this hard work:

> At that time, there was still water inside the field, which made it difficult to operate the pedal by foot. Moreover, it was hard (*xinku*), because you had

> to pedal continuously for four or five hours. In addition, the water was very hot because of the sun. Even before this, during the plant protection period, there were annoying leeches, snakes, and insects, and the field was dirty because of the fertilizer: human and animal excrement, which you spread with your bare hands! It was terribly bitter at the time (*ku de budeliao*)!
> The rice kernels were carried to a drying floor. There, they had to be constantly turned over with the help of a tool like a rake without teeth. Afterwards, the remaining, unwanted straw and weeds were removed with a harrow. Subsequently, the dried rice kernels were transported into a large communal granary, where they were later rationed according to the number of household members.
> Before the rice could be eaten, it had to be husked. Initially, we did this with a type of mill. Later we used powered machines. Finally, the husked rice had to be winnowed by a winnowing machine to blow out the remaining husks and dirt. Because the rice was still not clean enough, a flat bamboo sieve was used for winnowing.

A round bamboo winnowing basket served a similar purpose. This was held with two hands, performing specific shaking movements:

> During this procedure the bamboo basket had to be simultaneously shaken and slowly turned. If you master this technique, the rice is collected on one side of the basket, and the unwanted dirt is filtered out on the other side. (Interview, 13 February 2011, from fieldnotes.)

While they had apparently managed to thresh the rice, although with considerable physical discomfort, Mr. Zhao admitted that he had never learnt to master the winnowing basket technique. In fact, as ethno-archaeologist John C. Whittaker has noted, '[d]eceptively simple tools such as baskets and trays may rely heavily on specialized skills for their effective use' (Whittaker 2014, 135).

Answering my question of how the Zhao couple had learned to cultivate rice, they responded that they had just 'copied other people' (*genzhe bieren zuo*) and had 'watched the People's Commune'. On the one hand, they did not see anything difficult about rice cultivation. In the rather deprecating tone of many urbanites, they stated that 'it is all physical labour (*tili huo*)'. On the other hand, their experiences of weeding and winnowing show that rice cultivation is not very easy. Instead, it requires skills which need practice and long-term engagement with the material and the environment (Sigaut 1994; Ingold 2006).

It is obvious that, with the introduction and dissemination of mechanized harvesting and processing technologies, the corpus of related knowledge has been transformed and, to some extent, become redundant, as long as new technologies are available. There has been a move away from the rich embodied knowledge needed for seemingly simple tools, towards the different knowledge required to operate machines, the workings of which are only fully understood by their engineer designers. At the same time, the introduction of harvesting and processing machines is another factor that has set labour free for migrant work, creating a situation in which migrants lack agricultural knowledge, such as migrant worker Xiao Chen from Anhui, who stated that her mother-in-law could use a winnowing basket properly, which she was unable to do herself (personal interviews, 26 February 2011 and 5 March 2011).

Culinary rice knowledge

We also need to take into account the consumption of rice plants, because rice consumption is an important aspect of attributing value to the paddy field resource. Two-thirds of the Chinese population eat rice as their main staple food, especially in southern China (GRiSP 2013, 106). Farmers commonly retain not only administrative, but also imagined ties to it, as well as feelings of belonging to the land. Conventionally, culinary rice knowledge has been held by women, with bodies and food the two central media through which this knowledge is communicated. Village women know how to prepare specialities such as sweet rice balls, rice tofu, fried rice crackers, rice chips, sticky rice balls or rice noodles. Some of these are consumed only on special dates in the agricultural calendar, like New Year or the Spring Equinox (*chunfen*). Moreover, the women know which variety of rice to use for which dish, and can recognize the right variety using their visual and tactile senses.

Special knowledge is connected to fermented rice, which is called 'rice wine' (*mijiu* or *tianjiu*) and is eaten with chopsticks. This rice wine was the first thing I was offered when I arrived. It is believed to be good for health, and traditional Chinese medicinal substances may also be added to it. This suggests a wider knowledge about health and the body, connected to food preparation. This medical knowledge also became obvious when Yuemei and I were offered salty rice porridge when we suffered from severe dysentery after eating industrially-processed food from a nearby market during an engagement ceremony.

The distribution of this knowledge has not changed substantively within Green Water Village. Nevertheless, I observed that in Shanghai Mr. Wu took over providing family meals in his restaurant, showing that he, too, had acquired some cooking skills. Mr. Wu's wife assumed the task of cooking again when at home. The couple joked that during his time in Shanghai Mr. Wu had become a Shanghainese man – the stereotype of Shanghainese men being that they cook, which is said to set them apart from other men in China whose wives always prepare the food. In fact, the sent-down Mr. Zhao mentioned above always made the food on the numerous occasions I visited his family.

The diet and the cooking utensils available have changed and, with this, so has the related knowledge. According to Wu's (2010) description, people ate brown rice three times a day for a long time. Poor farmers ate brown rice soup in the morning, steamed brown rice at noon, and a morsel of brown rice in the evening. Even poorer farmers had to supplement their diet with small particles of rice husks, strips of sweet potato, vegetables and wild herbs. Moreover, as poor people often lacked salt and oil, a common way to prepare rice dishes was simply to steam the rice with vegetables on top, accompanied with roasted chillies (Wu 2010, 174).

Today, this situation has changed, with Wu claiming that the practice of consuming steamed rice with vegetables and chillies has become fashionable in local restaurants (ibid.). Generally, my observations suggest that Green Water villagers now have sufficient food. On an everyday basis they commonly eat steamed white rice with fried vegetables, occasionally noodles or porridge, as well as some local pork or fish. Migrants, who do not have their own cultivated rice, eat bought rice instead. Moreover, the spread of electricity and rice cookers has created other options for preparing food, and decreased dependence on fuels such as rice straw. Meanwhile, grain consumption in China has also decreased in general, in favour of increased vegetable and animal product consumption (NBSC 2019, sec. 6-4).

In economic terms, the importance of these changes to the resource of paddy fields and food production is that migrant work – in both the short and medium term – has rendered villagers (especially migrants) independent from cultivating and preparing their own food. This has given them the option of purchasing their food, including industrially-processed produce that require less preparation knowledge and time. Nevertheless, continuous cultivation is still essential to preserve paddy fields as a resource in the long term.

The issue of the socio-cultural value of the resource paddy field remains, though. Some rice specialities are more important than others in terms of health and local notions of the body. Moreover, food practices are generally a crucial means of identity-construction (Ohnuki-Tierney 1995; Mintz and

Du Bois 2002; Oxfeld 2017). Therefore, this socio-cultural realm of rice must also be taken into consideration when exploring how farmers manage the paddy field resource under migration pressure. Here, culinary knowledge goes beyond mere technical food preparation abilities, and questions of rice production go beyond economic calculations.

Wider plant use

More than just the grains of the rice plant are made use of, so the corpus of knowledge about its usage is not just limited to women and cooking. Rice husks and, particularly, rice straw are important by-products of rice cultivation. Rice husks can have many uses. The last remains of old houses dating back about a century attest that, in past times, rice husks mixed with clay were an integral building material for village houses, although constructing with bricks has rendered this use of husks obsolete. Moreover, husks can be used to feed animals and for fertilization, however, their importance here has lessened with the introduction of industrial fodder and chemical fertilizers respectively.

Rice straw is even more multifunctional. In contemporary house construction, fresh concrete is covered with rice straw to prevent it from cracking, while in the past, the old clay-brick houses' roofs were covered with rice straw (Wu 2010, 247). It is also used to make doormats or prayer mats, as flooring in oxen and pig barns, as string to tie vegetables or rice seedlings, to make hats, brooms, or ropes. Moreover, it serves as a base to ferment tofu, to build fires or, in the past, to make straw shoes. People can purchase sleeping mats made of rice straw at the market and, in 2011, most people slept on these kinds of mats. This seems to be changing, however, as I observed that industrially-produced mattresses now form part of a 'modern' dowry. Other straw products, too, are increasingly being replaced by industrial products – in Yuemei's words: 'People don't want rice straw anymore, they can buy products, they have the money' (personal interview, 23 January 2011).

This alludes to the fact that migrant work, which has made villagers relatively affluent compared to the past, and spurred their imagination of the good life, is not only affected by socio-technical transitions, but also affects them. Furthermore, it points to complex linkages in the socio-technical system of rice farming, and to the multifaceted consequences of changes within this system. When purchasing industrial products, it no longer matters whether rice straw is lost during the harvest by using a combine harvester, since nowadays, the remaining rice straw in the field is simply burned anyway. However, an unwanted side effect of this is that the frogs in the field die during this process. According to some villagers, these amphibians were key to pest management, so their loss necessitates the use of more pesticides. From the perspective of the corpus of rice straw knowledge, as the consumption of rice straw decreases, the related knowledge is reducing in importance, and this knowledge is shifting from being held by farmers towards being held by industry and product designers.

Rice rituals

Rice is also used in ritual practices. As Fredrik Barth has demonstrated, rituals are a powerful medium of knowledge transmission (Barth 2002). On the one hand, rituals, which were an important way of representing and transmitting knowledge in pre-1949 China, and which were banned as superstitious in collective times, have been reviving in the post-reform household system of knowledge transmission.

Three examples illustrate this revival and the deeper social meaning of rice. First, a female custodian of a temple in another village in Anren County showed me how she regularly places some cooked rice and noodles on the heads of the stone lions at the temple entrance, while reciting a sutra. Second, prior to the family dinner on New Year's Eve the ancestors were offered rice wine, followed by rice and toppings. Third, during a funeral in another nearby village, a chair was placed next to the coffin, on which was a plate with five grains, including rice. It is assumed that the Divine Farmer Shennong had taught humans how to cultivate the five grains – rice, two types of millet, beans and wheat (Yang and An 2008, 70). Villagers assume that the ghost of their deceased sits down on the chair. The five grains are supposed to scare off other ghosts. Peng Mu (2008), who conducted fieldwork on the world of the deceased in neighbouring Chaling County, confirms that uncooked rice plays a role in various exorcising rites. Her informant, a Taoist priest, explained that this is because the rice is connected to the Divine Farmer Shennong and is therefore believed capable of warding off ghosts. Peng lists several other instances of rice used in funeral practices, including tossing rice into a coffin and placing a bag of uncooked rice on a coffin near the deceased's head, along with an axe (Peng 2008, 124).

On the other hand, some rituals have lost their importance with the introduction of new farming technology and schedules. These relate especially to the rice customs that are more technical and directly connected to rice cultivation and the farmers' calendar. Wu (2010, 165-166) lists a range of local festivals, some of which are linked to rice cultivation. According to my host Zhou Wenlu, only some of these still exist, e.g. *kai yangtian men*, 'opening the door of the seedbed' and *chang xin*, 'tasting the new [i.e. freshly harvested rice]'. Moreover, he still remembers the vanished custom of 'burning of the seedbed paper' from his childhood in the 1960s and 1970s mentioned above (personal interview, 22 January 2011).[12] In contrast, Grandpa Zhou and his family claim that these customs only existed prior to 1949, that only his parents practiced them as described by Wu (2010), and that all of them have disappeared today (personal interview, 03 February 2011). This is also a common narrative in the county gazetteers. In any case, it is interesting that people today continue expressing some memory of these ritual practices, which suggests that they must indeed have been a powerful means of conserving and transmitting knowledge.

While the extension of the corpus of knowledge in line with new technologies has weakened the importance of some farming rituals, the less technical, spiritual knowledge around rice and the paddy field resource nevertheless remains central. It touches upon pivotal questions of life and death. This is an instance where knowledge and skills clearly transcend technical abilities, reaching into the realm of the socio-cultural continuity of households and their patrilines.

Agricultural deskilling and extended knowledge repertoires

Taken together, in the past decades rural China has witnessed a complex transformation of the system of knowledge transmission. The outcomes of this transformation may at first sight appear somewhat contradictory, including deskilling on the one hand and extending repertoires of knowledge on the other hand. It is therefore important to distinguish analytically between different levels, especially with regard to individual and distributed skills, as well as practical and discursive aspects. Depending on the specific focus, the outcomes will differ.

On the one hand, the transformation obviously creates several challenges. As Sigrid Schmalzer convincingly shows with regard to farmers in the collective system of the 1950s to the early 1980s, the issue of deskilling has emerged, although this is not straightforward. She asserts that the vision propagated by the state was one of skilling farmers, and in certain fields, e.g. hybrid rice, a number of farmers did receive some training. Nevertheless, at the same time, a developmentalist narrative prevailed that discriminated against existing knowledge forms. Accordingly, techniques that required

12 Xiao Chen from Anhui also mentioned *kai yangtian men*.

more skills were replaced by technologies introduced from the outside, such as new seed varieties or farm chemicals, as soon as the economic situation allowed for such substitutions (Schmalzer 2016, 126-127).

Nowadays, in addition to the ongoing rapid transformation, or even loss of agricultural skills due to the quick adoption of post-Green Revolution technologies, there is also the issue of long-term migration. There is no doubt that young migrant farmers lack the kinds of individually embodied practical farming skills that are acquired through daily long-term engagement with the paddy fields, their soil, water, plants, and farm tools. What is happening here can be understood as a contemporary Chinese version of 'agricultural deskilling' (Braverman 1974; Stone 2007). Importantly, deskilling in Braverman's sense does not necessarily entail the actual loss of embodied skills, rather the downgrading of those skills that have become obsolete in the industrial workplace. There is a clear parallel with this in China, where – as exemplified by the young migrant nurse Caixia – farming and rural life are commonly denigrated. This antipathy is also fostered by the formal school education system, by the media and global consumerism (White 2012, 11-12). Moreover, Braverman highlights the shift of control over knowledge. When manual techniques are replaced by machines or hybrid seeds, this shifts the control of knowledge out of the farmers' hands, into the scientists' hands (Stone 2007). While this has implications for the future preservation of rural households' field resources, this situation also has wider consequences. As Ben White notes: 'Thinking about youth, farming and food raises fundamental questions about the future, both of rural young women and men, and of agriculture itself' (White 2012, 19).

On the other hand, the transformation of the knowledge system also provides new opportunities for farmers, both in the city and in relation to farming. With regard to city life, the fact that migrants acquire new skills and networks has been widely noted.[13] Nevertheless, these specific migrant skills have rarely been explored in depth. Some of the younger migrants I interviewed had learned skills which were unrelated to farming. For example, Caixia and Yuemei both went to university, entering the fields of health care and insurance respectively. They became what Western policy makers, paying special attention to educational achievements, commonly call 'highly skilled migrants'. In contrast, I could observe that older migrants, who had spent thirty or forty years in the countryside prior to migration,

13 See, for instance, Li (2006, 177); Fan, Sun, and Zheng (2011, 2167); Yuan and Niehof (2011); Meng (2014, 44); and Chen (2015, 116).

built upon their previously-acquired skills and the social organization of those in the city.

For instance, Xiao Chen had grown up in an environment of austerity that constantly required ingenuity to cope with everyday life in the countryside. Working as a cleaner in Shanghai, she creatively applied her skills at finding low-cost solutions and repairing whatever broken things came her way, declaring, 'at home, we do it like this!' Meanwhile, Mr. Wu's family relied upon, adapted, and pooled their skills in their street restaurant to make a living in Shanghai. Although occasional negotiations about the distribution of knowledge occurred, in the city the family basically continued to follow the customary inside/outside dichotomy to organize female and male labour and knowledge. Accordingly, the female family members mainly worked inside the restaurant. Drawing on manual skills acquired earlier at home, e.g. through making cloth shoes, they performed tasks requiring great manual skills, such as placing particularly brittle or slippery vegetables on skewers with great dexterity and speed. Meanwhile, the male family members commonly took on the tasks outside the restaurant. These included making more distant deliveries, using a motorcycle to drive across the city to the wholesale market, and transporting heavy bags of goods back to the restaurant. In a way, these tasks reflected the fact that, in the countryside the use of agricultural machinery as well as carrying things such as heavy bags of fertilizer was commonly considered a task for the men. Moreover, knowledge about a special mix of spices was kept as a trade secret and was primarily transmitted from and to male relatives, preferably of the same patrilineage, which is a customary way of organizing knowledge within rural Han Chinese families (Kaufmann 2011; 2016).

These examples show that there are several layers or sediments of skills which migrant farmers build upon, develop, and use in different, sometimes highly individualized ways. In any case, both younger and older migrants used part of the income they earned from their skills in the city to support farming activities back home. In addition, some interviewees used their urban skills upon their return to the village, such as when Yuemei's father Zhou Wenlu used his construction skills to build a new house at home. Others, such as Caixia's mother, temporarily made use of her newly acquired entrepreneurial skills to open a small restaurant at home. In these ways, they could mitigate some of the challenges described above.

With regard to the farming skills used in the countryside, those Green Water villagers who had grown up almost exclusively with manual techniques were still alive, and some could develop new skills, for instance learning to operate the combine harvester. From a more holistic point of view that does not focus on individual skills but takes into consideration distributed skills, therefore, in the 2010s the village society contained a broad range of distributed knowledge. I suggest that this has, in fact, contributed to enlarging farmers' socio-technical spaces for manoeuvre, in view of coping with the conflicting pressures of farming and migration.

References

ACGCC, 安仁县志编纂委员会 Anren Xianzhi Bianji Weiyuanhui (Anren County Gazetteer Compilation Committee). 1996. 安仁县志 *Anren Xianzhi* (*Anren County Gazetteer*). Beijing: Zhongguo Shehui Publishing.

Altieri, Miguel A. 1987. *Agroecology: The Scientific Bases of Alternative Agriculture*. Boulder: Westview Press.

Anagnost, Ann. 1997. *National Past-Times: Narrative, Representation, and Power in Modern China*. Durham: Duke University Press.

Barth, Fredrik. 2002. 'An Anthropology of Knowledge'. *Current Anthropology* 43 (1): 1-18.

Bergstedt, Cecilia. 2016. *Cultivating Gender: Meanings of Place and Work in Rural Vietnam*. Copenhagen: NIAS Press.

Bossen, Laurel. 1994. '中国农村妇女什么原因使她们留在农田里 Zhongguo noncun funü: Shenme yuanyin shi tamen liu zai nongtian li? (Chinese Rural Women: What Keeps Them Down on the Farm?)'. In 性别与中国 *Xingbie Yu Zhongguo* (*Gender and China*), edited by Xiaojiang Li, Hong Zhu, and Xiuyu Dong, 128-154. Beijing: Sanlian Shudian.

—. 2011. 'Reproduction and Real Property in Rural China: Three Decades of Development and Discrimination'. In *Women, Gender and Rural Development in China*, edited by Tamara Jacka and Sally Sargeson, 97-123. Cheltenham: Edward Elgar Publishing.

Braverman, Harry. 1974. *Labour and Monopoly Capital: The Degradation of Work in the Twentieth Century*. New York: Monthly Review Press.

Bray, Francesca. 1984. *Agriculture*. In *Science and Civilisation in China*, Vol. 6, Part 2, edited by Joseph Needham. Cambridge: Cambridge University Press.

—. 1994. *The Rice Economies: Technology and Development in Asian Societies*. Reprint. Berkeley: University of California Press.

—. 2004. 'Rice, Technology, and History, the Case of China'. *Education About Asia* 9 (3): 14-20.

—. 2008. 'Science, Technique, Technology: Passages between Matter and Knowledge in Imperial Chinese Agriculture'. *The British Journal for the History of Science* 41 (3): 319-344.

—. 2013. *Technology, Gender and History in Imperial China: Great Transformations Reconsidered*. New York: Routledge.

Bray, Francesca, Peter A. Coclanis, Edda L. Fields-Black, and Dagmar Schäfer, eds. 2015. *Rice: Global Networks and New Histories*. New York: Cambridge University Press.

Chang, Te-Tzu. 2000. 'Rice'. In *The Cambridge World History of Food*, edited by Kenneth F. Kiple and Kriemhild Coneè Ornelas, 132-149. Cambridge: Cambridge University Press.

Chen, Chen. 2015. 'Rural-Urban Migration and Circularity in China: Evidence from Anhui Province'. PhD diss., University of California Los Angeles.

Chevalier, Alexandre, Elena Marinova, and Leonor Peña-Chocarro. 2014. 'Factors and Issues in Plant Choice'. In *Plants and People: Choices and Diversity Through Time*, edited by Alexandre Chevalier, Elena Marinova, and Leonor Peña-Chocarro, 3-14. Oxford: Oxbow.

Chi, Ch'ao-ting. 1936. *Key Economic Areas in Chinese History*. 2nd ed. New York: Paragon Reprint Corp.

Croll, Elisabeth J., and Huang Ping. 1997. 'Migration For and Against Agriculture in Eight Chinese Villages'. *The China Quarterly* 149: 128-146.

CZ, 郴州地区民间文学集成编委会 Chenzhou Diqu Minjian Wenxue Jicheng Bianweihui (Editorial Board of the Comprehensive Collection of the Folk Literature of Chenzhou District), ed. 1988. 中国民间谚语集成湖南卷: 郴州地区分卷 *Zhongguo minjian yanyu jicheng Hunan juan: Chenzhou diqu fen juan* (*Hunan Volume of the Comprehensive Collection of Chinese Folk Proverbs: Section Volume of Chenzhou District*). Chenzhou: N.p.

Eyferth, Jacob. 2009. *Eating Rice from Bamboo Roots: The Social History of a Community of Handicraft Papermakers in Rural Sichuan, 1920-2000*. Harvard: Harvard University Press.

Fan, C. Cindy, Mingjie Sun, and Siqi Zheng. 2011. 'Migration and Split Households: A Comparison of Sole, Couple, and Family Migrants in Beijing, China'. *Environment and Planning* 43: 2164-2185.

Gaetano, Arianne M., and Tamara Jacka, eds. 2004. *On the Move: Women and Rural-to-Urban Migration in Contemporary China*. New York: Columbia University Press.

Gao, Qi-jie, and Chuan-hong Zhang. 2010. 'Agricultural Technology Extension System in China: Current Situation and Reform Direction'. *Management Science and Engineering* 2 (4): 47-58.

Geertz, Clifford. 1963. *Agricultural Involution: The Processes of Ecological Change in Indonesia*. Berkeley: University of California Press.

Gilbert, Erik. 2015. 'Asian Rice in Africa: Plant Genetics and Crop History'. In *Rice: Global Networks and New Histories*, edited by Francesca Bray, Peter A. Coclanis,

Edda L. Fields-Black, and Dagmar Schäfer, 212-228. New York: Cambridge University Press.

GRiSP, Global Rice Science Partnership. 2013. *Rice Almanac*. Los Baños: International Rice Research Institute.

Hanks, Lucien M. 1972. *Rice and Man: Agricultural Ecology in Southeast Asia*. Chicago: Aldine.

He, Congzhi, and Jingzhong Ye. 2014. 'Lonely Sunsets: Impacts of Rural-Urban Migration on the Left-behind Elderly in Rural China'. *Population, Space and Place* 20 (4): 352-369.

Ingold, Tim. 2000. *The Perception of the Environment: Essays on Livelihood, Dwelling and Skill*. London: Routledge.

—. 2006. 'Walking the Plank: Meditations on a Process of Skill'. In *Defining Technological Literacy: Towards an Epistemological Framework*, edited by John R. Dakers, 65-80. New York: Palgrave Macmillan.

Jacka, Tamara. 1997. *Women's Work in Rural China: Change and Continuity in an Era of Reform*. Cambridge: Cambridge University Press.

—. 2006. *Rural Women in Urban China: Gender, Migration, and Social Change*. London: Routledge.

Jepson, Anne. 2014. 'Gardening and Wellbeing: A View from the Ground'. In *Making and Growing: Anthropological Studies of Organisms and Artefacts*, edited by Elizabeth Hallam and Tim Ingold, 147-162. Farnham: Ashgate.

Kaufmann, Lena. 2011. *Mala Tang – Alltagsstrategien ländlicher Migranten in Shanghai*. Wiesbaden: Harrassowitz.

—. 2016. 'Everyday Strategies of Rural Migrants: Assembling Skills in Mala Tang Production'. *Journal of Chinese Dietary Culture* 12 (2): 63-112.

—. 2019. 'Staying Behind, Moving On: Women, Technology and Migration in Contemporary Rural China'. *ZANTHRO*, Zurich Anthropology Working Paper No. 8: 1-20.

Kolb, Raimund Th. 2003. 'Reis'. In *Das grosse China-Lexikon*, edited by Brunhild Staiger, Stefan Friedrich, and Hans-Wilm Schütte, 618-622. Darmstadt: Wissenschaftliche Buchgesellschaft.

Kovach, Michael J., Megan T. Sweeney, and Susan R. McCouch. 2007. 'New Insights into the History of Rice Domestication'. *Trends in Genetics* 23 (11): 578-587.

LaFleur, Robert André. 2020. 'Round and Square: China's Lunar Calendar 2020'. 2020. Accessed 10 January 2020. http://robert-lafleur.blogspot.com/.

Lave, Jean, and Etienne Wenger. 1991. *Situated Learning: Legitimate Peripheral Participation*. Cambridge: Cambridge University Press.

Li, Yuyu. 2006. 'The Impact of Rural Migration on Village Development: A Comparative Study in Three Chinese Villages'. PhD diss., Johns Hopkins University.

Liu, Jieyu. 2017. 'Ageing in Rural China: State, Family and Gendered Care Responsibilities'. In *Handbook of Welfare in China*, edited by Beatriz Carrillo, Johanna Hood, and Paul Kadetz, 285-296. Northampton: Edward Elgar Publishing.

Marchand, Trevor H.J., ed. 2010. 'Making Knowledge'. *Special Issue, Journal of the Royal Anthropological Institute* 16 (May).

Marchand, Trevor H.J. 2014. 'Skill and Aging: Perspectives from Three Generations of English Woodworkers'. In *Making and Growing Anthropological Studies of Organisms and Artefacts*, edited by Elizabeth Hallam and Tim Ingold, 183-201. Farnham: Ashgate.

Meng, Xiangdan. 2014. 'The Feminization of China's Agriculture: A Sociological Analysis'. PhD diss., Wageningen University.

Mintz, Sidney W., and Christine M. Du Bois. 2002. 'The Anthropology of Food and Eating'. *Annual Review of Anthropology* 31 (1): 99-119.

Murphy, Rachel. 2004. 'Turning Peasants into Modern Chinese Citizens: "Population Quality" Discourse, Demographic Transition and Primary Education'. *The China Quarterly* 177: 1-20.

—. 2006. 'Citizenship Education in Rural China: The Dispositional and Technical Training of Cadres and Farmers'. In *Chinese Citizenship: Views from the Margins*, edited by Vanessa L. Fong and Rachel Murphy, 9-26. London: Routledge.

NBSC, National Bureau of Statistics of China. 2019. *China Statistical Yearbook 2019*. Beijing: China Statistics Press.

Needham, Joseph. 1974. 'The Nature of Chinese Society: A Technical Interpretation'. *University of Hong Kong Gazette* XXIII (5): part 2.

Netting, Robert McC. 1993. *Smallholders, Householders: Farm Families and the Ecology of Intensive, Sustainable Agriculture*. Stanford: Stanford University Press.

Ohnuki-Tierney, Emiko. 1995. *Rice as Self: Japanese Identities through Time*. Princeton: Princeton University Press.

Oxfeld, Ellen. 2017. *Bitter and Sweet: Food, Meaning, and Modernity in Rural China*. Oakland: University of California Press.

Peng, Mu. 2008. *Shared Practice, Esoteric Knowledge, and Bai: Envisioning the Yin World in Rural China*. PhD diss., University of Pennsylvania.

Perdue, Peter C. 1987. *Exhausting the Earth: State and Peasant in Hunan, 1500-1850*. Cambridge, MA: Harvard University Press.

Qi, Xing. 1986. *Traditionelle Feste und Gebräuche in China*. Beijing: Foreign Language Press.

Rawski, Evelyn Sakakida. 1972. *Agricultural Change and the Peasant Economy of South China*. Cambridge, MA: Harvard University Press.

Roberts, Kenneth D. 1997. 'China's "Tidal Wave" of Migrant Labour: What Can We Learn From Mexican Undocumented Migration to the United States?' *International Migration Review* 31 (2): 249-293.

Schippers, Thomas K. 2014. 'Agricultural Practices: Change and Stability'. In *Exploring and Explaining Diversity in Agricultural Technology*, edited by Annelou van Gijn, John C. Whittaker, and Patricia C. Anderson, 339-341. Oxford: Oxbow.

Schmalzer, Sigrid. 2016. *Red Revolution, Green Revolution: Scientific Farming in Socialist China*. Chicago: University of Chicago Press.

Schweizer, Thomas. 1989. *Reisanbau in einem javanischen Dorf*. Köln: Böhlau.

Sigaut, François. 1994. 'Technology'. In *Companion Encyclopedia of Anthropology*, edited by Tim Ingold, 420-459. London: Routledge.

Skeldon, Ronald. 1996. 'Migration from China'. *Journal of International Affairs* 49 (2): 434-455.

Smerdel, Inja. 2014. '"Training Oxen Meant Training for the Children"'. In *Exploring and Explaining Diversity in Agricultural Technology*, edited by Annelou van Gijn, John C. Whittaker, and Patricia C. Anderson, 258-268. Oxford: Oxbow.

Sternfeld, Eva. 2009. 'Biologischer Anbau "Made in China."' *EU-China Civil Society Forum Hintergrundinformationen* 12: 1-12.

Stone, Glenn Davis. 2007. 'Agricultural Deskilling and the Spread of Genetically Modified Cotton in Warangal'. *Current Anthropology* 48 (1): 67-103.

Sweeney, Megan, and Susan R. McCouch. 2007. 'The Complex History of the Domestication of Rice'. *Annals of Botany* 100 (5): 951-957.

VMZ, Ethnographic Museum of the University of Zurich. 2017. 'Skills and Aging: New Trajectories in Anthropology of Technology'. 9 May 2016. Accessed 16 February 2018. https://www.musethno.uzh.ch/de/museum/newsarchiv/Seed-Money-Grant.html.

White, Ben. 2012. 'Agriculture and the Generation Problem: Rural Youth, Employment and the Future of Farming'. *IDS Bulletin* 43 (6): 9-19.

Whittaker, John C. 2014. 'Threshing Processes and Tools – Exploring Diversity in the Past: An Introduction'. In *Exploring and Explaining Diversity in Agricultural Technology*, edited by Annelou van Gijn, John C. Whittaker, and Patricia C. Anderson, 133-135. Oxford: Oxbow.

Wilken, Gene C. 1987. *Good Farmers: Traditional Agricultural Resource Management in Mexico and Central America*. Berkeley: University of California Press.

Wittfogel, Karl August. 1957. *Oriental Despotism: A Comparative Study of Total Power*. New Haven: Yale University Press.

Wu, Chunyuan 伍春元, ed. 2010. 益相史话 *Yixiang shihua* (*Historical Narrative of Yixiang*). Beijing: Jiuzhou Press.

Xin, Liangjie, Yuzhi Fan, Minghong Tan, and Luguang Jiang. 2009. 'Review of Arable Land-Use Problems Inpresent-Day China'. *Ambio* 38 (2): 112-115.

XT, 湘潭市民间文学集成编委会 Xiangtan Shi Minjian Wenxue Jicheng Bianweihui (Editorial Board of the Comprehensive Collection of the Folk Literature of Xiangtan City), ed. 1988. 中国谚语集成湖南卷: 湘潭市分卷 *Zhongguo yanyu jicheng Hunan juan: Xiangtan shi fen juan* (*Hunan Volume of the Comprehensive Collection of Chinese Proverbs: Section Volume of Xiangtan City*). Xiangtan: N.p.

Yang, Lihui, and Deming An. 2008. *Handbook of Chinese Mythology*. New York: Oxford University Press.

Yuan, Juanwen, and Anke Niehof. 2011. 'Agricultural Technology Extension and Adoption in China: A Case from Kaizuo Township, Guizhou Province'. *The China Quarterly* 206: 412-425.

Zapata, Lydia. 2014. 'Exploring Diversity in the Past: An Introduction'. In *Plants and People: Choices and Diversity Through Time*, edited by Alexandre Chevalier, Elena Marinova, and Leonor Peña-Chocarro, 16-19. Oxford: Oxbow.

Zhang, Jiayan. 2014. *Coping with Calamity: Environmental Change and Peasant Response in Central China, 1736-1949*. Vancouver: UBC Press.

3 Reference Models for Transmitting Knowledge

Abstract
This chapter describes one specific verbal medium of paddy field knowledge transmission, farming proverbs, discussing the role these proverbs play at the nexus of rice farming and rural-urban migration in China. Based on two anthologies of oral vernacular literature, the chapter asserts that these agricultural maxims provide evidence for the transformation of farming technology and the system of knowledge transmission. In addition, it argues that, first, the strength of these sayings lies precisely in their flexibility, which has made them a platform of knowledge negotiation between peasants and the state; and, second, that these proverbs have the potential to serve as a back-up resource for retaining paddy field knowledge.

Keywords: China, oral knowledge transmission, rice-farming proverbs, negotiation of agricultural knowledge, rural-urban migration, farmer-state relationship

Yuemei and I were walking back to Green Water Village with Teacher Yang. He was a slim young man with glasses, wearing sweat pants and trainers. His ears and hands had turned red because of the cold. We had met him in the county seat of Anren and he accompanied us to the village to visit the local school, where Yuemei and he were planning some activities to encourage and support the primary school children. As we approached the village, my gaze fell on a newer red brick house (see Figure 4). Somebody had painted eight white characters on one of its walls, visible to everyone entering the village from the main road. The road was muddy, covered with melting snow and puddles. Matching the season, the theme of the characters – a proverb – was winter, too. In a rhythmic, almost rhyming parallel structure, it read:

Kaufmann, Lena, *Rural-Urban Migration and Agro-Technological Change in Post-Reform China.* Taylor & Francis Group 2021
DOI: 10.5117/9789463729734_CH03

Figure 4 A proverb painted on a wall: 'People shouldn't relax in winter, and the fields shouldn't waste in winter'

Photograph taken by the author

人不冬闲 *ren bu dong xian*
田不冬荒 *tian bu dong huang*
'People shouldn't relax in winter, and the fields shouldn't waste in winter'.

As it related to paddy fields, the proverb sparked my interest, so I asked about it. Teacher Yang illustrated the meaning by referring to his own situation. He was a primary school teacher from the area around Green Water. He had grown up cultivating rice, however, when he finished his studies he had immediately started working as a teacher. He now lived in nearby Heshi Township with his nuclear family, keeping 1.4 *mu* of rice fields in his home village. He explained that – mainly due to the weeds – paddy fields 'have to be used every year, not planting [rice] doesn't work'. Due to his profession, he was too busy for farming and he lived far away from his fields. To preserve them and maintain the fertility of the soil in spite of his absence, he lent his fields to 'other people' (*bieren*). He told me that the fields needed to be constantly used, which is also why rape seed is cultivated in winter. He stated that, for these very reasons, the proverb reminded villagers not to let their fields lie fallow in winter (personal interview, 21 January 2011).

This proverb, which sounded like a practical guideline with a moral undertone, fascinated me by the way it conveyed vital knowledge about the paddy field resource in a brief, yet beautiful form. Indicating a verbal layer of vernacular knowledge transmission, it triggered my interest to find out more about local farming proverbs. Such proverbs are, in fact, short mnemonic texts. They constitute one piece in the whole mosaic of knowledge transmission. This chapter aims to focus attention on this communicative medium (Barth 2002).

Drawing on two anthologies of oral vernacular literature,[1] it discusses the role these proverbs play in transmitting, negotiating and retaining paddy field knowledge. The proverbs stem from a major state-supported mass movement of oral literature collection in the 1980s – a crucial moment in the transformation of the knowledge system. This was an effort organized by university institutes as well as local folk literature associations and gazetteer bureaus with the help of volunteers, to collect and document Chinese vernacular rhyming sayings, folksongs and storytelling at the moment of their anticipated disappearance due to the ongoing popularity of television and modern media. Each county and city collected volumes for internal publication (*neibu*), from which separate province volumes would later take the representative or best texts, sayings and songs to be included into what would become a national series of Chinese popular literature (Flitsch 2002).

Proverbs comprise part of the repertoire of knowledge that farmers can draw on to deal with their paddy fields under various circumstances. Since they comprise a sort of reservoir of encoded knowledge, I suggest that they have the potential to give farmers specific advice, or act as resource to fall back on at the transitional moment when knowledge is being transformed.

As the French Sinologist and Durkheim student Marcel Granet observed in the 1930s, Chinese proverbs had been an important form of conceptual expression as far back as in ancient China (Granet 1934, 54). As China scholar Ingo Schäfer notes, such proverbs are a collectively possessed good that is, at least potentially, shared within society. They are 'language material' (*Sprachmaterial*), i.e. fixed imprints of readily available constructions which the Chinese language keeps in stock. Transmitted over centuries, this cliché-like stock of language forms transmits experiences and images,

1 The two anthologies come from the areas of Xiangtan and Chenzhou in Hunan Province. Xiangtan is not far from my field site, but the related anthology (XT 1988) is more comprehensive than the one from my field site Chenzhou (CZ 1988). The former lists 1002 farming proverbs, in contrast to the latter, which only contains 197. The latter were selected from more than 40,000 proverbs collected altogether, from all the counties and cities in Chenzhou (ibid., 194).

typifying descriptions, feelings, emotions, patterns of thought, insights, maxims and conceptual models. It is used to describe relationships between the world and humans, nature and society, and between human beings, as well as historical processes, conditions and world views, 'in which the most diverse sides of being and consciousness find their reflection' (Schäfer 1983, 67).[2] The usage of such proverbial language forms 'evokes a familiar horizon of opinions and judgements, establishes commonalities, and presents identifications' (ibid., 69).[3] Accordingly, this language form has been the customary way of transmitting patterns of thought and worldviews in China for centuries (ibid., 114).

It is not my intention here to trace or prove their historicity, instead I acknowledge that proverbs are an amorphous communicative medium. Being a stock of more or less fixed oral language constructions, their strength and resilience lies precisely in their flexibility to change and adapt, whilst remaining a medium that enjoys a certain authority, because it is easily depicted as a form of tradition. Because of this flexibility, the proverbs literally illustrate the transformation of the knowledge system as described in Chapter 2. I argue that, due to this flexibility, proverbs are a complex medium that may also be used to negotiate knowledge, political and moral values. Well aware of the value and potential of this communicative medium, the Chinese Communist Party (CCP) has undertaken considerable efforts to document and preserve proverbs. Moreover, it has used this particular form of communication that farmers are familiar with and understand well, to communicate new scientific and moral-political knowledge. In fact, as Schäfer notes, Mao Zedong's speeches and writings abound with proverbs and many made their way onto banners and buildings as slogans (Schäfer 1983, 42) – a common practice in contemporary rural China as well. In this way, proverbs have also become a political medium in the People's Republic of China (PRC), serving as a platform for negotiation between state scientific and farmers' knowledge.

Transmitting farming knowledge through proverbs

For centuries, various societies have used proverbs as teaching tools to convey moral values and social skills (Mieder 2004, 146). To a Western public

2 In German: 'in denen [...] die verschiedensten Seiten des Seins und Bewußtseins ihre Widerspiegelung finden' (Schäfer 1983, 67).

3 In German: 'ruft [...] einen vertrauten Horizont von Anschauungen und Urteilen wach, stellt Gemeinsamkeiten her, bereitet Identifikationen vor' (Schäfer 1983, 69).

which rarely uses vernacular oral literature, proverbs may appear to play only a complementary role in the transmission of farming knowledge. Yet, especially in oral societies, proverbs have a central role, being *the* mnemotechnic means for transmitting values and expert knowledge (Schippers 1992, 103). China is well known for its long history of literacy. Nevertheless, orality continues to be important, specifically in rural areas and among the elderly and female farmers, who have not received formal education.[4] For Chinese people an oral residue (Ong 2002) of knowledge transmission remains through a living and vibrantly developing oral literature, especially in the first decades of the PRC. Proverbs are a source of memorizing and verbally participating in knowledge at different levels of abstraction – e.g. political, metaphorical, moral, social or technical, or as situated humour.

In everyday life, before the arrival of modern scientific farming, Chinese farming proverbs were the only farming formulas available. Li Liqing, who edited the Chenzhou Volume of the *Comprehensive Collection of Chinese Folk Proverbs* (CZ 1988, 5), dedicates a considerable section of the volume's preface to agricultural proverbs. He claims that these farming proverbs are '"semi scientific" (knowing that something is true, but not knowing why)'.[5] Li asserts that every farmer in Chenzhou, where Green Water Village is located, knows these proverbs. This is '[b]ecause in the past farmers relied on Heaven to eat'.[6] If they failed to pay close attention to the changes of nature – with the help of the proverbs – they would unquestionably face a dreadful fate. He further acknowledges that farming proverbs have been of paramount importance to China's agricultural production for several millennia, as they have always had a guiding function, containing rich details about production and life experiences, and serving as a means of education and knowledge transmission. Li also recognizes their new, political meaning, i.e. the function of 'turning the [Chinese Communist] Party's guidelines and policies into the conscious actions of the masses' (CZ 1988, 5).[7]

4 From a worldwide perspective, China has a high literacy rate. Only about 5 percent of the Chinese population over 15 years old are illiterate. With 7.52 percent of Chinese women over 15 counting as illiterate, the proportion is higher among women, however (NBSC 2019, sec. 2-15), and it is even higher in rural areas and among the elderly. Moreover, rural inhabitants who only have a basic command of reading and writing are not counted as illiterate in these statistics. Oral literature thus remains important.

5 In Chinese: '"半截子科学",（知其然，不知其所以然）' *ban jiezi kexue, (zhi qi ran, bu zhi qi suoyiran)* (CZ 1988, 5).

6 In Chinese: '因为过去时代，农民靠天老爷吃饭' *yinwei guoqu shidai, nongmin kao tian laoye chifan* (CZ 1988, 5).

7 In Chinese: '使党的方针、政策，化为群众的自觉行动' *shi dang de fangzhen, zhengce, huawei qunzhong de zijue xingdong* (CZ 1988, 5).

Proverbs remain a powerful means of teaching and learning in contemporary rural China, as I observed in Green Water Village when they surfaced as codified knowledge.[8] Often, they only make sense in context and can only be fully understood by knowledgeable insiders. For instance, the following proverb on harvesting only makes sense in the context of using a sickle.

九熟十收， 十熟九收。	*Jiu shu shi shou,* *shi shu jiu shou.*	When 90 percent [of the rice] is ripe, you harvest 100 percent, when 100 percent is ripe, you harvest 90 percent. (XT 1988, 248)

Using a sickle, the grain is cut in a way that means over-ripe rice kernels would fall to the ground, because they were attached too loosely to the plant. Not all of the grain would make its way to the granary. If using a combine harvester, which cuts and collects the ripe grain directly, this knowledge becomes obsolete. This not only hints to the challenges of reinterpreting agricultural proverbs for new scientific, social, and political conditions (see Schmalzer 2016, 103-109). It also shows, once again, how Barth's (2002) faces of knowledge help us to understand the way in which media interact dynamically, rendering some proverbs unintelligible when confronted by technological change.

As encoded knowledge mnemonics, proverbs are not guidelines to acting, but rather codes to understand ways of doing in the moment of performance, in the sense of sentient ecologist Tim Ingold (2000). Reflecting on learning processes and distinguishing between knowledge and information, Ingold states that, rather than merely accumulating information, '[o]ur knowledgeability consists [...] in the capacity to situate such information, and understand its meaning, within the context of a direct perceptual engagement with our environments' (Ingold 2000, 21). It is obviously not possible to learn rice farming through the medium of proverbs alone. The Green Water migrant woman Yuemei confirmed this, talking modestly about her own rice farming skills:

> When I was little, I copied my parents and planted rice, I did quite a lot of fieldwork, and my parents explained quite a few things about rice farming to me, therefore I understand a bit more [about rice farming] than other people. (Email exchange, 17 July 2013.)

8 In fact many scholars of rural China mention the use of such proverbs, for instance Meng (2014, 77); Steinmüller (2013, 98); and Oxfeld (2017, 34, 36), although only few pay closer attention to this medium as a way of agricultural knowledge transmission, e.g. Schmalzer (2016, 108).

Yuemei was therefore invaluable in helping me to decipher some of the proverbs. Nevertheless, even she admitted: 'We have heard some [proverbs] so many times that we know them off by heart, I also know some of them. However, there are other phrases that I don't understand' (ibid., 16 July 2013). Whether this was due to her long migration experience (ten years at that point), or to the rapid technological changes since she left farming, remains unclear. As Yuemei's example shows, farming knowledge can only be fully acquired in context. Likewise, information contained in the proverbs can only properly be understood by a person who has had sensual experience of the context, gained from engaging with their environment through touch, taste, smell, seeing or hearing (Ingold 2000, 21). Proverbs can thus become optimal mnemonic aids and may also serve as teaching tools, in the form of codes. To better understand how this works, it is useful to consider the form of the proverbs and then briefly reflect on their content.

Proverb form

Investigating the form of Chinese agricultural proverbs enables us to appreciate their resilience and effectiveness in transmitting encoded content, including moral values or new information. Since the Chinese language has a limited phonetic inventory, the language contains a particularly high number of homophones, which is especially useful for creating rhymes. Rhymes, in turn, are excellent mnemonic aids.

There are relatively clear parameters when it comes to form, which also implies that the content of the proverbs may change. Like the proverb painted on the wall in Green Water Village, most of the proverbs have a similar form and structure, consisting of two parallel sentences of the same length, separated by a comma. Each sentence is between three and seven characters long. The two parts usually have a uniform syntax. Often, the last character in the first sentence rhymes with the last character in the second sentence. The following structure is particularly common (rhyming characters and homonyms are in bold):

XXX**X**, XXX**X**. (4 + 4 characters)

The following are examples of proverbial instructions on the proper handling of rice seeds and the right times for weeding:

盐水浸种， 苗不生虫。	*Yanshui jin* ***zhong,*** *miao bu sheng* ***chong.***	If you immerse the seeds in salt water, pests won't grow on the seedlings. (XT 1988, 219)

小暑大暑， 快把草除。	*Xiaoshu da**shu**,* *kuai ba cao **chu**.*	When *xiaoshu* (Lesser Heat), and *dashu* (Greater Heat) approach, quickly get on with weeding. (XT 1988, 209)

Interestingly, as Schäfer notes, Mao Zedong often used the same structure as a stylistic feature of his writings. This was a way to ensure that his audience could easily understand, focus on and remember his main points (Schäfer 1983, 73-76). Chinese policy makers today still rely on similar linguistic structures to popularize new policies. For example, in 2015 Premier Li Keqiang coined a slogan for the framework of the 'Made in China 2025' industrial strategy, which includes innovation of agricultural equipment. His eight-character slogan – *dazhong chuangye, wanzhong chuangxin* – aims for 'mass innovation and entrepreneurship' (Central Government of the People's Republic of China 2017).

However, in the collections of farming proverbs from Hunan, the following structures also appear frequently:

XXXX**X**, XXXX**X**. (5 + 5 characters)
XXXXXX**X**, XXXXXX**X**. (7 + 7 characters)
XX**X**, XX**X**; XX**X**, XX**X**. (3 + 3; 3 + 3 characters)

A few proverbs only consist of one sentence, which is usually seven characters long. Occasionally, sayings contain a list, or they comprise two sentences with varying, nevertheless rhyming, structures.

In contrast to Indo-European languages such as English, the Chinese language is able to convey substantial amounts of content in a few words. Therefore, a few characters are sufficient to succinctly transmit a range of information, and the proverbs are short enough to make them easy to memorize and recite in everyday life. Similar to the 'Song of the 24 Solar Terms' (see Appendix III), the rhymes, rhythm, and parallel sentence structure all facilitate their memorization. Grammatically, the structure is simple. Most proverbs only contain nouns and adjectives and/or verbs. Some even consist merely of a sequence of nouns. This, too, assists in remembering and understanding them.

It is interesting to view these oral proverbs (*yanyu*) in relation to the idiomatic written expressions of traditional scholars (*chengyu*). According to Schäfer, both written and oral forms of expression are products that have been generated collectively. However, oral proverbs often have a regional character, are more similar to the spoken language, and somewhat more flexible in their form. In contrast, being close to the written classical Chinese language, *chengyu* have a more fixed choice of words, syntax and grammar. Moreover, they are tied to the four-character form. Despite these differences, there is also a connection between written *chengyu* and oral *yanyu*. Accordingly, sometimes parallel written and oral versions of proverbial expressions exist. Moreover, many old written *chengyu* that remain in use today can be found in classics such as the *Shiji*, i.e. the history of ancient China by Sima Qian, or in the Confucian classics, both of which were compiled more than two millennia ago. Nevertheless, many of these *chengyu* originated from orally transmitted fables and *yanyu*. An example is the seemingly modern written idiom *li ling zhi hun* (commonly translated as 'to lose one's head through material greed'), which Mao Zedong used in order to criticize the Soviet Union's revisionism. Nevertheless, the *Shiji* explicitly indicates the oral origin of this expression (Schäfer 1983, 44-46, 50).

Stylistically, the proverbs are modest in comparison to the written Chinese language. The usage of *chengyu* in written literature denotes a sophisticated style and reflects the academic knowledge acquired by the writers. In a way, the farmers' proverbs may be perceived as their oral counterparts, standing for the practical knowledge acquired by farmers. Despite their apparent simplicity, many of the farmers' proverbs are not only instructive, but also highly melodious and witty, bearing evidence of a particular vernacular style. This style is reflected in the common usage of analogies, among other techniques, as in the following example:

田平如镜，	*Tian ping ru jing,*	The field as level as a mirror, the mud as
泥烂如浆。	*ni lan ru jiang.*	soft as paste. (CZ 1988, 161)

Likenesses are drawn, for example, with regard to seasons, cultivation steps, and water levels. Moreover, there are analogies between cultivation tasks and everyday objects outside the realm of farming. These give a special meaning and emphasis to what is being said about rice cultivation. Sometimes, they carry a moral subtext, for example when water for the field is compared to the vital importance of breast milk for humans. In sum, the distinctive form of these proverbs makes them an optimal communicative medium for memorizing and transmitting farming knowledge orally.

Proverb contents: a fall-back repertoire

In view of the paddy field predicament in particular, and rapid technological modernization in general, these proverbs have a special potential for keeping knowledge at hand. This potential is further understood by considering what type of knowledge lends itself to being transmitted in this way, and what other function the proverbs play for Chinese farmers. For these purposes, I have classified the proverbs in the two collections into three overlapping categories. While I provide some examples below, I list about 150 more farming proverbs in Appendix IV, to provide a more detailed overview of the range of rice farming proverbs and their encoded knowledge. As proverbs comprise a repertoire that is widely distributed in society, it is difficult to answer the question of personal repertoires. To my knowledge, there is no study concerning which individual farmer knows which proverbs. Therefore, we just have to take these proverbs as a resource that farmers in the twenty-first century can potentially access. Measuring the number of proverbs about a certain topic reveals which knowledge is suitable for communication via proverbs, which knowledge is regarded as important and central, and which is particularly complex. For example, there are copious proverbs about fertilizing and irrigation. According to Schmalzer (2016, 105), in the collective era the state also had a particular interest in collecting farmers' knowledge about fertilizing.

The first and most numerous category contains locally specific facts and rules that are based on and reveal experience (see Appendix, A.1-11). It provides highly detailed, practical instructions concerning the individual steps in rice farming, as in the following three examples. The first proverb is an instruction for producing 'pit fertilizer' (*dangfei*) by fermenting different substances, including excrement (see Santos 2011, 494-495). The second informs farmers how to transplant different varieties and crops correctly, while the third explains how to irrigate the seedbed properly:

沤凼肥，冇得巧， 一层土来一层草， 常灌水来常翻搅。	*Ou dangfei, mao de* ***qiao****, yi ceng tu lai yi ceng* ***cao****, chang guan shui lai chang fan* ***jiao****.*	If you soak 'pit fertilizer', there is no skill [i.e. this is the only way to do it, there is no other skilful solution]: after a layer of soil comes a layer of grass, water it often and turn it over and stir it often. (XT 1988, 229)[9]
粘插一把， 糯插三根， 棉花苗，打单身。	*Nian cha yi ba, nuo cha* ***san gen****, mianhua miao, da* ***danshen****.*	Transplant a handful of sticky *nian*-rice, transplant three roots of *nuo*-rice, plant cotton seedlings separately. (XT 1988, 233)

9 The dialectism 冇得 (pronounced *mào dái*) is equivalent to the standard Chinese 没有 *meiyou* ('not have', 'there is not').

日灌夜排扯露水， 晴晒秧田浅灌水。	*Ri guan ye pai che lu* ***shui,*** *qing shai yangtian qian guan* ***shui.***	Water during the day, drain the water at night and haul in the dew, dry the seedbed in the sun when it is sunny and irrigate it with shallow water. (CZ 1988, 163)[10]

The knowledge in this first category is communicated explicitly. It concerns the seasons, fields, seed varieties and selection, sowing and cultivating seedlings, transplanting, plant protection, fertilizing, weeding, pest management, irrigation, harvesting, and mixed cultivation. The proverbs describe in detail which cultivation steps should take place at which time, the water level which is suitable for each stage of plant growth, or the type of fertilizer which is appropriate for each type of field and crop.

The second category of proverbs concerns skill (see also A.12-14). There are less of these proverbs than those in the first category. On the one hand, the proverbs in this category centre on farmers' knowledge at a meta-level. They reveal that farm work is indeed understood as a craft, which cannot be learnt in a short period of time. Moreover, it is even more difficult to master than other crafts. Occasionally, single aspects that are perceived as being particularly difficult are emphasized, such as the cultivation of seedlings. In other cases, they stress which types of knowledge are especially important for farmers, such as knowledge about the seasons and the right moment to undertake each cultivation task. The following saying is an example of proverbs that address the skill of sensing the right moment:

打铁要看火候， 作田要抢时候。	*Da tie yao kan huo****hou,*** *zuo tian yao qiang shi****hou.***	In forging ironware you have to look at the crucial moment of temperature, in farming you have to seize the moment. (XT 1988, 207)

On the other hand, this category also contains proverbs that refer to aspects of farming knowledge more implicitly and subtly. This is often embodied knowledge, as in the following instructions on how to harvest, thresh and winnow rice, as well as plough properly at each stage of tillage:

割禾要轻， 打禾要稳。	*Ge he yao qing, da he yao wen.*	To cut the grain you have to be gentle, to thresh the grain you have to be stable. (CZ 1988, 165)

10 Here 扯 *che* means 'to haul' or 'to foster'; the dew is perceived as positive.

重打轻扬， 粒粒进仓。	*Zhong da qing **yang,** li li jin **cang.***	If you hit hard and winnow gently, all the grain will enter the granary. (XT 1988, 249; see also CZ 1988, 165)
头道深， 二道浅， 三道象洗脸。	*Tou dao **shen,** er dao **qian,** san dao xiang xi **lian.***	The first round [of ploughing must be] deep, the second round shallow, the third round is like washing your face. (XT 1988, 215)

Two topics are central to the proverbs in this category: the working body and handling agricultural tools. The first provides insights about the body as a tool, and about the knowing body, especially hands and legs or feet. The second theme reveals facets about the skilled handling of tools. In addition, bodily hardship, strain, effort and diligence are highlighted regularly, and it is stated that rice farming is not only about skill, but also about physical endurance and perseverance, as the following two adages suggest:

一粒粮食一滴汗。	*Yi li liangshi yi di han.*	One grain of rice, one drop of sweat. (XT 1988, 207)
换土如换金， 全靠手脚勤。	*Huan tu ru huan **jin,** quan kao shou jiao **qin.***	Changing the soil is like changing money, everything depends on hard-working hands and feet. (XT 1988, 216)

While the first proverb could also be used in the form of a joke, to express consent or sarcasm, as a metaphor or exaggeration, this category of proverbs is, at the same time, a way of rendering tacit knowledge and skill more explicit. It is one of many facets this oral medium affords.

The third, and smallest category (see also A.15) is rather heterogeneous and stands apart from the two others. I have categorized it as a group because it relates to the intensification of rice farming. While migration is never mentioned in these proverbs, they relate to the knowledge that is crucial when taking decisions about labour allocation, which is important for migration decisions. The following three proverbs remind farmers about the rewards of intensifying rice production:

田土是个宝， 越耕越是好。	*Tian tu shi ge **bao,** Yue geng yue shi **hao.***	Field and soil are a treasure, the more you plough, the better it is. (XT 1988, 214)

田间管理如绣花， 工夫越细越到家。	*Tianjian guanli ru xiu**hua**, gongfu yue xi yue dao**jia**.*	To care about the fields is like embroidery, the more thorough the effort, the more perfect it is. (XT 1988, 243)
修好塘坝， 本小利大。	*Xiu hao tang **ba**, ben xiao li **da**.*	If you repair the ponds and dykes well, this will entail small capital and large benefit. (CZ 1988, 160; XT 1988, 224)

In contrast, the two proverbs below warn farmers about the consequences of not caring for their fields properly. This is relevant knowledge when it comes to taking migration-related land-use decisions.

衣服不洗要脏， 田不耕种要荒。	*Yifu bu xi yao **zang**, tian bu geng zhong yao **huang**.*	If you don't wash your clothes, they become dirty, if you don't plough and plant your fields, they become barren. (XT 1988, 214)
田荒三年是草， 土荒三年是宝。	*Tian huang san nian shi **cao**, tu huang san nian shi **bao**.*	If a [paddy] field lies barren for three years it is [full of] weeds, if the soil [used for dry farming] lies barren for three years it is a treasure. (XT 1988, 216)

Schippers notes that there are numerous proverbs in the instances where farmers have the possibility to take practical action (Schippers 1992; 2014). His comments about proverbs related to the timing of farm work in southern France also seem valid for rice farming proverbs from Hunan. Schippers bases his hypothesis on the observation that, in southern France, there are plenty of proverbs about the busy times of sowing and harvesting, with virtually none about other times of the year. In addition, he discerns that pastoralists who follow a regular working cycle have more working proverbs than farmers. This leads him to suggest that there is 'a circular causality between local folk knowledge and the possibilities of performing concrete action on crops and animals'. By adhering to the idea of a repertoire of knowledge, he highlights how proverbs reflect the modalities of actions which pastoralists and farmers respectively 'can undertake concretely to prevent accidents or "worries"'. Referring to a set of proverbs about good and bad years, he concludes that the folk knowledge which farmers need to master, 'not only expresses the mental and social "domestication" of the various agricultural times but also of the unforeseen and the unpredictable events that characterize every new yearly cycle' (Schippers 2014, 79-80).

From this perspective, proverbs are indeed a repertoire of communicated knowledge that farmers can draw on in response to diverse situations. This becomes especially clear in those proverbs from Hunan about time, particularly the seasons. These proverbs relate to the issue of planning, an aspect that deserves closer attention, because it indicates the strategic agency of rice farmers. The following are examples of proverbs that focus on time:

种是金，土是银，错过节气无处寻。	*Zhong shi jin, tu shi yin, cuoguo jieqi wuchu xun.*	Planting is gold, the soil is silver, but if you miss the solar terms you have nowhere to search. (XT 1988, 207)
迟动三天手，减少半年粮。	*Chi dong san tian shou, jianshao ban nian liang.*	If you get to work three days late, the grain [to harvest] is reduced by six months. (XT 1988, 208)
惊蛰早，清明迟，春分犁田正当时。	*Jingzhe zao, qingming* ***chi****, chunfen li tian zheng dang* ***shi****.*	*Jingzhe* (Awakening of Insects) is too early, *qingming* (Clear and Bright) is too late, *chunfen* (Spring Equinox) is exactly the right time for ploughing the fields. (XT 1988, 209)

The sheer number of proverbs about time reveals that knowledge about the right season and the perfect moment – Kairos – are regarded as crucial in rice farming. Moreover, they suggest that time management and foresight are crucially important. This foresight extends to at least one cultivation per year – even longer when practices of seed selection or the provision for drought years are included.

As anthropologist Jan Patrick Heiss demonstrates in his case study of a Manga village in Niger, the planning of fieldwork is specific and only explicit to a certain extent. This is because not every step may be scheduled in detail. Therefore, knowledge about possible situations, their processes and related actions have to be included in the planning stage (Heiss 2003). This is also the case in rice farming. Many of the proverbs may be seen as references to preventive knowledge about potential scenarios and how to respond to them, for example with regard to strategies for minimizing risks when planting several varieties of rice. It is mainly these proverbs about the seasons and selecting the right moment that implicitly remind farmers that foresight is an important precondition for their actions. In this sense, the proverbs not only prompt farmers about important cultivation steps and measures, but also provide them with a repertoire of wide-ranging potential blueprints for action. This is what makes them so valuable, not least in the current situation of transition.

Educating the masses

In view of their potential, it is no coincidence that proverbs have also become a political medium for educating farmers. As Schäfer asserts, even before the communist victory, during the Yan'an period of the late 1930s and early '40s, Mao Zedong reflected intensively on language. For Mao, finding a proper linguistic form and style that the masses could easily absorb and understand was a central requirement for realizing a Chinese Marxism. Mao therefore turned to the language of the 'masses' (*qunzhong*), which he perceived as being particularly rich, vivid, and expressive with regard to real life. He formulated the task of learning from the language of the masses – mostly peasants – in order to develop a language that would appeal to precisely these masses. Proverbial language, in particular, played a special role in this process (Schäfer 1983, 12-13, 21, 112).

Mao deliberately adopted proverbial language forms to convey his ideas. These could be either ancient proverbs – whether in their original form or adapted – or self-created proverbs, written or oral. Often, he started or finished an idea or sentence with proverbs, or aligned several in a row. Two examples of his usage of proverbial forms are the above-mentioned written *chengyu* 'to lose one's head through material greed' and the oral proverb *bu ru hu xue, yan de hu zi*, which literally translates as 'how can you catch tiger cubs without venturing into the tiger's den?' and corresponds to the English 'nothing ventured, nothing gained' and which appears in Mao's philosophical essay *On Practice* (1965 [1937]). In this and numerous other cases, Mao Zedong used language forms that stemmed from the life-world of his audience, reinterpreted them and placed them into new contexts, feeding them back to his audience (Schäfer 1983, 33-39, 90-92).

Mao used this familiar and potentially shared form of communication to popularize his ideas and give them a 'Chinese flavour' (*chinesischen Geschmack*). Proverbial language helped him to translate Marxist theories into less foreign and abstract concepts that were more accessible, digestible and intelligible to a wide audience (ibid., 69, 81). Using proverbial forms allowed him to tie them in with a rich cultural heritage of historical worldviews and speak directly to the experience horizons of his audience, for example by 'associatively evoking complexes of historical experiences of subjugation and exploitation' (ibid., 33).[11] Such collective 'language images' (*Sprachbilder*) had an emotional effect on his audience, who could easily decipher their figurative meaning. Mao used proverbs to either link a new thought to a well-known historical formula or transmit new content through such a formula. Referring to autochthonous experiences and insights helped to lend authority and emphasis to what he said, confirm his ideas, illustrate the content, prove his theories, and reformulate his thoughts in a memorable way. At the same time, the older images which the proverbs evoked also entered into developing the new ideas (Schäfer 1983, 33-34, 71, 81).

The examples provided by Schäfer suggest that educating farmers, the Communists' core constituency, was mainly pursued in view of propagating moral-political values. However, such ideological propaganda went hand in hand with more practical issues. One example was the introduction of Green Revolution farming methods in the 1960s and 1970s. As Schmalzer notes, however, it was quite a political and ideological challenge to value and refer to old farmers' knowledge on the one hand, while rejecting 'traditional knowledge' for the sake of 'scientific knowledge' on the other. In practice, collectors and extension agents found themselves in a situation that sometimes required the reinterpretation of farmers' knowledge in order to match it with the presiding farming ideal of the time (see Schmalzer 2016, 103-107). While today the tension regarding the evaluation of traditional and scientific knowledge remains, the current context of the farmers' paddy field predicament poses yet more new challenges. It is therefore interesting to take a closer look at folk literature collection efforts in the reform period.

Textualizing vernacular knowledge

The early experiences of collecting folk literature and using it as form of propaganda in the 1930s set the stage for the Communist Party's future collection projects in the 1980s (Flitsch 2002, 223). However, the reform

11 In German: 'assoziativ Komplexe historischer Erfahrungen der Knechtung und Exploitation wachrufen' (Schäfer 1983, 33).

period collection efforts also had a new dimension. In addition to using oral literature for educational purposes, the systematic national collection efforts of the 1980s implied a desire to salvage this oral heritage. In the PRC, however, it is not seen as contradictory to both safeguard folk literature and use or adapt it for contemporary educational purposes.

One of the reform period collection projects was the *Santao Jicheng* project from the 1980s. The project combined the desire to salvage material which was anticipated would soon be lost in the incipient era of modernization, with educational purposes. It therefore has a special place within the CCP tradition of collecting folk literature. With regard to safeguarding folk material, the project was implemented at precisely the threshold of the major socio-economic and technological transition outlined in the preceding chapters. This was no coincidence. It was, as Mareile Flitsch (2002) describes, at this time in the early 1980s that folklorists became aware of what had vanished. This was at a moment when they had just been rehabilitated, after folkloristic activities had been banned during the Cultural Revolution. Scholars sensed the rapid and enormous changes to come and, therefore, saw the need to document folk literature comprehensively and systematically, before it was gone forever. This view was shared by the folklorists who had not been allowed to train students during the Cultural Revolution and were now growing old. Thus, the Research Association of Folk Literature and Folk Art, which the scholars belonged to, applied to undertake a national project which would systematically document folk literature – the *Santao Jicheng* project (ibid., 224-225).

Once approved by the government, this project was initiated by the Ministry of Culture's National Commission for Nationality Affairs and the Chinese Association of Folk Literature in 1984. Under the leadership of a national editorial committee, funded by the state, and informed by clear guidelines, in 1985 the *Santao Jicheng* project embarked on the comprehensive collection of folk literature at every administrative level. The aim was not only to preserve, but also to disseminate folk literature. Between 1984 and 1990 alone, about 2 million people engaged in the collection process, collecting 1,840,000 folk tales, 3,020,000 folk songs, and 7,480,000 proverbs. The collected and edited material began being published from the end of the 1980s onwards, in the counties, regions and cities of every Chinese province. This resulted in the step-by-step publication of the serial *Zhongguo minjian wenxue santao jicheng*. These Three Comprehensive Collections of Chinese Folk Literature – in short *Santao Jicheng* – comprise the three genres of folk tales, folk songs, and proverbs (ibid., 225-226). The two anthologies from Hunan Province I draw on (CZ 1988; XT 1988) have their roots in the *Santao Jicheng* project.

In view of the transformation of knowledge described in Chapter 2, it is now time to reevaluate the collection efforts of this project. On the one hand, as explained below, as ethnographic sources the proverbs presented in this book illustrate farmers' knowledge and the transformation of this knowledge. On the other hand, they also provide glimpses into the CCP's practices of educating farmers. In this regard, the prefaces of the two Hunan anthologies describe in detail that those proverbs that were perceived as good and educational were included in the publication, while others have been rejected. What is more, the collected material now appears as a treasure of farming formulas that may assist the future transmission of farming knowledge, as well as offer a medium to cling on to and adapt in the moment of agro-technological and social transformation.

Negotiating knowledge and farmer-state relationships

Proverbs are not only a means of propaganda and education, they also provide a platform for the negotiation of knowledge. The following four proverbs are prime examples of how the CCP used proverbs to impose moral and technical knowledge onto farmers. In view of the vigorous promotion of Green Revolution agriculture outlined in Chapter 1, it appears reasonable to assume that the state had some influence in crafting these proverbs. The first focuses on 'scientific cultivation':

科学种田， 越种越甜。	*Kexue zhong* ***tian,*** *yue zhong yue* ***tian.***	If you farm scientifically, the more you plant, the sweeter it gets. (XT 1988, 243)

The second, rather slogan-like proverb addresses the issue of mechanization:

要想农业发展快， 必须实现机械化。	*Yao xiang nongye fazhan kuai,* *bixu shixian jixiehua.*	If agriculture is to develop quickly, you have to realize mechanization. (XT 1988, 243)

The third saying refers to chemical fertilizer:

种田肥当家， 巧施氮磷钾。	*Zhong tian fei dang* ***jia****,* *qiao shi dan lin* ***jia****.*	In farming, fertilizer rules the roost, skilfully apply nitrogen, phosphorus, and potassium. (CZ 1988, 162)

Finally, high-yielding rice varieties are mentioned:

麻婆崽多， 短禾谷多。	*Ma po zai* ***duo****,* *duan he gu* ***duo****.*	The pockmarked woman has many children, the short rice [i.e. semi-dwarf high-yielding varieties] has many grains. (CZ 1988, 159)[12]

These four proverbs, which promote scientific and mechanized rice cultivation, are extremely interesting. They retain the old form in order to codify new knowledge. As the saying goes, they are like 'new wine in old wineskins'. They point towards a flexibility of the medium of proverbs that Schäfer also observes in Mao's use of *chengyu* (see Schäfer 1983, 57-67).

In a way, the proverbs can also be seen as an interesting interface when it comes to negotiating the farmer-state relationship. The proverbs have various, interrelated faces, which make them so flexible. These range from serving as memorates of technical farming details, as guidelines for potential action, and as a medium of political and moral education. While the state may use them to transmit scientific knowledge, farmers might use the very same proverbs in a sarcastic or joking manner, for example to express some scepticism or disregard for the knowledge the state was trying to impose.

What is certain is that farmers are not passive adopters of the encoded knowledge promoted by the government. Just as they adopt and adapt the farming technologies introduced by the state strategically, they also adopt and adapt the codes and knowledge encoded in the proverbs, incorporating their own experiences with these new technologies into their available repertoire of knowledge. In the twenty-first century, the proverbs are therefore much more than just simple sayings about folk wisdom. Along with other resources such as technologies, labour, knowledge, and skill, the proverbs

12 The legendary Qing Dynasty 'pockmarked woman' is best known for her still widely-appreciated tofu dish from Sichuan Province called *mapo doufu*. Using her image in this proverb thus has both a traditional and a positive connotation.

are now part of the socio-technical resources that farmers have for finding and negotiating complex solutions to manage their paddy fields. The next two chapters show what such solutions look like in practice.

References

Barth, Fredrik. 2002. 'An Anthropology of Knowledge'. *Current Anthropology* 43 (1): 1-18.

Central Government of the People's Republic of China. 2017. '李克强谈"中国制造2025": 从制造大国迈向制造强国 Li Keqiang tan "Zhongguo zhizao 2025": Cong zhizao daguo maixiang zhizao qiangguo (Li Keqiang on "Made in China 2025": From a Manufacturing Country towards a Manufacturing Power)'. Accessed 9 January 2020. http://www.gov.cn/guowuyuan/2017-08/10/content_5216727.htm.

CZ, 郴州地区民间文学集成编委会 Chenzhou Diqu Minjian Wenxue Jicheng Bianweihui (Editorial Board of the Comprehensive Collection of the Folk Literature of Chenzhou District), ed. 1988. 中国民间谚语集成湖南卷: 郴州地区分卷 *Zhongguo minjian yanyu jicheng Hunan juan: Chenzhou diqu fen juan (Hunan Volume of the Comprehensive Collection of Chinese Folk Proverbs: Section Volume of Chenzhou District)*. Chenzhou: N.p.

Flitsch, Mareile. 2002. 'Mündliche Überlieferungen und die Problematik ihrer Aufzeichnung in der VR China'. *Oriens Extremus* 43: 221-236.

Granet, Marcel. 1934. *La pensée chinoise*. Paris: Albin Michel.

Heiss, Jan Patrick. 2003. *Zur Komplexität bäuerlicher Feldarbeit in Afrika: Eine Fallstudie in einem Manga-Dorf (Niger)*. Münster: Lit.

Ingold, Tim. 2000. *The Perception of the Environment: Essays on Livelihood, Dwelling and Skill*. London: Routledge.

Mao, Zedong. 1965. *On Practice*. Beijing: Foreign Languages Press.

Meng, Xiangdan. 2014. 'The Feminization of China's Agriculture: A Sociological Analysis'. PhD diss., Wageningen University.

Mieder, Wolfgang. 2004. *Proverbs: A Handbook*. Westport: Greenwood Press.

NBSC, National Bureau of Statistics of China. 2019. *China Statistical Yearbook 2019*. Beijing: China Statistics Press.

Ong, Walter J. 2002. *Orality and Literacy: The Technologizing of the Word*. 2nd ed. London: Routledge.

Oxfeld, Ellen. 2017. *Bitter and Sweet: Food, Meaning, and Modernity in Rural China*. Oakland: University of California Press.

Santos, Gonçalo. 2011. 'Rethinking the Green Revolution in South China: Technological Materialities and Human-Environment Relations'. *East Asian Science, Technology and Society* 5 (4): 479-504.

Schäfer, Ingo. 1983. *Populäre Sprachformen und politische Argumentation: Zur Funktion der Idiomatik in den Schriften Mao Zedongs*. Frankfurt am Main: Haag + Herchen.

Schippers, Thomas K. 1992. 'Le proverbe comme forme de transmission de savoirs: Quelques réflexions à propos d'un corpus de proverbes météorologiques'. In *Les voies de la parole*, edited by Jean-Noël Pelen and Claude Martel, 99-105. Aix-en-Provence: Université de Provence.

—. 2014. 'Seasonal Variations in Crop Tending and Folk Knowledge in Southern France'. In *Exploring and Explaining Diversity in Agricultural Technology*, edited by Annelou van Gijn, John C. Whittaker, and Patricia C. Anderson, 79-80. Oxford: Oxbow.

Schmalzer, Sigrid. 2016. *Red Revolution, Green Revolution: Scientific Farming in Socialist China*. Chicago: University of Chicago Press.

Steinmüller, Hans. 2013. *Communities of Complicity: Everyday Ethics in Rural China*. New York: Berghahn.

XT, 湘潭市民间文学集成编委会 Xiangtan Shi Minjian Wenxue Jicheng Bianweihui (Editorial Board of the Comprehensive Collection of the Folk Literature of Xiangtan City), ed. 1988. 中国谚语集成湖南卷: 湘潭市分卷 *Zhongguo yanyu jicheng Hunan juan: Xiangtan shi fen juan* (*Hunan Volume of the Comprehensive Collection of Chinese Proverbs: Section Volume of Xiangtan City*). Xiangtan: N.p.

4 Technological Choice in the Wake of Migration

Abstract
This chapter analyses the strategic technological decisions which Chinese rice farmers take to manage their farmland in a context of rural-urban migration. Based on ethnographic field research, it does so mainly through the example of one left-behind woman and her choice of harvesting technologies. Proposing a repertoire perspective on technological choice, the chapter sheds light on the diverse socio-technical factors behind such decision-making. It argues against a linear perspective of technological development, showing why it makes sense for farmers to simultaneously draw on a repertoire of old and new technologies, rather than simply opting for mechanization in order to compensate for the migrated labour. This also provides additional insights into the complex relationship and causalities between agricultural technology and migration.

Keywords: China, agricultural decision-making, mechanization of harvesting technology, rural-urban migration, socio-technical system, repertoires of technology

Approaching Green Water Village for the first time, I accompanied Yuemei, who was coming home after a year of migrant work in Beijing. In Green Water, she would meet her left-behind mother Mrs. Luo, her father, the construction worker Zhou Wenlu, and her two younger siblings who had also migrated, to celebrate the approaching New Year of the Rabbit. Squashed into the loading space of a three-wheeled autocycle, the main road led us directly through rice fields. The rice had already been harvested and I wondered why long stalks were sticking out of the ground in some fields, while they had been cut short in others (see Figure 5). Yuemei told me that this was related to the choice of harvesting technology, depending on whether a combine harvester or a sickle had been used.

Kaufmann, Lena, *Rural-Urban Migration and Agro-Technological Change in Post-Reform China*. Taylor & Francis Group 2021
DOI: 10.5117/9789463729734_CH04

Figure 5 Fields harvested with a combine harvester (left) and a sickle (right)

Photograph taken by the author

Upon arrival in Green Water, we were warmly welcomed by Mrs. Luo, who offered us fermented rice and other food. At nightfall, she told me to share the bed with Yuemei. There we slept – Yuemei and I head to toe – under the same cover on a thin, hard, compact yet airy mattress made of rice straw.

Mrs. Luo was a short, slim woman with a big smile. Her long black hair was tied into a loose ponytail that she had tucked under a pink knitted hat. As the village houses had no heating, in and around the house she wore brown stripey trousers, a dark blue jacket with red sleeve protectors, and colourful slippers that she had crocheted. Talking with her about harvesting, I was astonished to discover that she actually used both harvesting technologies in her fields, cutting some of her rice by hand with a sickle and some with a machine, rather than opting for just one of these two competing technologies.

Whenever I walked through the village, I paid special attention to the farming technology in use. Some tools were simply leant against houses, others were kept behind them, stored in sheds or attics. I was amazed to come across a whole potpourri of diverse technologies, ranging from stone mortars to motorized ploughs. I was struck by the way that people safeguarded them, keeping them all, even if they had stopped using them. For example, Mrs. Luo's neighbour had not disposed of her threshing tub, even though she used a motorized threshing machine. Mrs. Luo's brother, who lived in a nearby village, kept a manual rice mill and stones to husk rice manually. Meanwhile, Mrs. Luo's husband Zhou Wenlu took his rice to a fellow villager who owned an electric mill. I was intrigued to find out more about their choices and, as I inquired further, I found that the reasons behind choosing a particular technology were complex and, in many cases, intimately related to the phenomenon of migration. For Mrs. Luo, it was a central part of the arrangements she and her family had figured out to manage the household's fields in the face of the out-migration of her husband, her daughter Yuemei and the other two grown-up children.

Drawing on these insights, this chapter sets out from the basic assumption that there is a close relationship between technological change and social change such as the migration phenomenon. I suggest that the villagers' practices stand in contrast to the narrative conveyed by local gazetteers and statistics, which draws a rather linear picture of technological development. As Francesca Bray notes, such an understanding of technological development is also equivalent to the common-sense Western model of technological progress. It is one marked by capitalist criteria of efficiency, such as mechanization and economies of scale, which we have all internalized at school and which, for a long time, has tended to portray societies who do not implement these technologies as inferior or, in the case of China, as a failure (Bray 1994, xiv; see also Sigaut 1994, 435). However, my ethnographic research reveals a picture that, in reality, is more complex, less monocausal and linear. In this picture – a snapshot of Green Water's 'history of technology-in-use' (Edgerton 2007, xi) – farmers are not passive recipients of the new technologies promoted by the government. They are social actors who consciously choose, evaluate, and use different technologies, based on their available resources and to suit their individual circumstances.

When looking at the intersection of farming technology and migration, mechanization has generally attracted the most scholarly attention. It is now widely recognized, by scholars from various disciplines, that there is a close relation between mechanization and migration, and that the labour-saving capacities of mechanization can be a powerful way for rural households to diversify their livelihoods, by enabling some family members to take up migrant work in the cities (see, e.g., Rigg, Salamanca, and Thompson 2016).

In Green Water Village, mechanization is also often used in this way. However, it is intriguing that mechanization only partially substitutes manual technologies – despite being so vigorously promoted by the Chinese government and welcomed by many farmers for sparing them from tedious farmwork. I suggest that considering the issue of technological choice and migration from the perspective of the anthropology of technology may shed some light on the issue. It helps to explain why farmers use competing technologies simultaneously to protect their fields, and why they retain and store other obsolete tools. Overall, it provides some insights into the complexity of everyday life decisions and the projects that villagers pursue.

Following Barth (2002), I understand farming technologies, in their socio-technical dimension, as a medium in which paddy field knowledge is stored and transmitted as a vital part of the villagers' repertoire of knowledge for dealing with the paddy field resource. As archaeologist Helena Knutsson notes, farm tools can be used as mnemotechnic resources, improving memory 'by storing information in material objects and their treatment' (Knutsson 2014a, 278). In investigating the earliest introduction of farming to the not yet farming-focused environment of Scandinavia (about 4000-2000 B.C.E.), she claims that the new tools that were introduced can actually be seen as a 'handbook of farming practices', i.e. 'a kind of physical manual with attached narratives, which ensured the success of the [farming] enterprise' (Knutsson 2014b, 310). The idea of agricultural technology as a 'farming manual' is useful for understanding the role that particular technologies play in the specific layout and transmission of the overall system of farming knowledge. In Green Water Village and other rice farming areas, while well-known and long-used tools retain knowledge, new tools introduced into this system are mnemotechnic resources that represent new solutions for particular problems in rice farming.

Technologies do not stand alone. They are are 'imbued with meaning, acquired, transmitted, and performed in a social context' (Smerdel 2014, 286). In order to be effective as knowledge stores, they require a practitioner to use them with the help of their skilled body. Anderson and colleagues speak of 'muscle memory', referring to agricultural skills that have gradually become incorporated in farmers' bodies (Anderson et al. 2014, 5). As the aspect of the skilled body has already been discussed in Chapter 2, I will not refer to it explicitly below. Nevertheless, I conceive the body as well as the wider community of practice in which knowledge about tools is developed, practiced, and transmitted as integral elements of the technologies described. In view of this, it is useful to distinguish between technologies and techniques. Drawing on Francesca Bray, I understand technology broadly as 'social-material networks or systems, including sets of techniques and equipment, but also trained personnel, raw materials, ideas and institutions' (Bray 2008, 320). In this chapter, I mainly focus on farm tools and machines as two specific types of technology, and less on 'techniques', although I do include some in my descriptions when they compete with or complement certain technologies. 'Techniques' denote 'the skilled practices that go into the material production of knowledge as well as the production of artefacts' (ibid.). In practice, all technologies require certain techniques of handling, and all techniques have a level of interaction that includes components such as the body or a tool.

What is more, technologies and techniques are never merely technical, but always also socio-cultural productions (Lemonnier 1993b). It is therefore highly relevant to look at technological choices in order to better understand Chinese households' strategies for coping with their predicament. According to Pierre Lemonnier, the term 'technological choices' highlights 'the sorting of possibilities on which the development of a technical system is de facto based' (Lemonnier 2012, 301). It refers to the process of selection as well as to its results. Which farming technologies are used in practice depends on various factors. While Lemonnier adopts a long-term perspective that spans several millennia, stating that technological choice mostly happens unconsciously and unintentionally (ibid.), I focus on farmers' more immediate situations and argue that their choice is clearly strategic. Moreover, it is not necessarily exclusive. Together, the repertoire of technologies can provide solutions to different socio-technical problems that have occurred before and may potentially reoccur again, in the form of, for instance, a shortage of fuel, electricity or cash, or a wave of return migration due to an economic crisis.

Tilling with power ploughs and oxen

One brief example of strategic choice relates to tilling technology. The main tool used in Green Water Village for deeper tillage is the plough, either pulled by oxen or operated with motors. Moreover, there is a whole range of other tillage tools, which mostly serve for surface operations. Most households around Green Water use a power plough. While in 2011 they did not always own one, they could rent one from other villagers at the cost of 90 to 100 Yuan (about 13 to 14 USD) per *mu*. Farmers from nearby Paishan County reported paying only half that price. In any case, only one or two hours were needed to plough one *mu* with this machine (see Figure 6).

In contrast, as several Green Water villagers explained to me in 2011 and 2017, an ox would take a whole day to plough the same amount of land. Moreover, an ox needs to be tended and grazed, and the farmer needs to pay attention to it, making sure that the animal does not eat or

Figure 6 Field preparation with a power plough

Photograph taken by Zhou Yuemei

trample on other farmers' crops (see also Perdue 1987, 133). Nevertheless, it remains useful. First, some fields are so 'deep and soft' that a machine would get stuck in them, so they can only be ploughed with the help of cattle. Second, it is cheaper to use cattle than rent a machine. Besides, oxen serve as a form of insurance, since they can be sold for cash in times of need. Moreover, their dung can be used as manure, as stated in the following proverb: 'The dung of an ox fertilizes three fields of seedlings' (*yi tiao niu de fen, san miao tian de fen*) (XT 1988, 230). Oxen manure is cheaper, and better quality, than chemical fertilizer. These are some of the reasons why some households continued to raise and use oxen, either alone or in combination with tilling machinery. Moreover, they also continued to use hoes, an indispensable and multifunctional tool. While it would be possible (although labour-intensive) to use only a hoe for field preparation, in the 2010s the Green Water villagers employed it in spaces that were inaccessible to oxen and machines, such as the corners of a field. In making their choices, they thus considered a whole range of factors. Among these, financial and time considerations were especially important when it came to migration, as in the case of Yongcai's household, who stopped raising oxen when the three sons migrated.

Harvesting with sickles and combine harvesters

A second example of strategic choice is even more complex and concerns harvesting technology. Harvesting technology is particularly apt for looking at strategic choices, because the harvest represents one of the two peak seasons in rice farming (see Chapters 2 and 5 for planting). In order not to lose any of the harvest, the activity has to be carried out quickly, i.e. within a time frame of only ten days. In this context, the combine harvester is especially appealing for migrant households who lack labour. However, the sickle also retains some advantages and – as the case of Mrs. Luo shows – it is not a simple choice of either/or. Rather, the two technologies present elements in a repertoire from which to choose, according to varying household circumstances.

Sickles are the conventional harvesting implement in Green Water Village. Today's sickles are industrial products (see Figure 7). Since about 2006, sickles with wooden or plastic handles have been produced, which may be purchased for less than five Yuan (about 0.7 USD) at the local market. According to Grandpa Luo, a part-time carpenter, these last for about five years, two years longer than hand-crafted sickles (personal interview, 24 January 2011). Harvesting with a sickle is performed by grabbing a bushel of the ripe rice plant with the left hand, while the right hand cuts the bushel several centimetres above the ground, making a movement close to the ground swiping from right to left, towards the body. Thus, the upper part of the body is bent towards the field, and the knees are slightly bent (see also Bray 1984, 335).

For an in-depth understanding of the use and choice of the sickle, it is useful to draw on the comparative insights of François Sigaut. According to Sigaut, using a sickle has been the dominant harvesting technique in Europe, Asia and North Africa for many centuries. In his analysis, Sigaut highlighted important connections between the tool and the harvested product. These also apply to Green Water Village and contribute towards rendering the choice of harvesting technology so complex. In his descriptive classification, or 'technical lineage' (*lignée technique*) of harvesting tools, the sickle-technique is the eighth of nine techniques Sigaut outlined. Each technique involves a certain 'way of action' (*mode d'action*) (pulling out, picking up, beating, stripping off, breaking off, cutting through pressure, cutting through friction, cutting through friction with a launched tool) to obtain a certain 'product' (*produit*) (whole plants, grains and spikelets, spikes and panicles, a handful of stalks, an ensemble of stalks) (see Sigaut 1991, 33, 37).

Figure 7 A hand-made sickle (above) and an industrially-produced sickle (below)

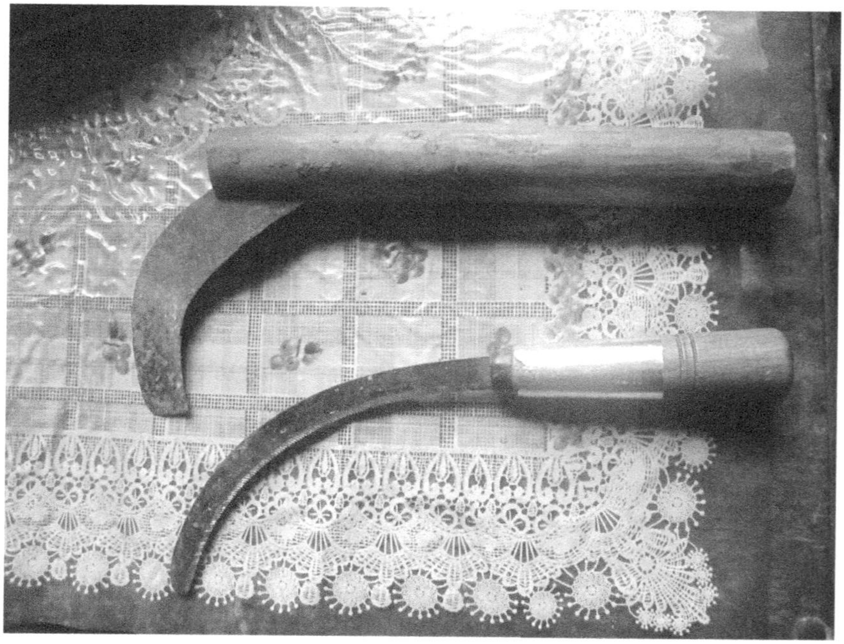

Photograph taken by the author

According to this classificatory scheme, the sickle collects a 'handful of stalks' (*une poignée de tiges*) held in one hand by 'cutting' (*couper*) them with the other hand 'through friction' (*par friction*). This technique is particular with regard to the stalks, because it implies that the cut ears of the grain are not cut off, but they rest on the stalks.[1] On the one hand, this is safer for the cutter, as it grants a security distance between tool and body. On the other hand, keeping part of the stalk on the harvested grain enables collecting, transporting, and processing the harvested grain in characteristic sheafs (Sigaut 1991, 33, 37). The sickle is therefore seen as the technological solution to harvesting the straw together with the grain, implying that the straw is valuable enough to deserve the extra effort in threshing and transport (Anderson and Sigaut 2014, 90).

Moreover, there is a linkage between tool and farm animals. In this regard, Sigaut suggested some interesting connections between the usage of a sickle and the usage of animal power. He asserted that, unlike a harvesting knife or manual harvesting, the sickle can cut considerable amounts of grain at

1 For the strategic advantages of using a finger knife to cut just the grain, see Miles (1979) and Bray (1984, 330).

once. This bigger amount implies higher requirements for the labour needed for transportation and threshing, leading him to conclude that it would seem to only be economical to use sickles in combination with (draught) animal power. At the same time, it valorizes the straw produced, making it usable as fodder and litter for the animal (Sigaut 1991, 41). The latter certainly used to be the case in Green Water Village.

Sickles also have implications with regard to gender and the social organization of sickle-harvesting knowledge. Sigaut claimed that, while the more nimble-fingered harvesting techniques such as finger knives exclusively required women's labour, sickles is done by various groups. Harvesting with a sickle is typically a woman's task in non-Mediterranean Europe and India, but a man's task in the semi-arid areas of Morocco, Spain, central Asia and north China (ibid., 41-42). In southern China's Green Water Village, harvesting with a sickle is performed across age and gender boundaries. Yuemei recalls harvesting from her school years in the 1990s, when children were sent to the harvested field to glean, i.e. collect left-over rice panicles and kernels by hand.[2] This was an aspect of the 'part-work and part-study system' (*qin gong jian xue*), comprising both educational activity and contributing to society. Generally, however, she added that men, women, and children (including herself) all had to help harvesting with a sickle, and nobody was spared from the hectic job of cutting the grain (personal and text conversations, 23 January 2011 and 24 November 2016). In view of enskilment, this may be seen as an important step in ensuring that children acquire the skills for rice farming. The introduction of the combine harvester has certainly transformed this practice.

Combine harvesters have only recently been introduced to Anren County through a county government initiative. According to the *Anren County Gazetteer*, rice was harvested mechanically in Anren for the first time in October 1998. At that time, the County Department of Agriculture had invested more than 100,000 Yuan (about 14,130 USD) to purchase a combine harvester from Zhejiang Province. More than 3000 people are reported to have watched the spectacle of a demonstration by this exciting new vehicle. In 2000-2001 the department brought in six more machines, setting up teams to provide machine-harvesting services for local farmers for 50 Yuan (about 7 USD) per *mu*. Next came 30 new combine harvesters in 2003, processing 2490 hectares, which corresponds to about 10 percent of the total arable area (ACGCC 2011, 284, 300-302). In addition, several privately-owned combine harvesters are kept in the area. Hunan has a relatively high number of

2 According to Sigaut (1991, 41), gleaning is typically a woman's task in Asia.

Figure 8 The only combine harvester in Green Water

Photograph taken by the author

combine harvesters – a machine which is not common in every Chinese rice growing area – compared to other Chinese provinces. With around 130,000 combine harvesters in 2018, the province ranks sixth nationally (NBSC 2019). In fact, between 2000 and 2018 alone, there has been a more than forty-fold increase in combine harvesters in Hunan (HPBS 2019, sec. 12-8).

The farmers from Green Water confirmed that combine harvesters were a very recent innovation. Only the richest farmers could afford them – there was just one machine in Green Water Village, which Hugen's family had purchased in 2007 (see Figure 8). In 2011, Hugen was a stout and confident villager in his early thirties, practising seasonal migration to raise his two children, one each from his current and former wives. He also owned the largest house, and his household was said to be the richest in Green Water. Hugen stated that the combine harvester had cost 50,000 Yuan (about 7060 USD) and that his family had saved up for eight years to buy it (personal interview, 1 February 2011).

Ever since Hugen's family bought the combine harvester, Green Water villagers had been using it for cutting and threshing. They paid Hugen up to 100 Yuan (about 14 USD) per *mu*, a considerable cost that not everybody was willing or able to afford. Hiring manual harvesting services to replace the missing labour would be even more expensive (see Chen 2016), which is why some farmers continued to rely on the household labour they had available.

Despite its high cost, the combine harvester wrought considerable time and labour savings. Hugen explained that it took about ten minutes for him to harvest one *mu*. Only one person was needed to run the machine, with a second one walking behind it to put the rice into bags. In contrast, four people would take about one day to harvest one *mu* manually. This is an important factor when it comes to making technological choices when faced with off-farm migration. Renting the services of a combine harvester means that migrant workers do not need to return for the busy harvesting season. Besides, it has enabled villagers such as left-behind women to pay somebody else for the task of harvesting, rather than performing it themselves or engaging in labour exchange (personal and text conversations in 2011 and 2016).

With regard to the division of labour and knowledge, it is notable that combine harvesters in Hunan are operated exclusively by men, sparing women, children, and migrants from much of the harvesting work. Mrs. Luo's husband Zhou Wenlu sees this in relation to the complex and uneven physical features of the terrain, which render it difficult to operate the machine. In contrast, he states that it is easier to drive a combine harvester in the flat terrain of northern China, which is why women there can also drive a combine harvester (text conversation, 29 April 2017). Here it is worth noting that certain new skills are needed to drive a combine harvester, and that the new technology and the skills to operate it are attributed only to men, which points towards the transformation of the knowledge system, including the issue of deskilling discussed in Chapter 2.

Choosing harvesting technologies

Regarding the choice between a sickle and a combine harvester, both technologies ensure the cultivation and, hence, the protection of the rice field. At the same time, each of them has certain advantages and disadvantages that farmers weigh against each other. My aim here is not to provide an exhaustive list of factors around decision making, but rather to suggest that the choice is indeed complex and goes beyond simple economic reasoning.

Labour and costs are the two central factors Green Water villagers take into account when choosing between a sickle or a combine harvester. They are also interconnected when it comes to opting for mechanization in the context of migration. On the one hand, migrant remittances – as well as migrants' skills to earn them – enable those remaining in the village to afford harvesting services, as in the case of left-behind woman Mrs. Luo, or to invest in purchasing machines, like combine harvester owner Hugen. On the other hand, the need to replace migrated labour through labour-saving technology is obvious. This is related to the double effect of migration on agriculture through missing labour and increased cash income through remittances, which has been widely documented in studies examining the impact of migration on agricultural production (see, e.g., Davis, Carletto, and Winters 2010).

Reality is more complex, however, and farmers not only consider the financial and human capital available for making technological choices. When it comes to harvesting, the choice of method is generally related to the following factors: the desired part of the plant, including the possible use of the stems; plant morphology; field density; and soil type (Anderson and Sigaut 2014, 92). Moreover, different techniques and technologies affect each other, because farming is a system (Sigaut 1991). Whittaker compares this mutual influence to interdependent organisms in a given system. From this perspective, technologies 'occupy a particular functional niche within their social and technical environment, interacting and sometimes competing with other technologies' (Whittaker 2014a, 355). His analogy helps to better understand the competitions and dynamics between different technologies. In Green Water Village in 2011 both sickles and the combine harvester co-existed. This is, because a 'technology [here the sickle] survives as long as it maintains a competitive edge – technical, economic, or even social – over technologies with similar functions [here the combine harvester]' (ibid.).

The combine harvester's main competitive edge, in view of the missing emigrated labour, is certainly its labour-saving capacity. In contrast, sickles are not only a low-cost implement, but they also cut the plants closely to the ground and therefore have the competitive advantage of producing rice straw. This is in contrast to the long, standing rice straw that the combine harvester leaves behind on the field, as described in the introduction of this chapter. That straw is lost and cannot be used for other purposes (see Chapter 2). Although increased cash incomes through migration have rendered farmers more independent from the product of rice straw, they still need to decide and weigh up if this independence is affordable or makes sense.

Within the sphere of farming, one of rice straw's main purposes is to produce fodder for the oxen. Because oxen are first and foremost kept as ploughing animals, a change in harvesting technology directly affects the choice of ploughing technology. This is an inverse instance that perfectly confirms the close connection between the sickle harvesting technique and the use of draught animals observed by Sigaut (1991, 41). Moreover, regarding the connection between instructing animals and instructing children mentioned in Chapter 2, the influence of harvesting technology on ploughing technology will eventually have consequences for the transmission of rice farming skills in many other realms besides harvesting.

In addition, the choice of a different harvesting technology also means that, as ever more grain is harvested, farmers will need to depend on new means of carrying, threshing and processing this higher volume (ibid.). There is a clear connection to the harvesting method applied prior to threshing and, more generally, to the whole *chaîne opératoire* or 'operational sequence' of the entire agricultural process (Leroi-Gourhan 1964; Anderson 2014). Aware of how rice farming machines affect the use of other machines, Oshiro (1985, 328) has termed this 'technical linkage', and singled out this linkage as one decision-making factor in purchasing harvesting and other farm machinery in Japan.

The combine harvester – a machine that combines the tasks of harvesting and threshing – also illustrates that the shift in threshing technology is affected by a shift in harvesting technology. Nevertheless, the combine harvester is singular, because it only produces the grain. Here it becomes evident that, like the choice of harvesting technology, the method chosen for threshing is strongly connected to the product desired, e.g. grain, straw in whole stems, broken straw; for human or animal consumption; or as a building material. Moreover, as Whittaker shows, the choice is linked to many other interrelated factors, e.g. social, environmental, crop-specific, and technological (see Whittaker 2014b).

Besides these more complex considerations, there are a whole range of further practical factors that come into play when choosing between a sickle and a combine harvester. First, as Grandpa Zhou stated, the amount of rice planted was a factor in the decision: if farmers cultivated little rice (an option favoured by many left-behind household members), i.e. less than one *mu*, they generally preferred to use a sickle for harvesting (personal interview, 28 January 2011). One-to-three *mu* of rice fields, explained Zhou Wenlu however, was a good size for using a combine harvester (text conversation, 29 April 2017). A smaller cultivated area could be managed with manual techniques in the given time frame, even with few people. At the same time, the human harvesters did not have to bear the financial burden of machine harvesting.

Second, because not every plot was suitable for machine harvesting, the size of the field played a role. In Hugen's words, if a field was too small, the combine harvester 'cannot enter' it. This echoes the findings of other scholars on the relation between farm size and mechanization. For example, Tian et al. (2015, 1249) state that, in nearby Poyang, Jiangxi Province, a combine harvester is used on larger and flatter plots, which then also has an influence on crop choice (rice or cotton). An awareness of this relationship between mechanization and farm size has led policy makers in China and elsewhere to conclude that bigger farms are a necessary precondition for modernizing farming (see Rigg, Salamanca, and Thompson 2016, 119). However, despite Green Water villagers' use of machines on bigger plots, in the overall picture of the resilience of smallholder or family farms, and in the ongoing debate about whether mechanization actually fosters bigger farm sizes or supports small farms, my findings point towards the latter. Green Water villagers use mechanization to ensure cultivation and, hence, to protect fields for the family; but generally empirical findings on the issue remain inconclusive (ibid., 125-126).

Third, the shape of a field and the characteristics of the soil are important. For instance, Mrs. Luo used a sickle in some parts of her field, and a combine harvester in other parts of the same field. This was because the machine would get stuck in the mud in certain places. This was also confirmed by Grandpa Zhou and by the combine harvester's owner, Hugen, who explained that he would not drive his combine harvester into a field if there was too much water in it and the soil was too muddy (personal interviews in January-February 2011).[3]

Overall, the complexity of factors that come into play outlined here – which provide merely a glimpse into the issue of technology choice – contribute to explaining why scholars have not yet been able to clearly determine the relationship between migration and the adoption of agricultural mechanization, despite numerous efforts to do so (see Rigg, Salamanca, and Thompson 2016, 125). Other factors not mentioned here may involve further practical issues, such as the way a crop is sown (i.e. broadcast or transplanted) and the way it ripens (evenly or at different stages) (Bray 1984, 322, 335), the availability of subsidies for machinery and extension services, as well as socio-cultural factors. For example, mechanization frees up fragile left-behind elderly people to offer their labour in exchange, or gives other households with several members the possibility of offering their labour in return for cash or other favours. Other examples are influences based on gender and age (see Song 1998; Yuan and Niehof 2011), political factors, or the use of certain technologies as markers of status or identity (Lemonnier 1993a, 18-19). Furthermore, in reviewing the factors around purchasing combine

3 See also Bray (1994, 56).

harvesters when they first reached Japan in the 1970s, Oshiro (1985) finds that critical 'time windows' for accomplishing certain tasks, time saved which can be used for other income-generating activities, loans, reduction of heavy work, desire for ownership, social values (the feeling of being a modern farmer), village dynamics (having an equal status with others), and technical linkages all played a role.

The association of certain techniques and technologies with 'backwards' or 'modern' farming practices also arose in Green Water village. For example, in comparing stone mills and mechanized mills, Grandpa Zhou held the opinion that the stone mill was an 'underdeveloped machine' (*bu fada de jiqi*) and he preferred to mill his rice mechanically rather than manually (personal interview, 28 January 2011). Similarly, the villagers perceived combine harvesters and other machines as 'modern'. This deprecation shows clear parallels to the perception of traditional toilet and fertilizing techniques as backwards, as mentioned by Santos (2011, 497), and to the pervasive narrative of linear technological development underlying the local gazetteers. This hints at the discursive level of technology adoption, which is not necessarily congruent with actual practices. In fact, it seems that in practice Green Water farmers' perceptions of 'modern' and 'backward' farm implements only had a minor impact on their choices. Other factors around mechanization weighed heavier, said Luo Baowen – for example, that a machine would get through the work 'very quickly' and 'doesn't make [you] too tired', or that it 'sets free the labour force' (*jiefang laodong li*) (personal interview, 22 January 2011). Nevertheless, the hint at modernity shows how pervasive the narrative of farmers and backwardness in contrast to science and modernity is, persistent to a degree that not just state agents, but even farmers themselves refer to it (Schmalzer 2016, 108-109).

Technological choice from a repertoire perspective

Chinese rice farmers now have at their disposal several generations of farm implements. Each of these technologies provides solutions for particular problems in dealing with the paddy field resource under varying circumstances. In adopting certain agricultural technologies, farmers weigh up multifaceted practical factors – not just considerations of the available labour and financial capital. Technological choices involve numerous factors, ranging from the technical-ecological to the socio-cultural (Lemonnier, 1993b). Moreover, because farming is a socio-technical system, the outcomes of particular choices often affect other spheres of the system, which renders the choice even more complex.

Here, however, I would like to shift the focus: away from these distinguishing decision-making factors, towards highlighting the repertoire character of the knowledge and skills inscribed in these technologies that underlie technological choice; and away from discourse towards practice. In fact, as Lemonnier notes, 'it seems that societies choose between a number of possible technical solutions', some of which may at first sight appear illogical in terms of their material achievements, but nevertheless follow their own socio-cultural logics (Lemonnier 1993a, 16). Such a repertoire perspective contrasts with the standard narrative of linear development, in which 'new' replaces 'old', as presented in official discourse and partly also in farmers' discourse on 'backwards' and 'modern'.

In this regard, a look at actual agricultural practices reveals that Green Water villagers in fact make use of many 'non-synchronous'[4] technologies simultaneously, which also occurs in other farming systems around the world (see van Gijn, Whittaker, and Anderson 2014). In Green Water Village, it is most obviously exemplified by the simultaneous use of the oxen and 'power plough', the sickle and the combine harvester, as well as by the coinciding practices of transplanting and direct seeding. From a repertoire perspective it becomes clear that farmers are taking up technologies strategically in a way that best fits their current circumstances and constraints. This also partly explains why some seemingly old technologies persist, despite strong government and economic incentives to replace them with newer ones.

Finally, a repertoire perspective also sheds light on causalities. With reference to the long-standing debate about the causal relationship between agro-technological change and population size (Malthus 1798; Boserup 1965), the case of Green Water Village contributes to highlighting the complexity of factors revealed by post-Boserupian research, including ecological, political-economic, and social factors (see Stone 2001). Regarding the more immediate question of whether new technologies cause migration through freeing up labour, or whether migration causes the adoption of new technologies because of the need to replace labour, and through the availability of remittances to invest in technology, my data suggest that the situation is, in fact, multicausal. In the first instance, the introduction of labour-saving technologies in combination with the introduction of the household system in the 1980s certainly set free millions of farmers. In the

4 The concept of non-synchronicity goes back to the German Marxist philosopher Ernst Bloch (1885-1977). For more about its origin and more recent adoptions, see Flitsch (2008, 270).

twenty-first century, these technologies have become an established part of farmers' household strategies that can be used strategically to enable part of the household to pursue migrant work in the city – or abandoned again according to complex socio-technical logics.

References

ACGCC, 安仁县志编纂委员会 Anren Xianzhi Bianji Weiyuanhui (Anren County Gazetteer Compilation Committee). 2011. 安仁县志 *Anren Xianzhi, 1989-2003* (*Anren County Gazetteer, 1989-2003*). Changsha: Hunan Renmin Publishing.

Anderson, Patricia C. 2014. 'Trampling the Crops with Animals'. In *Exploring and Explaining Diversity in Agricultural Technology*, edited by Annelou van Gijn, John C. Whittaker, and Patricia C. Anderson, 138-140. Oxford: Oxbow.

Anderson, Patricia C., Annelou van Gijn, John C. Whittaker, and François Sigaut. 2014. 'The Dimension of Tools, Skills and Processes: Exploring Diversity'. In *Exploring and Explaining Diversity in Agricultural Technology*, edited by Annelou van Gijn, John C. Whittaker, and Patricia C. Anderson, 3-15. Oxford: Oxbow.

Anderson, Patricia C, and François Sigaut. 2014. 'Introduction: Reasons for Variability in Harvesting Techniques and Tools'. In *Exploring and Explaining Diversity in Agricultural Technology*, edited by Annelou van Gijn, John C. Whittaker, and Patricia C. Anderson, 85-92. Oxford: Oxbow.

Barth, Fredrik. 2002. 'An Anthropology of Knowledge'. *Current Anthropology* 43 (1): 1-18.

Boserup, Ester. 1965. *The Conditions of Agricultural Growth: The Economics of Agrarian Change under Population Pressure*. London: Allen & Unwin.

Bray, Francesca. 1984. *Agriculture*. In *Science and Civilisation in China*, Vol. 6, Part 2, edited by Joseph Needham. Cambridge: Cambridge University Press.

—. 1994. *The Rice Economies: Technology and Development in Asian Societies*. Reprint. Berkeley: University of California Press.

—. 2008. 'Science, Technique, Technology: Passages between Matter and Knowledge in Imperial Chinese Agriculture'. *The British Journal for the History of Science* 41 (3): 319-344.

Chen, Chunhong 陈春红. 2016. '晾晒难 收割难 水稻收获两痛点如何破解 Liangshai nan shouge nan shuidao shouhuo liang tongdian ruhe pojie (Drying Is Difficult, Harvesting Is Difficult: How to Break the Two Sorrows of Harvesting Paddy Rice)'. 中国农机化导报 *China Agricultural Mechanization Herald*. 30 August 2016. Accessed 9 September 2016. http://www.camn.agri.gov.cn/Html/2016_08_30/2_1842_2016_08_30_30001.html.

Davis, Benjamin, Gero Carletto, and Paul C. Winters. 2010. 'Migration, Transfers and Economic Decision Making among Agricultural Households: An Introduction'. *The Journal of Development Studies* 46 (1): 1-13.

Edgerton, David. 2007. *The Shock of the Old: Technology and Global History since 1900*. Oxford: Oxford University Press.

Flitsch, Mareile. 2008. 'Knowledge, Embodiment, Skill and Risk: Anthropological Perspectives on Women's Everyday Technologies in Rural North China'. *East Asian Science, Technology and Society* 2 (2): 265-288.

van Gijn, Annelou, John Whittaker, and Patricia C. Anderson, eds. 2014. *Exploring and Explaining Diversity in Agricultural Technology*. Oxford: Oxbow.

HPBS, Hunan Provincial Bureau of Statistics. 2019. *Hunan Statistical Yearbook 2019*. Beijing: China Statistics Press.

Knutsson, Helena. 2014a. 'Blades as Messengers of Agriculture: A Case Study from Scandinavia'. In *Exploring and Explaining Diversity in Agricultural Technology*, edited by Annelou van Gijn, John C. Whittaker, and Patricia C. Anderson, 278-285. Oxford: Oxbow.

—. 2014b. 'The Complex Art of Changing Lifestyles on the Verge of the Neolithic'. In *Exploring and Explaining Diversity in Agricultural Technology*, edited by Annelou van Gijn, John C. Whittaker, and Patricia C. Anderson, 295-310. Oxford: Oxbow.

Lemonnier, Pierre. 1993a. 'Introduction'. In *Technological Choices: Transformation in Material Cultures Since the Neolithic*, edited by Pierre Lemonnier, 1-34. Oxford: Oxford University Press.

—., ed. 1993b. *Technological Choices: Transformation in Material Cultures Since the Neolithic*. London: Routledge.

—. 2012. 'Technology'. In *The Oxford Handbook of Linguistic Fieldwork*, edited by Nick Thieberger, 298-316. Oxford: Oxford University Press.

Leroi-Gourhan, André. 1964. *Le geste et la parole I: Techniques et langage*. Paris: Albin Michel.

Malthus, Thomas. 1798. *An Essay on the Principle of Population*. Reprinted in 1986. Harmondsworth: Penguin Classics.

Miles, Douglas. 1979. 'The Finger Knife and Ockham's Razor: A Problem in Asian Culture History and Economic Anthropology'. *American Ethnologist* 6 (2): 223-243.

NBSC, National Bureau of Statistics of China. 2019. 'National Data: Annual by Province'. Accessed 6 January 2020. http://data.stats.gov.cn/english/easyquery.htm?cn=E0103.

Oshiro, Kenji K. 1985. 'Mechanization of Rice Production in Japan'. *Economic Geography* 61 (4): 323-331.

Perdue, Peter C. 1987. *Exhausting the Earth: State and Peasant in Hunan, 1500-1850*. Cambridge, MA: Harvard University Press.

Rigg, Jonathan, Albert Salamanca, and Eric C. Thompson. 2016. 'The Puzzle of East and Southeast Asia's Persistent Smallholder'. *Journal of Rural Studies* 43: 118-133.

Santos, Gonçalo. 2011. 'Rethinking the Green Revolution in South China: Technological Materialities and Human-Environment Relations'. *East Asian Science, Technology and Society* 5 (4): 479-504.

Schmalzer, Sigrid. 2016. *Red Revolution, Green Revolution: Scientific Farming in Socialist China*. Chicago: University of Chicago Press.

Sigaut, François. 1991. 'Les techniques de récolte des grains: Identification, localisation, problèmes d'interprétation'. In *Rites et rythmes agraires: Séminaire de recherche*, edited by Marie-Claire Cauvin, 31-43. Lyon: GDR-Maison de l'Orient.

—. 1994. 'Technology'. In *Companion Encyclopedia of Anthropology*, edited by Tim Ingold, 420-459. London: Routledge.

Smerdel, Inja. 2014. 'Concluding Remarks'. In *Exploring and Explaining Diversity in Agricultural Technology*, edited by Annelou van Gijn, John C. Whittaker, and Patricia C. Anderson, 286. Oxford: Oxbow.

Song, Yiching. 1998. '"New" Seed in "Old" China: Impact of CIMMYT Collaborative Programme on Maize Breeding in South-Western China'. PhD diss., Wageningen University.

Stone, Glenn Davis. 2001. 'Theory of the Square Chicken: Advances in Agricultural Intensification Theory'. *Asia Pacific Viewpoint* 42 (2-3): 163-180.

Tian, Qing, Daniel G. Brown, Lin Zheng, Shuhua Qi, Ying Liu, and Luguang Jiang. 2015. 'The Role of Cross-Scale Social and Environmental Contexts in Household-Level Land-Use Decisions, Poyang Lake Region, China'. *Annals of the Association of American Geographers* 105 (6): 1240-1259.

Whittaker, John C. 2014a. 'Some Principles of Technological Decline: The Case of the Tribulum'. In *Exploring and Explaining Diversity in Agricultural Technology*, edited by Annelou van Gijn, John C. Whittaker, and Patricia C. Anderson, 355-356. Oxford: Oxbow.

—. 2014b. 'Threshing Processes and Tools – Exploring Diversity in the Past: An Introduction'. In *Exploring and Explaining Diversity in Agricultural Technology*, edited by Annelou van Gijn, John C. Whittaker, and Patricia C. Anderson, 133-135. Oxford: Oxbow.

XT, 湘潭市民间文学集成编委会 Xiangtan Shi Minjian Wenxue Jicheng Bianweihui (Editorial Board of the Comprehensive Collection of the Folk Literature of Xiangtan City), ed. 1988. 中国谚语集成湖南卷: 湘潭市分卷 *Zhongguo yanyu jicheng Hunan juan: Xiangtan shi fen juan* (*Hunan Volume of the Comprehensive Collection of Chinese Proverbs: Section Volume of Xiangtan City*). Xiangtan: N.p.

Yuan, Juanwen, and Anke Niehof. 2011. 'Agricultural Technology Extension and Adoption in China: A Case from Kaizuo Township, Guizhou Province'. *The China Quarterly* 206: 412-425.

5 Land-Use Strategies

Abstract
This chapter describes the land-use and land-arrangement strategies that Chinese migrant and left-behind rice farmers use to manage their farmland and off-farm migration. These include both social and technical strategies – only some of which accord with state expectations – such as leaving behind family members, building houses on farmland, using labour-saving technologies, switching from rice to cash crops, or even abandoning fields. Using specific household cases, the chapter demonstrates how peasants draw on a wide repertoire of available resources to handle their situation. Shedding light on the logics behind these decisions, it argues that, in taking seemingly technical agricultural decisions, farmers are in fact pursuing various long-term and short-term projects that best match their fluctuating current and anticipated future household situation.

Keywords: China, socio-technical household strategies, land-use arrangements, rural-urban migration, migrant-left-behind nexus, intensive and de-intensive rice farming

During my stay in Green Water Village I learned that making strategic use of farming technology was only one way to preserve paddy fields under conditions of missing labour due to migration. Mrs. Luo and her family drew on a whole repertoire of strategies, consisting of proven techniques as well as seemingly experimental, or even drastic measures.

In fact, the entire living arrangement of Mrs. Luo's household was a strategic response to cope with the paddy field predicament, and not an easy one. As mentioned before, Mrs. Luo's husband Zhou Wenlu was a migrant in his early fifties, working for a construction company that moved to different sites across the country every year. This slim and earnest man with a suntanned, beardless face had been the first to leave the household. That was around the beginning of the reform period in the late 1980s and early 1990s, when his three children – Yuemei, her younger sister and brother

Kaufmann, Lena, *Rural-Urban Migration and Agro-Technological Change in Post-Reform China*. Taylor & Francis Group 2021
DOI: 10.5117/9789463729734_CH05

– started primary school. Having only received basic schooling himself, but being fond of studying and reading books to learn autodidactically about things such as medicinal herbs, he longed to give his children a good education. Thus, he succumbed to the pressure of paying for their education. In addition, he hoped to provide his son Pengyu with a sufficient material basis to find a good wife who would eventually bear grandchildren and take care of himself and Mrs. Luo in their old age. Zhou Wenlu therefore put up with the burden of leaving the family alone. He accepted the tiring working conditions common on many construction sites: the seven-day working schedule, the nights in crowded containers, where men from all over China speak to each other in different dialects, squeezed into bunk beds, with the only private space consisting of a bed slot divided from the rest of the room by a mosquito net.

Once the three children had grown up, they also left the village in search of the good life. Yuemei, the eldest, was the first child to leave in the early 2000s. She studied diligently and passed the difficult entrance examination for a Beijing university. Although her sister Linjie did not manage to gain a place at university, she followed Yuemei to Beijing anyway, where Yuemei financially and organizationally supported Linjie's vocational training instead. One year the two sisters had felt lucky to see their father more frequently, when his company had also been working in Beijing. Meanwhile their little brother, Pengyu, followed the cohort of younger village men. He went south to Guangdong Province, where he found a job in mining, just as other fellow villagers had done before him. This job was facilitated by Yuemei, who had graduated by then and begun working in a German company, and so was able to pay for the digger operating training he required.

Meanwhile, the somewhat frail Mrs. Luo remained all by herself in Green Water. This was not her native village (*niangjia*), but the village she had moved to when she married Zhou Wenlu. At that time, mobile phones were not yet in general use to keep in touch regularly and, during the first years of Zhou Wenlu's migration, there were not even telephone lines in the village.[1] Thus, Mrs. Luo had to wait for the Spring Festival until she would finally see her husband and children again. The rest of the year, she usually lived on her own, taking care of the household's fields, and trying her best to maintain the rice cultivation. Only after Yuemei and her sister consecutively married other migrant workers and each gave birth to a baby in Beijing, did Mrs. Luo leave the countryside for the first time, visiting her daughters in order to

1 Telephone lines were laid in 2002 and mobile phone communication enabled in 2003 (Wu 2010, 246).

help them out. She soon returned to Green Water, however, because she felt ill, which the family attributed to her body not being used to city life. Back home, she continued looking after the fields. Occasionally she visited her aged parents, who lived close by. More frequently, she spent time working alongside the other women left behind in Green Water.[2]

Based on brief case studies such as this one of Mrs. Luo and her family, this chapter looks at the land strategies of migrants and their left-behind household members, which include both land use and land arrangement. I investigate farmers' strategic decisions between intensive and de-intensified rice cultivation, in view of their available socio-technical resources. I argue that land-use decisions are not simply the application of different techniques and technologies on production decisions. Instead, farmers are, in fact, pursuing larger 'projects' (Ortner 2006; Farquhar 2006). One of these projects is certainly the long-term preservation of the paddy field resource, thereby retaining an important social and material safety net. Other projects range from finding a marriage partner, ensuring security in old age, continuing the patriline, affording their children's education, safeguarding their own health or, more generally, getting the best out of both the rural and the urban world in search of the 'good life'.

It is useful to apply a repertoire perspective to the strategies described in this chapter, as this allows for a comprehensive analysis that goes beyond simply examining individual strategies, as is common in the literature. Most of the strategies take the form of concrete 'agricultural practices' (Schippers 2014a, 339). At their base lies a repertoire of knowledge and skills for dealing with paddy fields in diverse circumstances. This includes particular techniques and technologies to protect the paddy field resource, to transform fields into other valuable resources or, in rare cases, even to allow it to deteriorate. Moreover, a repertoire perspective adds the necessary historical depth. Drawing on Schippers (2014a), I suggest that many of the agricultural practices that make up the repertoire of knowledge span a longer time frame, containing knowledge that has been accumulated, tested, and adapted in local society on a long-term basis.

Table 2 provides a simplified overview of twelve land-use strategies that the investigated farmers from Hunan and Anhui pursued in view of their paddy field predicament.[3] Moreover, it lists these strategies along with the

2 Based on personal observations and multiple conversations with Mrs. Luo's family, 2010-2016.

3 Increased animal husbandry and fishery as well as growing grains for animal feed instead of human food seem to be further important strategies (see OECD 2005, 52-54; Huang 2016). Except for one small pig farm, I did not observe these in my field sites, however.

Table 2 Overview of the villagers' land-use strategies

	Strategy	Actors	State position
1	Leaving behind family members	Migrants with paternal parents	Indirectly promoted
2	Renting out the fields	Migrants and those left behind in need of labour	Tolerated, to some extent promoted
3	Seasonal return	Migrants working nearby or in flexible conditions	Allowed
4	Mutual help	Able-bodied left-behind people with the necessary network	Allowed
5	Hiring labour	Left-behind people who can afford it	Allowed
6	Labour-saving technologies	Left-behind people who can afford it	Promoted
7	Direct seeding	Left-behind people in need of labour	Promoted if in connection with modern field management practices that ensure sufficient grain yield
8	Single-season rice	Elderly and infirm left-behind people in need of labour	Tolerated
9	Abandoning the fields	Migrants and left-behind people in need of labour	Forbidden
10	House construction	Sons of marital age and their parents	Forbidden if on farmland
11	Dry fields	Left-behind people in need of labour	Tolerated
12	Cash crops	Left-behind people in need of labour and cash	Tolerated, revenue source for local governments

main actors involved and the position of the state, which represents the major structure in which farmers operate. Farmers use some strategies to pursue their own projects in line with state expectations. This is the case with those strategies that allow farmers to sustain intensive rice farming (Table 2, 1-7). These are generally tolerated or encouraged by the state. In practice, however, the position of the local and the central state may differ somewhat. Generally, though, even though countermeasures are not always enforced, the state is rather oppositional to those strategies that entail a de-intensification of rice farming (Table 2, 8-12). This is because the state has its own projects of grain sufficiency, agricultural productivity and, more generally, national sovereignty and stability.

For analytical purposes, I present the strategies here as being distinct and modular. In reality, however, farmers usually employ several different strategies at the same time, adapting them according to changing circumstances. Moreover, the differentiation between migrant and left-behind actors is not so clear-cut in practice, as most strategies involve both parties and the categories themselves are fluid. Even though some may seem simple, it should be kept in mind that every strategy is the result of complex decision-making processes and of balancing different, not always easy options.

Some of these strategies may seem surprising because they do not contribute to preserving the paddy fields. For centuries rice production in China has followed a logic that aimed primarily at intensifying rice production and cultivating sufficient rice to feed growing populations (Rawski 1972, 140). Recently, this has changed, and not only in China, where today's farmers follow new logics of using land and allocating labour (Rigg, Salamanca, and Thompson 2016, 128). Thus, in light of farmers' household circumstances and, from their personal perspectives, these strategies do make sense, even when they conflict with state imperatives.

Sustaining intensive rice farming

The following strategies allow farmers to sustain intensive rice farming despite part of their households being away from the fields.

Leaving behind family members

One strategy for migrants is to leave some family members behind to cultivate the fields. This is the most efficient way to ensure that fields stay in their hands, preserving the paddy field resource for the whole family, including returning migrants. This was a common strategy among my interview partners from Hunan and Anhui provinces. Econometric and human geographic studies confirm that it works well, especially where there are elderly individuals who can be left behind (Carter and Yao 2002; Xie and Jiang 2016). In view of patrilocality, these are preferably the husband's parents, as in the case of Grandpa Zhou. As explained in my introduction, he stayed behind with his wife and disabled mother while his son Zhou Wenbao, his daughter-in-law and two granddaughters, Lanying and Lanxiang, all migrated. If there are no elderly members to leave behind, however, it is more difficult to pursue this strategy, as in the case of the restaurant owner Mr. Wu from Anhui. His parents had died, while he had migrated to Shanghai with his entire household. In need of labour in the city to run the business, there were no other close household or family members left to take care of his fields in his absence. He therefore followed a different strategy, i.e. renting out the fields.

However, some migrants leave behind family members other than paternal parents (see Fan and Wang 2008, 211-214). This was the case of left-behind Mrs. Luo, whose situation was described at the beginning of this chapter. Her case is quite unusual though, because women of her age – in their early fifties – often migrate, too. The family had decided that Mrs. Luo's frail health impeded her from migration. In this way, Mrs. Luo could simultaneously benefit from and protect the safety net provided by the paddy field resource.[4]

Renting out the fields

Another common land-arrangement strategy is to lease or lend the fields to other villagers. This is usual for migrants and left-behind people who are unable to cultivate all the family fields by themselves. It was pursued by Green Water villagers, as well as my interviewees from Anhui, and is also widely described in the literature on farming and migration.[5] This strategy entails continuing intensive rice farming and, hence, preserving the paddy field resource. In this way, skilled family labourers are replaced by other skilled labourers rather than by labour-saving technologies or techniques, although combining land rental with a change of technologies is also possible and common.

It seems that in China there has been a transition from more in-family arrangements in the early 1990s towards more land rentals in the 2000s. In fact, this would correspond to a shift in migration patterns, with the increasing participation of women and migrating couples since the new millennium (Gaetano and Jacka 2004; Zhou 2005; Bossen 2011). As more able-bodied family members migrate, households increasingly need to make external arrangements for their paddy land.

The amount of land rented out is considerable. According to a survey of 525 migrant households in Hunan Province, in the late 1990s, more than half of the farmers had rented out land, corresponding to 28.7 percent of the total contracted area (Fang 1998, 171). This proportion is almost as high

4 Here it would be particularly interesting to investigate how exactly such complex decisions are negotiated among household members, and how these various aspects are weighed against each other in the long run. This would be the subject of a future study, however.

5 See, for instance, Li (2006); Fan and Wang (2008); Tilt (2008); Jin and Deininger (2009); He and Ye (2014); Nguyen, Rigg and Derks (2015); and Xie and Jiang (2016).

as the average amount of Chinese land rented out in 2016 (Glenn and Yao 2016). Such land rentals are legally sanctioned, to enable villagers to adjust to the increasing mismatch between allocated land and labour (Heerink et al. 2007, 794).

The following migrant case provides an insight into making the decision to rent out paddy land. I got to know Mr. Li, the brother-in-law of restaurant owner Mr. Wu, in 2008 during a field trip to rural Anhui. Mr. Li and his family had a house on the hill there, with an ox in the shed, chickens and a guard dog. The view was impressive, a wide-reaching landscape of paddy fields crossed by a fast-flowing river. Mr. Li told me he double-cropped rice on an area of five to seven *mu*, much more than what was generally cultivated around Green Water. He proudly showed me around his house, pointing out the various rice processing technologies and making me touch the cotton and rice he had harvested and stored in bags in the attic. With a large smile, he asked me to take photographs of him and his animals. At that moment, Mr. Li struck me as one of the few farmers who had firmly determined to stay.

Surprisingly, the next time I met Mr. Li and his wife was in Shanghai in 2011. He had joined Mr. Wu's restaurant business as his apprentice. Mr. Li and his wife had 'contracted' (*chengbao*) their fields out to other people in return for rent. Mr. Li's grown-up children had also migrated and his elderly father, who had stayed, had to take care of his somewhat confused old mother who was no longer able to look after herself. Alongside these changing care relationships among family members, which are challenged by the ageing population, changing family norms, and migration, the migration also meant there was nobody left to take care of the fields properly.[6]

Left-behind farmers also pursue the land-rental strategy. Mrs. Luo's case shows that having to care for small grandchildren in a split household, in which the middle generation has migrated, can exert additional pressure on

6 As the ageing population is creating new challenges for traditional care arrangements (Buch 2015, 279), China's rural elderly are also finding themselves in a difficult situation (see, e.g., Yan 2003; He and Ye 2014). To receive the necessary care from their children, many grandparents put considerable effort into caring for their grandchildren (Cong and Silverstein 2011; Santos 2017). Migration additionally challenges and restructures care relationships (Alber and Drotbohm 2015, 3-4). Whilst in China sons used to be the primary care providers for their parents together with their wives, migrated sons now commonly provide only financial contributions, while many daughters step in to provide direct daily care for their natal families (see, e.g., Liu 2017, 292). In Mr. Li's family this was not the case, however, since his sister had also migrated. Nevertheless, I could observe in several cases that migrated women, including Mr. Li's sister, cared deeply about their natal families by providing financial care and food gifts for their parents, even more than for their in-laws, thereby affirming 'relatedness and belonging' (Alber and Drotbohm 2015, 2).

labour constraints. In 2015, Mrs. Luo had to lease the household's paddy fields out for one agricultural year because, together with her daughter-in-law, she had to care for two newborn grandchildren at home. Moreover, just before the busy agricultural season, she had to assist her own daughter with her baby in Beijing. Renting out the fields thus ensured that they were protected and the grandchildren were cared for, which guaranteed the continuity of the patriline as well as ensuring there would be people around to care for the grandparents in their old age.

While modalities differ in terms of renting arrangements, generally this strategy does not seem to be a way to earn additional income. Some farmers I interviewed gave away their land for free or in exchange for rice straw; others received some cash in return. The cash amount seems negligible, however (Jin and Deininger 2009, 633). Referring to farmers from central-eastern Hunan, Li Yuyu (2006) found that they did not receive anything, but actually had to provide something to the people who farmed their fields. This is certainly context-specific. In 2004 taxes were still being levied, grain prices were low, and there were other viable economic options, due to the village's proximity to Xiangtan City. Nevertheless, it underlines the importance and necessity for Chinese rice farmers to protect the value of their paddy field resource, at whatever cost (see also Pieke 2002, 8). Generally, according to Article 37 of the Chinese Law on Rural Land Contracts, when farmers subcontract their land, the two parties should sign a written contract (NPC 2002). However, in practice, this does not always seem to be the case.

Renting out your fields requires trusting the people you are leasing them to. As sociologist Niklas Luhmann asserts, a premise for trust is risk, to which trust is the solution. Moreover, trust requires special social institutions, a precondition of which is 'familiarity' (Luhmann 1988, 94-97). As the land transfer market in rural China is only gradually becoming officially institutionalized, in 2011 migrating farmers could not rely on any legal framework that would formally secure their temporary land transfers during their absence. Thus, I suggest that the 'community of practice' is central here (Lave and Wenger 1991). As studies in the field of skilled practice have suggested, successful material production is based on the effective management of social ties – often grounded in relations of trust – whether within a given community or between different communities of practice.[7] Where the tenure situation is insecure and there are no formal written transfer contracts, migrant farmers have to turn to trusted members in their community of practice, usually relatives, to not only ensure that they can reclaim their

7 For the former see, e.g., Clifford Collard (2016). For the latter, see, e.g., Eyferth (2009).

fields when they return, but also that their fields are treated properly so their value is retained during their absence, which requires skilful handling.

In this respect it is interesting to briefly consider the farmers who rent the fields from others in exchange for some sort of formal remuneration. My Anhui and Hunan interlocutors explained that these were often rice farmers who either planted cash crops full-time, or planted large-scale rice commercially. They had sufficient household members at home or hired additional help during the harvesting period. Moreover, they were usually close fellow villagers who were mostly relatives and/or neighbours, given the patrilocal residence pattern. In other words, migrants leased their fields to members of their own community of practice, preferably to close and familiar individuals who they were related to and who they trusted.

The strategy of renting out fields is not a new one, although today's context and outcomes differ from the past. Between the eighteenth and early twentieth centuries it was common in Hunan for wealthy land-owners to rent their land to poorer tenants, receiving an initial silver deposit and a regular rent paid in kind (Rawski 1972, 121). In post-reform China, tenancy and ownership relations have been profoundly transformed. Nevertheless, when viewed in the broader context of previous centuries, the renting strategy can be said to form part of a long-term repertoire of knowledge for dealing with the paddy field resource. It is a strategy that is proving useful again for many farmers in the current migration context. A significant side effect in the contemporary context is that, by using this practice, farmers are implicitly facilitating the transition towards 'big household' (*dahu*) farming. Thus, farmers are actually acting in line with state objectives. Big household farming is part of a national policy framework which favours increasing agricultural productivity through large-scale commercial farming and ensuring that deserted farmland is used (Li 2006, 401; OECD 2014).

Seasonal return

Some migrants migrate seasonally and return in the peak season. This season is called *shuangqiang*, i.e. the 'double rush' of simultaneously harvesting the first crop and transplanting the second crop of rice.[8] Around Green

8 This refers especially to rice farming, because rice requires particularly high labour inputs during a short time window, to prevent the seedlings from withering during transplanting, and to deter the ripe rice, which is only fixed loosely to the plant, from falling to the ground. Due to the ripening times and temperature tolerance, both actions have to take place simultaneously, otherwise the second harvest will not mature in time before the winter.

Water Village, only a few people seemed to employ this strategy, although I also met some commuters. The practice generally requires proximity to a big city, which is not the case for Green Water Village.[9] Alternatively, it necessitates flexible migrant job arrangements, as in the case of Mrs. Luo's sister-in-law Zhou Meijuan and her husband from Paishan Township, who ran their own decoration business in the Guangzhou metropolis.

Hugen, the owner of Green Water's combine harvester, also commuted. He worked as an excavator operator in Guangdong Province, receiving a monthly income of 4000 Yuan (about 570 USD). Every summer he returned for four months to harvest the household's fields and to offer his harvesting services to fellow villagers. This guaranteed him higher returns than the excavation business. With his project of family building in mind, he certainly needed this money. Hugen had just married for the second time. Because he already had a son with his first wife, the child from this second marriage was born outside the legal birthrate in the framework of the so-called 'one-child policy'. Although this policy was effectively changed into a 'two-child policy' in 2016, in 2011 the second birth caused him some costs. Moreover, the baby was a girl and he had a preference for sons. Therefore, he was ready to pay another fine and to have a second, from his perspective hopefully male, child with his new wife. The money earned through commuting would allow him to provide his sons with houses, making good marriage matches for them, thereby also continuing the patriline and ensuring care for him and his wife in their old age. Against this background, the strategy of returning in the peak season had considerable advantages for Hugen, including the fact that it meant his paddy fields continued to be cultivated despite his own and his wife's migrant jobs.

By practicing such circular migration, he was also acting in line with state expectations. While migrants returning seasonally are able to profit from being productive in both the rural and the urban worlds, the state benefits from migrant workers' cheap contribution to China's growing industry, as well as from upholding and ensuring sufficient grain production.

Mutual help

Another way to deal with the labour shortages wrought by migration is to cooperate in the peak tasks of transplanting and harvesting. Mutual aid helps

9 The practice is reported from many places in and beyond China. Among others, see Murphy (2002); Li (2006, 224); De Brauw (2010); He and Ye (2014, 362); and Oxfeld (2017, 39).

to avoid expensive alternatives such as hiring labour or renting machines. Thus, cutting production costs enabled Mrs. Luo and her husband Zhou Wenlu to save up the money they needed to build a new house for their son.

While labour exchange has been common practice for centuries, the current rural exodus also poses new challenges for farmers in this respect. Many villagers who would normally have exchanged their labour have migrated, and those who remain in the village have a different demographic composition, being mostly elderly, women, ill people, and children. Therefore, this raises questions about new arrangements of labour exchange. For example, in previous times Mrs. Luo and Mrs. Zhang, the mother of Hugen who owns the combine harvester, would have relied on the help of household members, close relatives, and direct neighbours. As they have all left the village though, the two village women now help each other in the peak season, for instance, to transplant each other's fields (see Figure 3).

In the case of Mrs. Zhang, this was not a given. Being known as the richest householders in the village, Mrs. Zhang and her son Hugen were the only people who resided in a walled compound. Born in 1955, Mrs. Zhang was a woman who claimed that she had originally come from a city and been dispatched to Green Water Village. Shouting, 'Aiya, my fate (*ming*)! My fate is not good!' she was frustrated that her biological siblings, who had not been given away, enjoyed a better life in the city. Nevertheless, in the village she was still part of a prestigious family, her stepfather having been a cadre in the People's Liberation Army in 1948. She had therefore enjoyed five years of schooling, which was more than other village women of her age were given. Recently, she had simply reclaimed some fallow land from the village and established her own tea tree plantation.

Therefore, the other villagers rather disliked Mrs. Zhang and shunned her. Nevertheless, Mrs. Luo got along well with her. This may also be connected to their personal relationship. Even though the two families were not closely related, both women had nevertheless married into the same Zhou lineage and lived in the same village group. In their case, the basic arrangement was still similar to pre-migration times, where labour was exchanged for labour. The difference in the 2010s was that they had to seek out more distant relatives or fellow villagers who were available for labour exchange.

A prerequisite for engaging in labour exchange is that people are in the right physical condition to offer their own labour. For middle-aged Mrs. Luo and Mrs. Zhang this was still the case. However, it may prove difficult for old or infirm people to offer their own labour, especially if they are also responsible for looking after their grandchildren. As the 67-year-old left-behind farmer Huang Guixiang from Hunan explained:

> I couldn't handle it [cultivating the paddy fields every year anymore]. It is hot, and I am very tired. I have to thresh the grain and fetch them to granary. I can do thousands of grams at most. And I don't get much money out of it. The agricultural inputs cost a huge amount of money [...] I am too old. I don't want to exchange labor with others. Labor exchanging also requires a lot of work. And it is unbearably hot. Renting a harvester is much easier. (He and Ye 2014, 364.)

If exchanging labour is not feasible, as in Huang's case, left-behind farmers have to find other arrangements, as outlined in the following sections.

Hiring labour

Hiring labour is one way to mitigate the shortage of skilled labour without offering labour exchange, when confronted with porous local networks due to migration. This leaves some household members free to engage in other time-consuming activities, such as migrant work or child rearing. It also means they do not have to go beyond their own physical barriers, but can act in the best interests of their own health. A precondition for hiring labour, however, is being able to afford it. According to Mrs. Luo's daughter Yuemei, nowadays it has become common in Green Water Village to hire non-relatives in exchange for money, because people have less time but more financial capital due to migration (email exchange, 6 October 2016).

Hiring labour can involve paying for manual labour, machine labour, local or external labour. For example, He and Ye (2014, 363) report from various provinces that '[t]here was a relatively stable team of hired labour in every village, composed of villagers who have extra labour in the family or less land to till'. They add that there were fixed wages, which were adapted to meet the usual local labour costs. For example, 15 Yuan (about 2.12 USD) per day was paid in the low season and double that amount in the peak season. In addition, meals had to be provided for the hired workers. In contrast, Murphy (2002, 83) describes how the farmers she surveyed in Jiangxi Province were reluctant to hire fellow villagers. They preferred, if at all possible, to hire cheaper labour from nearby poorer villages. Similarly, but linked to mechanization, Li Yuyu (2006, 225) reports from Ya'ai Village in south-eastern Hunan that there were teams of rural labourers from poorer areas in Hunan who roamed from village to village, offering services with their planting and harvesting machines for a price. According to my interviews, Green Water villagers arranged this sort of hiring locally. This was possible because there were people who own the necessary machines in the village, such as Hugen with his combine harvester.

The practice of hiring labour raises questions regarding the growing disparities within rural society. For ideological as well as organizational reasons, it was not common in the Mao era. In the more distant past, however, hiring labour had been common practice. Yet, it was not so pervasive in Hunan, because families there practised extensive rice cultivation methods rather than the intensive ones that required more helpers (see Rawski 1972, 130). Moreover, as Francesca Bray points out, 'the high degree of skill required [in conventional manual paddy rice farming] has made it difficult even to substitute hired labour for family labour' (Bray 1994, 56). Along with the increased financial means gained from migration, the new farming practices and technologies seem to have rendered hiring labour easier, because the workers do not require the same degree of skill as in earlier times.

Labour-saving technologies

For left-behind household members who lack sufficient expertise or strength by themselves to cultivate their paddy fields once many of the other villagers have migrated, labour-saving technologies can provide a way out. If farmers own or borrow these technologies, this does not necessarily involve hiring labour. The technologies include mechanization and farm chemicals, but also rather unexpected technologies such as building cement ridges between the fields that save labour by preventing grass from growing on the ridge and having to be cut. Due to its particular connection with migration (see Chapter 4), mechanization is especially common in many places.[10] Nevertheless, only villagers with the necessary financial capital have the option of mechanization.

One example is the left-behind Xie couple from a nearby village in Longshi Township. The couple have two sons and two daughters. When we met, their youngest daughter Ying was studying international trade in the city of Shenzhen, Guangdong Province. The other three children were labour migrants. Making use of the remittances from three children, the Xies were able to continue double-cropping. This was optimal for preserving the paddy

10 For China and Vietnam see, e.g., Murphy (2002, 73); Li (2006, 225); van den Berg et al. (2007); De Brauw (2010); Lo and Chen (2011); Yuan and Niehof (2011); and Nguyen, Rigg and Derks (2015).

fields and, moreover, provided the couple with some of the money needed to support Ying's education. Ying explained that it was possible to obtain two rice harvests per year because her parents 'use machines' (*yong jiqi*) (personal interview, 4 February 2011). This included, for example, a small electric threshing machine that could be operated by a single person if the paddy bundles were piled up nearby. In contrast, the old foot-operated machines required four or five people: two for threshing and two or three to bring in the paddy bundles.

Even though nowadays mechanization entails entirely new and more effective machines such as the combine harvester, the idea of substituting the human labour that lies behind mechanization is an older one (see Bray 1994, 54). In this sense, even though the outcomes are more far-reaching today, this strategy can also be seen as stemming from a larger repertoire of knowledge. Nevertheless, the new machines doubtless require different knowledge than the techniques and the technologies they replace. Therefore, the strategic use of mechanization is a good example of how villagers draw on the extended repertoire of knowledge that has become available through the state's modernization efforts.

Direct seeding

One way to preserve paddy fields whilst decreasing your own labour input in the planting season is to switch from transplanting to direct seeding. While the term 'direct seeding' encompasses a range of different techniques, in Green Water it involves wet seeding in the form of broadcasting pregerminated seeds.[11] In 2011, this strategy was not practiced in Green Water Village and farmers only mentioned it with regard to other places in Hunan. In 2016, however, Mrs. Luo stated that 'a lot of people' in the village had switched to direct seeding. She claimed that they did this because the work is 'not so bitter, but relaxed' and because 'families don't have time' (video conversation, 8 February 2016).

While the lack of labour due to migration is one important decision-making factor, there are a whole range of other points that farmers have to take into consideration when opting for the technique of direct seeding. First, according to Hugen who owned the combine harvester and whose sister had resorted to this technique, the field must be properly levelled beforehand. In conditions of a lack of male labour, this usually meant needing a tractor.

11 For specifications and classifications of rice seeding techniques see Bray (1984, 252); Pandey et al. (2002); and IRRI (2015).

Second, said Hugen, it is problematic to use pesticides: because the plants are not evenly distributed, it is not possible to apply pesticides evenly. Third, there are more weeds in direct seeded fields than in transplanted ones (personal interview, 1 February 2011). Nevertheless, Mrs. Luo explained that this does not mean farmers have to apply more farm chemicals in direct seeding. This apparent contradiction between more weeds and equal amounts of farm chemicals is resolved by the possibility of spreading work over the rice plants' entire growing cycle. Since the rice growing period is less intense, this allows farmers to turn to laborious manual weeding techniques. Fourth, Mrs. Luo claimed that yields are higher in direct seeding. Fifth, according to her, only early-season rice (usually hybrid rice) can be direct seeded, as late rice (local *ziku mi*) needs to be transplanted (video conversation, 8 February 2016).

While it has been noted that direct seeding is connected to the new hybrid rice varieties that facilitate this technique (Murphy 2002, 85; Li, Xin, and Yuan 2009, 7), climate also plays a role. In view of the approaching cold season, which local rice plants cannot tolerate, transplanting rice saves time. This is because the second crop can already be growing in a nursery while waiting for the first crop that occupies the fields to mature and be harvested. Directly seeding two crops would mean that the second crop would not ripen before it got too cold for the seeds (van den Berg et al. 2007). Yet generally, directly seeded rice plants ripen earlier than transplanted ones, because they do not have to re-establish their root system after being pulled out of the nursery bed.

Finally, there are also financial considerations. Direct seeding requires more financial capital, because seeds are more expensive and the technique requires twice as many seeds as transplanting (IRRI 2016). The decision for or against direct seeding is therefore highly complex, especially when taking into account the many advantages of transplanting (see Chapter 2). Part of this complexity is due to 'technical linkages' (Oshiro 1985), since the technique of direct seeding will influence the choice of other techniques and technologies employed in the subsequent rice-growing process.

Direct seeding is a distinctive strategy, because using direct seeding instead of transplanting means substituting a newer technique with an older one. This stands in sharp contrast to the linear narrative of technological development and progress. The technique of direct seeding (broadcasting rice grains) was the earliest method of rice cultivation. The shift to transplanting only occurred in the late Han Dynasty (23-270 B.C.E.) (Chang 2000, 140-141). It is worth noting that, in the current context of rural exodus, the technique is undergoing a revival in Hunan and elsewhere.[12] It is also remarkable that the social and technical knowledge about this more than two millennia-old technique has been retained, even though transplanting has long been the dominant practice and Green Water farmers today do not seem to have ever personally practiced direct seeding. Still, in a way, knowledge about the practice has been remembered collectively. Following Schippers (1992; 2014b), I suggest that proverbs may have played a role here. These sayings can transmit knowledge that is not required in certain situations, but may be crucial in others. For example, the proverbs on the explicit advantages of careful transplanting quoted in section A.4 simultaneously convey implicit information about the disadvantages of direct seeding. Moreover, farmers seem to have retained 'tactile memory' (Harries 2017), which allows them to now rediscover certain logics inherent in the seeds.

De-intensifying rice farming

In contrast to the strategies described above, the following strategies clearly entail a de-intensification of rice farming.

Single-season rice

Cultivating single-season rice involves moving from two harvests to one single rice harvest per year. Along with the practice of renting out their fields, this was my interlocutors' most common strategy and one that is widely reported in the literature.[13] Using this technique, rice farming is de-intensified to a degree that matches the (lacking) resources of left-behind farmers, whether physical, technological, financial or in terms of skilled labour. It is a suitable strategy for the elderly people who are left behind by their migrant household members, who only cultivate enough rice for their own subsistence.

Not only Mrs. Luo followed this strategy, but also her neighbour Wenjun. Wenjun and his wife lived in Guangdong Province all year round, he working

12 For instance in India, Bangladesh, Malaysia, Thailand, Vietnam, South Korea, and Japan. For details see Pandey et al. (2002); Labrada (2003); Kamoshita et al. (2009); and Ogura, Sukuchan, and Narioka (2011).

13 Among others, it is mentioned by Li (2006, 197, 373); van den Berg et al. (2007); Zhang, Li, and Song (2014); and Tian et al. (2015, 1253-1254).

as an excavator operator and she in a shoe factory. They only returned to Green Water for New Year. Meanwhile, Wenjun's parents stayed at home in the village. His father tended the paddy fields and his mother cared for his two little daughters. Living alone with the grandchildren, his parents had switched to single-season rice, which was sufficient to feed the four of them (personal interview, 27 January 2011). In this way the household could retain the paddy fields and care for the grandchildren, whilst also benefitting from the incomes of two migrants, which were needed to pay for the children's education.

This move to single-season rice is reflected in local statistics. According to the *Anren County Gazetteer*, the area planted with early and late rice decreased slightly between 1989 and 2003, whereas the area planted with middle rice – an indicator of single-season rice – rose slightly (see ACGCC 2011, 286). While more recent local statistics are not available, in view of increased migration and evidence from my interviews, I expect that the area planted with middle rice must have continued increasing from then on. In line with this, the outputs of middle rice in Hunan Province have consistently grown, while the outputs of double-cropped early and late rice have decreased (HPBS 2019, sec. 12-11). In view of this, Hunan Province has recently taken measures to promote double-cropping and ensure grain supply. These include assigning professionals to build seedling raising greenhouses and raise rice seedlings and intelligent plants, subsidizing transplanting machines, and sending more than 11,000 cadres to provide technical guidance for rice farmers (Xinhua 2020).

While circumstances differ today, it is not the first time that a switch from double to single cropping has occurred in Hunan. In the early Qing Dynasty (1644-1911) farmers employed a similar strategy, possibly due to changing land-labour ratios, the need to restore irrigation facilities after a period of rebellion, and the lack of sufficient water in some places (Rawski 1972, 220). This strategy is another case that contradicts the narrative of linear technological development.

Abandoning the fields

A practice that was less common among my informants is actively abandoning fields, despite all the negative consequences for their value. In 2011 it was estimated that farmers in China were abandoning about two million hectares of farmland every year, mainly due to migration (see Gao 2011). This is an issue that has attracted much attention, not least because of its implications for food security.[14] Generally, due to the complex interaction between different local socio-economic and environmental circumstances, as well as factors of spatial interaction within China in general, and Hunan Province in particular, there are stark local variations in instances of land abandonment. The reported rates of perennial abandonment in southern Chinese rice growing areas range from 7 percent to more than one third of all fields (Li 2006; Zhang, Li, and Song 2014; Xie and Jiang 2016). The rates of seasonal abandonment are even higher, ranging between almost 30 and 90 percent in Hunan. In this respect, Anren County, where Green Water Village is located, has a low level of seasonal farmland abandonment (0.00-15.07%) (Yu et al. 2017, 8-9).

The scope of abandonment also varies due to the particular strategies followed by rural households. Strictly speaking, there are two different strategies that involve the abandonment of farmland. On the one hand, there is the abandonment of selected individual plots, which He and Ye (2014, 364) call 'cutting the size of farming'. This is practiced by left-behind household members to reduce their heavy workload. On the other hand, there is the complete abandonment of farmland, in the framework of migrant households managing their land during their absence (Xie and Jiang 2016).

My interlocutors did cut the size of the land they farmed. Discarding their bad-quality fields gave the Green Water villagers more time and labour to dedicate to protecting their good-quality fields, as well as their own bodies. Generally, this strategy implies that farmers have to choose carefully which fields to abandon, requiring a deep knowledge of the characteristics of every individual plot. Land allocation during the reform period has led to a high degree of land fragmentation, with Chinese farmers' fields being dispersed, on average, over 6.06 plots of varying quality (Heerink et al. 2007, 794). None of my interlocutors left their high-quality paddy fields lying fallow, but it was common to abandon plots that were low quality or difficult for people and machines to access. Based on my observations of abandoned fields, I discerned that the strategy of complete abandonment was more common in the nearby administrative seat of Longshi Township than within Green Water Village. In that market town, I noticed a number of desolate fields that were neither small nor marginal (see Figure 9).

While I was not able to interview farmers who had completely abandoned their land to gain further insights into their decision for doing so,

14 Most studies focus on the driving factors of abandonment as well as the impacts, e.g. on food security or on the environment in different regions of the world. See, e.g., literature reviewed in Qin (2010); Zhang, Li, and Song (2014); and Yu et al. (2017).

Figure 9 An abandoned paddy field in Longshi Township

Photograph taken by the author

economists and geographers have singled out various factors that facilitate abandonment. These include the immature formal land rental market, labour shortages due to migration, the relatively good working situation of some migrants in the city, and specific plot characteristics. Moreover, a farmer's age has been identified as a decision-making factor, while gender has not. It was found that middle-aged migrants with rich farming experience felt close bonds to the land and agriculture, and were less likely to abandon their fields. In contrast, the young – usually migrants with little or no farming experience – and the old – because of their physical limitations and inability to carry out heavy farming tasks any longer – were more likely to abandon some of their land. The gender of left-behind farmers did not play a role in this decision, because they were either old, ill or tired (Xie and Jiang 2016; Zhang, Li, and Song 2014; Yu et al. 2017). In contrast, having more family members at home, specfically those older than 64, may also counteract land abandonment (Xu et al. 2017).

Either way, abandoning land runs counter to the state's legal framework, although its implementation differs across the regions. If farmland is abandoned for two consecutive years, it can be reclaimed by the original contracting entity (Xie and Jiang 2016, 264). The related fears of villagers have been reported from some parts of China, suggesting a stricter enforcement of the law (Roberts 1997; De La Rupelle et al. 2008, 28; Jin and Deininger 2009, 633). In Green Water Village farmers stated that people who abandoned their fields should have to pay a fine. This measure did not seem to be rigorously enforced or bear much influence on the villagers' decisions, however. In my interviews, Green Water villagers did not mention either this policy, nor any fear of losing their land. I was told that farmers in the village would only lose their land-use rights upon a *hukou* transfer, i.e. when registering themselves elsewhere. In view of this, and aware that, even if they wanted to, it was still almost impossible to obtain an urban *hukou* from big cities like Shanghai or Beijing, many Green Water Villagers were willing to remain registered in the village to retain their land-use rights (see also Andreas and Zhan 2016; Chen and Fan 2016).

With regard to land abandonment, in general, it seems that the central government has focused on providing farming incentives rather than imposing penalties in the last few years. The direct subsidies for agriculture and the other measures that aim to encourage farming described in Chapter 1 have not fully counteracted land abandonment, however. This may be one of the reasons why the Chinese Ministry of Agriculture addressed the issue in 2016 by announcing clearer guidelines for land transfer rights (Glenn and Yao 2016).

Overall, my findings on land abandonment suggest that farmers' decisions are based on an interplay of multiple complex factors taking place at different levels. Policies are certainly one important factor. Moreover, the fact that farmers choose which particular plots to discard, and that migration is not forced suggests that land abandonment is, in fact, often a conscious calculation. I therefore also view land abandonment as a strategy, rather than a lack of choice. In Green Water, this strategy was, however, not the preferred option. In view of a long-term repertoire of knowledge, in which farming knowledge is memorized collectively and transgenerationally in agricultural practices (Schippers 2014a, 339), it seems possible that earlier experiences of the outcomes of land abandonment in Hunan (see Perdue 1987, 72; Li 2006, 36) have been incorporated into today's farmers' decisions.

House construction

There is another strategy of using farmland to build a house on (see Figure 10). This strategy is illegal (NPC 2002). It contributes to the loss of irrigation ponds and facilities, and arable land. The latter has been a major problem in post-reform China. It has been estimated that, between the mid-1980s and the mid-1990s, about 6 percent of China's total cultivated area was

Figure 10 The foundations of a house under construction on a former paddy field

Photograph taken by the author

lost. Rural housing was estimated to account for about 5 to 6 percent of the total lost farmland towards the end of the last millennium (see Sargeson 2002, 928). The pace of this loss of arable land only began slowing down sharply in 2006, which is officially portrayed as a response to the national preservation policy of a '1.8 billion *mu* farmland redline' (*shi ba yi mu gengdi hongxian*) (see Chien 2015, 68).

Since the reform period, rural house construction has increased rapidly in China. Between 2000 and 2018 alone the per capita housing floor space of rural residents in Hunan Province has more than doubled, from 30.92 to 63.57 square metres per person (HPBS 2019, sec. 1-8). This phenomenon has attracted the attention of journalists, policy makers, and scholars. They have been particularly interested in analyzing the reasons why migrant workers build seemingly wastefully expensive, large, and often underoccupied houses in the countryside, which, moreover, sacrifices their scarce farmland.

While the issue has been seen as a response to market forces and insecurities about rural land entitlement,[15] it is in fact more than merely an

15 See, for instance, Feder et al. (1992); Oi and Walder (1999); and Ho (2001), cited in Sargeson (2002, 929).

economic or policy issue. Anthropologists and social scientists studying the link between migration and houses in China and elsewhere have comprehensively stressed the socio-cultural aspect of this construction. They have shown that house building is not simply a matter of investing migrant remittances. Instead, they have emphasized the complex socio-material meanings behind house building and the particular appearance of these new homes.[16]

With regard to rural China, social scientists have viewed the wave of house construction in connection with specific social and demographic aspirations. These include enhancing an individual's attractiveness to marriage and business partners; increasing family harmony by providing more privacy and space; enabling young couples to have more spatial and financial independence from their parents and in-laws; giving daughters-in-law who move into these new houses more space for personal agency; displaying a family's social 'face' (*mianzi*) and wealth; and securing support for the elderly in their old age by earning their children's gratitude.[17] In short, building a house serves in various ways the projects of sons, daughters-in-law, and parents, as well as, more generally, the whole family's endeavours.

Migrant work offers a promising way to fulfil these costly aspirations. Li Cuiping, the wife of restaurant owner Mr. Wu, asserted that her family had enjoyed higher social standing in the village community since they had constructed a multi-storey house. They often talked about it and planned to retire there in their old age, together with their son Fengfeng once he had grown up and got married. They proudly described how large it was and the consumer goods they had bought for it, such as a television, DVD player, and washing machine. This had only been possible through migrant work, as new houses are expensive. For example, according to Green Water combine owner Hugen, it costs over 100,000 Yuan (about 14,130 USD) to build a new house. His house, the most impressive in the village, cost more than twice that amount. My interlocutors from Anhui mentioned similar prices. These are exorbitant amounts of money for most rural families, considering that even a rather basic new house costs about 15 times the per capita annual net income in rural Hunan (HPBS 2014, sec. 20-25). Domestic construction has therefore also been described as a driver for migration, and it has been suggested that migrant remittances are foremost invested

16 For recent studies see, e.g., Miller (2008); Dalakoglou (2010); Levin and Fincher (2010); Walsh (2011); Chen (2015); and Pauli (2015).
17 These aspects are mentioned by Sargeson (2002, 942); Yan (2003, 154, 178-179); Chu (2010, 38); and Chen (2015, 119).

in house building.[18] Meanwhile, Chen (2015, 116) found that owning a new large house also encouraged migrants to return to the countryside and circulate. This is even true for the younger generation of migrants who lack agricultural skills and who might be expected to have less interest in investments at home.

My own findings suggest that owning a new house is, indeed, also a central concern for Green Water villagers, particularly for young men and their parents. The high number of new brick, concrete and tiled two- or three-storey houses also indicates this. In Green Water Village, houses are also sometimes built on farmland and, against the legal regulations, land is even sold. In fact, illegal sales are frequent in China and local officials have been promoting land transactions because they will produce extra revenues for them.[19] While the reasons why villagers sell their land still need to be investigated, my observations on house construction generally echo social scientists' findings from other parts of China, namely that house construction is a means to enhance sons' immediate marriage prospects (e.g. Sargeson 2002, 942; Chen 2015, 119-120). In Green Water Village, there is a prevailing scarcity not only of farmland, but also of land to build on. Moreover, in times of comparatively high returns from labour migration, there is intense social pressure to own a new, tiled, multi-storey house made from concrete and bricks. In the patrilocal system, unmarried young men especially feel this pressure. As Green Water villagers repeatedly stressed, if all an unmarried man had to offer was an old clay house, it would be difficult for him to find anyone who would be willing to marry him and move in with him.

The need to find a way to attract marriage partners has been impacted by national birth planning policies that – due to a preference for sons – have resulted in a skewed sex ratio. In Anren County, where Green Water Village is located, 52.9 percent of the population in 2003 was male, compared to only to 47.1 percent female (ACGCC 2011, 47). This distorted ratio means that more men are competing for fewer potential brides. In addition, rural men are facing a situation in which there are fewer women available in rural areas due to the augmented marriage options that are emerging with increasing out-migration (Chen 2015, 120). This was the case for Yuemei, for example, who found a husband from northern Gansu Province in 2012 whilst living in Beijing.

18 For details see Murphy (2002, 91); Sargeson (2002); Li (2006, 380); and Fan, Sun, and Zheng (2011, 2176).

19 See, for instance, Sargeson (2002, 929-930); Chien (2015, 67); and He and Xue (2014, 127).

While some Green Water men have therefore married wives from other provinces, such as Sichuan or Guizhou, most of them have put their efforts into house construction as a strategy for finding a marriage partner. Yuemei's younger brother Pengyu, i.e. the son of Mrs. Luo and Zhou Wenlu, is a case in point. When I met this young man in Green Water Village in 2011, he was a migrant worker in his early twenties who would soon approach the age when he was expected to get married. His family members unanimously agreed that it would be difficult for him to find a wife with their old, modest, flat brick house, a building that consisted of a living-cum-sleeping room, a second sleeping room, a kitchen, and two sheds, all with dusty floors. The windows were cracked and there was no bathroom or running water. The toilet was communal and, to generate fertilizer, combined with the pigsty. In short, these were not the conditions that young women from Green Water aspired to, especially those who had migrated before and been exposed to urban standards and dreams.

In the previous year, Pengyu's family had therefore bought a piece of land from a paternal relative. According to their future neighbour, Mrs. Zhang, the plot had cost 20,000 Yuan (about 2830 USD). In the following years, they built a multi-storey house there. This strategy was successful, since the house is now occupied by Pengyu's wife, their three young children, and his parents, who look after the children. Pengyu himself did not earn enough from his job as an excavator operator in Guangdong Province to afford the plot and the house, however. Therefore, his elder sister Yuemei, who was not yet married at that time, helped to finance Pengyu's house. She perceived this as 'giving something back' to her parents who had done so much for her. In turn, Pengyu's father Zhou Wenlu used the construction skills he had acquired during years of related migrant work to oversee and help with the construction. For the father, this house-building project was a reason to return home temporarily. In 2019, he returned home again, after his third grandchild was born – this time the son they had been longing for.

Pengyu's case supports Chen's (2015, 121) argument that house construction may be seen as a rational household strategy in the context of competition for wives. When viewed from the perspective of attracting marriage partners, the seemingly wasteful expenditure, on top of sacrificing scarce farmland, makes perfect sense. This is especially so when considering that, after the migrant workers have departed, those left behind have, on average, more farmland for fewer agricultural labourers.

Therefore, in view of the issue of resource use, I suggest that using farmland as building land in Green Water Village entails transforming one valuable resource into another valuable resource. Similar to a paddy field, a house is a resource that can be used in the long term. It provides material security and forms part of the safety net in a migration context marked by uncertainty. Moreover, by promising to attract future marriage partners, house construction directly affects old-age care and the continuity of the patriline through new grandchildren. In addition, as Sargeson notes, if the house is constructed on geomantically favourable sites, this is seen to not only impact on a family's present situation, but also has implications for their future (see Sargeson 2002, 944). A house is therefore a crucial resource that may be at least as valuable as a piece of farmland. At the same time, while households still retain several plots for rice farming, the absence and economic activities of the migrant members relativize the immediate impact of the related farmland loss.

Dry fields

The last two strategies both involve crop choices. One strategy is to plant subsistence crops that compete with rice and grow in dry fields. Plant choice is a complex issue. It is related to an interplay of numerous factors, ranging from natural and technical to cultural, social and symbolic ones (Chevalier, Marinova, and Peña-Chocarro 2014, 4). In Green Water Village, too, the social context is particularly influential. Accordingly, in the current context of out-migration, one major decision-making factor is the amount of labour input required for specific crops. The subsistence crops planted in dry fields are much less labour-intensive than rice, which makes them a serious competitor to rice. The most common dry crops in Anren County are sweet potatoes and beans, followed by some maize and sorghum (ACGCC 2011, 287).

Mrs. Luo illustrates this strategy, too. Her case suggests that cultivating dry fields is a strategy of left-behind people who lack sufficient skilled labour to farm all the household's paddy fields. Like the strategy of abandoning some of their fields, which is also a long-term decision, farmers need to choose carefully which wet fields to transform into dry fields, however. Being aware of this, and drawing on her socio-technical crop knowledge, Mrs. Luo had transformed all of the family's third-quality fields into dry fields. Due to their poor quality, the rice yields from these fields would have been fairly low, while the inputs of labour and fertilizer would have been comparatively high. This strategy allowed Mrs. Luo to simultaneously protect the family's most valued paddy fields by dedicating more attention and time to them, while relieving herself from severe labour constraints after the departure of the other household members. Moreover, she could sell surplus vegetables on the market in Longshi Township.

This option does not seem to be so straightforward, however. As I mentioned in my introduction, former paddy fields are not particularly suitable for dry crop cultivation. This renders the choice more complex and leads farmers to also reclaim new land to plant dry crops on, instead of only using their paddy fields. Accordingly, the county-level crop statistics show that the area cultivated with dry crops in Anren County had almost doubled between 1989 and 2003, yet there was only a slight decrease in the area of paddy fields (ACGCC 2011, 287).

Nevertheless, available figures at the provincial level, which take into consideration the whole reform period and include more recent statistics, suggest that the area sown with grain (mostly rice) had decreased by almost one fifth from 1978 to 2018 (HPBS 2019, sec. 12-2). Along with other approaches described in this chapter, the strategy of planting dry crops seems to have contributed somewhat to this decrease. Accordingly, Anren County's grain crop statistics, in which rice, sweet potatoes, beans, maize and sorghum all count as grain crops, show that the area of sweet potato and bean farming had increased between 1989 and 2003. In the same period, the land used to grow rice had decreased (ACGCC 2011, 284-285). Besides, the shapes and ridges of most dry fields in Green Water bear evidence of their former usage as wet fields (see Figure 11). In spite of this, grain yields, including rice, have continued to rise (ACGCC 1996, 286; HPBS 2019, sec. 12-6). This is closely related to the increased use of post-Green Revolution technologies and is a reason why the local government does not seem to punish the Green Water farmers' strategies, but responds to them by promoting new technology instead.

In view of a long-term repertoire of knowledge, dry crop cultivation has formed part of rice farmers' strategies for a long time. Rice farming has always had different seasonal requirements with regard to the labour needed. This has rendered it not only possible, but also desirable to combine with other suitable economic activities, as a way to minimize risks and raise incomes (Bray 1994). What makes the dry crop strategy distinct in the current context, however, is that the dry crops mentioned above are now competing with rice in an unprecedented manner.

Cash crops

Another de-intensification strategy is to cultivate cash crops in the paddy fields. The income thus obtained is higher than that generated from rice. Therefore, cash cropping seems to be a good alternative even to migration

Figure 11 Zhou Wenlu and Mrs. Luo water their dry fields (former wet fields)

Photograph taken by the author

for some farmers from Hunan, Anhui, and elsewhere.[20] Moreover, it also offers the possibility of preserving wet fields when opting for irrigated cash crops such as lamp rush (see below).

Local statistics indicate that cash crops are of growing importance. Between 1989 and 2003, the most common cash crop by far in Anren County was rape seed, which occupied about two thirds of the cash crop area. Moreover, tea, fruit, tobacco, cotton, and groundnuts were also common.[21] In the same period, the proportion of cash crops in relation to the total cultivated area in Anren County grew from 11.61 percent to 19.3 percent. During this time, the area cultivated with rape seed almost doubled (ACGCC 2011, 287-288). This was due to vigorous government promotion since 1976, but was also related to farmers' personal strategies.

Rape seed was, I observed, a common cash crop in both Anhui and Hunan. This crop is mainly cultivated for its oil, which is sold at market as well as

20 This is not only in China, but also in other parts of Asia such as the Philippines or Vietnam, see, e.g., Song (1998); McKay (2005); van den Berg et al. (2007); De Brauw (2010, 135); and Tian et al. (2015).

21 Vegetables are not included, and appear in a separate table (see ACGCC 2011, 287-288). More recent local statistics are not available.

consumed at home for cooking. In 2011, Mrs. Luo cultivated seven *fen* (about 467 m^2) of her fields with rape seed and sesame, and one plot of five and a half *fen* (about 367 m^2) with rape seed and sweet potatoes.[22] Three main objectives motivated her and other villagers' decisions to plant rape seed. First, it brought in additional income, offering Mrs. Luo some independence from her migrated family members. Second, because the timing of rape seed cultivation did not interfere with the peak agricultural seasons, it was possible for her to cultivate it even in the absence of other household members. Third, being planted in the empty winter paddy fields, it did not interfere with rice and prevented fields from turning to seed, thus protecting their value.

While rape seed renders paddy fields more economically productive, prevents weeds, is a complementary crop to rice and is, therefore, welcomed by both farmers and the state, other cash crops also compete with rice. Cultivating them runs counter to the central government's aim of achieving national grain self-sufficiency. Nevertheless, from the villagers' perspective of protecting their paddy field resource, it is no problem to substitute one of the two rice seasons with a cash crop. However, it is important to retain at least some rice cultivation.

Here, a second example of cash cropping also comes from Mrs. Luo's household. In the late 1990s and early 2000s, her three children were still at home and Mrs. Luo and her husband had to pay for their schooling. Therefore, they decided to replace their second rice crop with edible lilies (*baihe, Lilium brownii* F. E. Brown var. *viridulum* Baker), a medicinal plant and cash crop. One key consideration in this land-use decision was the particular stage of their family lifecycle and its specific needs at that point. This echoes earlier findings from economic anthropologists of agriculture, who – in debating theories from Chayanov's (1966) cycle of family size – identified cycles in household resources and needs as important variables in land-use decisions (see Barlett 1980, 558-559).

While affording children's schooling and making up for the lack of the children's labour were short-term objectives, the parents were also pursuing long-term projects through this lily cultivation. Like my Anhui interlocutors, they believed that investing in their children's education would produce more employment options and higher incomes in the future, contributing to 'changing fate' (Obendiek 2016). This was not only seen as being beneficial for the children, but also an important strategy for guaranteeing care in the parent's own old age. At the same time, the family did not need to give up

22 Ten *fen* equal one *mu*. One *fen* equals 1/150 ha, i.e. it is about 66.67 m^2.

Figure 12 Lamp rush growing in a wet field

Photograph taken by the author

their paddy fields entirely, retaining them as an additional resource both for their old age and for their son Pengyu, who was expected to inherit the fields and continue the patriline.

The third example also involves a commercial crop that competes with rice: lamp rush (*dengxin cao, Juncus effusus*; see Figure 12). This grows in bundles, with characteristic stems that stick out from the wet ground like long green needles. The spongy cores can be sold as cushioning material. The crop is also sold to Japan for tatami mat production, as well as to other Asian countries (Zhi 2010; ACGCC 2011, 291). Lamp rush grows throughout the winter and spring, so replaces the early rice crop.

Cultivating lamp rush is mainly undertaken by old women and their husbands, who have been left behind by their migrated family members. Important factors that influence their crop choice are the short-term objective of earning an extra income to secure the immediate subsistence of those left behind while, at the same time, compensating for the lacking household labour and preserving the paddy fields in the long term. Lamp rush is an optimal crop in this regard because, like rice and unlike most other crops, it grows in a wet field.

This crop already has a certain history in the area, which is closely tied to the socio-technical and political contexts of previous eras. According to the *Anren County Gazetteer*, in Qing and Republican China, between the mid-seventeenth and mid-twentieth centuries, lamp rush used to be a major local cash crop. As its name suggests, the rush was previously used for lighting. Due to its perceived superstitious usage, which was condemned by the Chinese Communist Party, and the increasing replacement of candles and oil lamps by other lighting technology, its cultivation decreased in the 1950s (ACGCC 1996, 295). At the end of the 1980s, with de-collectivization, it regained importance, and Longshi Township became one of the three cultivation centres in Anren County. In 2002, its extensive promotion was part of an ambitious local government project and one of the first rural commercialization projects in the prefectural-level city of Chenzhou. By 2003, 7100 *mu* of lamp rush were planted in the entire county, 5100 *mu* of which was in Longshi Township, where Green Water Village is located (ACGCC 2011, 291).

Lamp rush cultivation is fairly profitable, the output value of the crop being more than double that of rice: up to about 4000 Yuan (about 565 USD) per *mu*, compared to up to 1200 Yuan (about 170 USD) per *mu* for one season of rice (Zhi 2010; ACGCC 2011, 291). Seventy-year-old Granny Li from neighbouring Paishan Township sold a bundle of peeled rush for three Yuan, which was then used to line coffins. Furthermore, the profitability and popularity of lamp rush is connected to the comparatively low input of farm chemicals and labour it needs. This makes it an important crop for old people, whose bodies are no longer strong and who lack the support of others. In a newspaper report by Zhi Fujing (2010), who interviewed various farmers from Longshi Township, a 60-year-old woman, Duan Xiaoliu from Fengnan Village, stated that rush is 'good to manage' (*guanli de hao*). My Green Water interviewee Zhou Wenxiang confirmed this: 'Rush needs neither fertilizer nor pesticides; this is why you earn more than with rice farming' (personal interview, 25 January 2011). According to farmer Zeng Shelian, who is also quoted in Zhi's report, spraying farm chemicals was not necessary in the past, but climate change has rendered insecticides mandatory. In 2010, fertilizer and insecticides were applied twice a year, amounting to 100 Yuan of capital investment per *mu* if the land is fertile. This is, nevertheless, still less often and costly than in rice farming, and renders the cultivation of lamp rush 'much more relaxed' (*qingsong duo le*) (Zhi 2010).

Duan Xiaoliu, the old woman interviewed by Zhi, further draws an analogy between lamp rush and the proverb, 'you don't have to worry about marrying off the emperor's daughter' (*huangdi nü'er bu chou jia*), meaning you do not have to worry about the cultivation and sale of lamp rush:

> The quality of our lamp rush here is good. First, we have a cultivation history of more than 300 years and everyone's cultivation techniques are good; second, the climate is suitable here. (Zhi 2010.)[23]

According to the report, farmers in Longshi Township have the custom that 'three *fen* of grassfield [i.e. a field planted with lamp rush] support father and mother' (*san fen caotian yang die niang*). This is why farmer Yang Shizai, more than 70 years old, claims never to have needed the support of his son, asking only that he cultivate some *fen* with lamp rush for him (Zhi 2010).

In contrast to lamp rush's cultivation, its processing is highly demanding. Processing the rush involves carefully and quickly peeling off the inner part of the plant with a small knife, ensuring that the long core is not allowed to break (see Figure 13). Not everybody is skilled at this. Out of several grandchildren gathered around Granny Li, demonstrating her skills to me, only one ten-year-old girl was able to imitate her movement successfully.

Overall, for migration-affected households, cultivating lamp rush implies a shift in labour in terms of time, space, and degree: from the labour-intensive peak seasons of rice farming to a more balanced need for labour throughout the year; from the outside rice-farming, to the inside processing of lamp rush; and from heavy to light labour. Hence, the potential of the labour of senior female family members can be fully tapped in lamp rush cultivation and processing. Here, farming decisions in the context of migration are clearly being taken in view of old women's manual skills. Lamp rush cultivation seems to match the ideal gendered spheres of knowledge distribution in rural Chinese society, which normatively differentiate between male/heavy/outside and female/manual/inside tasks (Jacka 1997; Bray 2013). This may also be one reason why the crop is generally accepted among the old villagers. This lamp rush example clearly demonstrates that skill is an important factor in agricultural decision making, and that a focus on skill provides valuable insights into farmers' decision-making strategies.

23 In Chinese: '我们这里的灯芯草质量好。一是我们这里有300多年种植灯芯草的历史，大伙种草技术好；二是这里的气候条件适合。' *Women zheli de dengxin cao zhiliang hao. Yi shi women zheli you 300 duo nian zhongzhi dengxin cao de lishi, dahuo zhong cao jishu hao; er shi zheli de qihou tiaojian shihe* (Zhi 2010).

Figure 13 Granny Li peels the dried lamp rush

Photograph taken by the author

While the move from rice to cash crops may appear at first sight to be a post-reform period phenomenon, farmers were already abandoning paddy fields in favour of cash crops in China in the sixteenth and seventeenth centuries (Rawski 1972, 50; Bray 2013, 80, 82). Around that same time, Jiangxi immigrants led a flourishing cash crop trade with new food produce such as sweet potatoes, tobacco, and sorghum in Hunan (Perdue 1987, 97). Therefore, even though the particular cash crops may have changed, switching from rice to cash crops is another example that contradicts the narrative of linear technological development. Hence, with regard to a repertoire of farmers' knowledge about different ways to manage paddy fields, this switch is evidently one proven option available to farmers that has regained its appeal in the current context of rural emigration, changing food consumption patterns, and overall rural policy.

Moreover, since the Chinese grain market still appears less liberal than that of other crops, the decision to change to cash crops is often viewed in the framework of market liberalization, in addition to emerging export opportunities (e.g. OECD 2005, 53). It is perceived as a rational move by farmers to gain more income. Due to offering increased tax revenues, this move may even enjoy the support and pressure of local governments (ACGCC 1996, 290, 297;

Tilt 2008, 197; Murphy 2006, 19; Steinmüller 2013, 105). However, my findings show that, while profit is certainly a concern for farmers, it is not the only consideration. They also make decisions based on the social organization of farming and the related knowledge and skills in a migration context.

More than linear, more than technical

The twelve possible land-use and land-arrangement strategies described above form part of a whole repertoire of decision-making agendas, especially since each household farms several plots of land. Some of these strategies, such as cash-cropping, abandoning fields, planting single-season rice or even using mechanization might, at first sight, appear to be new in view of the described context of the field preservation-migration predicament. However, many are actually rooted in previous centuries. As historian Evelyn Rawski argues, farming practices have to be seen in the context of the entire peasant economy. Accordingly, in Qing China (1644-1911) some Chinese farmers diversified their income-generating activities through, for example, handicrafts and other enterprises. This resulted in stagnating rice productivity and 'left the farmer[s] with neither the interest nor the inclination to plunge more deeply into rice culture' (Rawski 1972, 142-143). Today a similar situation of diversification is occurring, as farmers are expanding their activities through migration and other off-farm jobs.

Nevertheless, despite the de-intesification strategies described in this chapter, rice yields in China have been growing, enabled by post-Green Revolution technologies and market dynamics. This is also connected to a gradual geographic shift towards rice production in northern areas since 1949, which is in part related to climate change (Wang and Hijmans 2019). Today, the country's most northernmost province Heilongjiang actually has the highest rice output in China, followed by Hunan, which had rice outputs of 26,855,000 tons and 26,740,000 tons respectively in 2018 (NBSC 2019, sec. 12-10). However, the population has also been growing, so food security is still an issue.

At the local level, some decisions, such as constructing houses, abandoning or converting fields, have long-term consequences for the value of the paddy field resource. At the same time, there is an underlying dynamic inherent to many of these decisions, as they are adjusted, rejected or combined according to changing demographic or economic circumstances such as children needing education, migrating, getting married, and having children themselves, or household members falling ill and growing older.

Here, rural and urban China are intimately intertwined through decision-making, and strategically thought of by the staying and migrating members of the community of practice world as one and the same sphere. This is partly because family members make strategic decisions together, based on the contribution made by both migrants and those left behind, which impact on all of them. Yet, even where decisions are not taken together, they affect the whole rural-urban household and its resources. For example, in 2011 Hugen's wife Xi said she did not want to, but felt compelled to migrate again. At that time, their baby was only three months old. Feeling driven to migrate was due to a major conflict with Mrs. Zhang, her mother-in-law. Mrs. Zhang was furious when she found out that Xi was already married and, in addition, not yet officially divorced. This dispute forced Xi out of the house and into factory work in Guangdong. Mrs. Zhang, in turn, had to adapt her rice farming practices to be able to care for both the baby and the fields, carrying the little child along on her back wherever she went.

In summary, when examining the members of migration-affected rice farming households as dynamic agents, their actions appear highly intentional ways to pursue various short-term and long-term projects. Even though the structural conditions of the paddy field predicament are difficult and the transformations occurring to the knowledge system are profound, farmers nevertheless retain individual agency to act in their own best interests as far as possible within the limits of their options and the resources available.

References

ACGCC, 安仁县志编纂委员会 Anren Xianzhi Bianji Weiyuanhui (Anren County Gazetteer Compilation Committee). 1996. 安仁县志 *Anren Xianzhi* (*Anren County Gazetteer*). Beijing: Zhongguo Shehui Publishing.

—. 2011. 安仁县志 *Anren Xianzhi, 1989-2003* (*Anren County Gazetteer, 1989-2003*). Changsha: Hunan Renmin Publishing.

Alber, Erdmute, and Heike Drotbohm. 2015. 'Introduction'. In *Anthropological Perspectives on Care: Work, Kinship, and the Life-Course*, edited by Erdmute Alber and Heike Drotbohm, 1-20. New York: Palgrave Macmillan.

Andreas, Joel, and Shaohua Zhan. 2016. 'Hukou and Land: Market Reform and Rural Displacement in China'. *The Journal of Peasant Studies* 43 (4): 798-827.

Barlett, Peggy F. 1980. 'Adaptive Strategies in Peasant Agricultural Production'. *Annual Review of Anthropology* 9 (1): 545-573.

van den Berg, M. Marrit, Huib Hengsdijk, Joost Wolf, Martin K. van Ittersum, Wang Guanghuo, and Reimund P. Rötter. 2007. 'The Impact of Increasing Farm Size and Mechanization on Rural Income and Rice Production in Zhejiang Province, China'. *Agricultural Systems* 94 (3): 841-850.

Bossen, Laurel. 2011. 'Reproduction and Real Property in Rural China: Three Decades of Development and Discrimination'. In *Women, Gender and Rural Development in China*, edited by Tamara Jacka and Sally Sargeson, 97-123. Cheltenham: Edward Elgar Publishing.

Bray, Francesca. 1984. *Agriculture*. In *Science and Civilisation in China*, Vol. 6, Part 2, edited by Joseph Needham. Cambridge: Cambridge University Press.

—. 1994. *The Rice Economies: Technology and Development in Asian Societies*. Reprint. Berkeley: University of California Press.

—. 2013. *Technology, Gender and History in Imperial China: Great Transformations Reconsidered*. New York: Routledge.

Buch, Elana D. 2015. 'Anthropology of Aging and Care'. *Annual Review of Anthropology* 44 (1): 277-293.

Carter, Michael R., and Yang Yao. 2002. 'Local versus Global Separability in Agricultural Household Models: The Factor Price Equalization Effect of Land Transfer Rights'. *American Journal of Agricultural Economics* 84 (3): 702-715.

Chang, Te-Tzu. 2000. 'Rice'. In *The Cambridge World History of Food*, edited by Kenneth F. Kiple and Kriemhild Coneè Ornelas, 132-149. Cambridge: Cambridge University Press.

Chayanov, Aleksandr Vasilevich. 1966. *The Theory of Peasant Economy*, edited by Daniel Thorner, Basile Kerblay, and R.E.F. Smith. Homewood: Irvin.

Chen, Chen. 2015. 'Rural-Urban Migration and Circularity in China: Evidence from Anhui Province'. PhD diss., University of California Los Angeles.

Chen, Chuanbo, and C. Cindy Fan. 2016. 'China's Hukou Puzzle: Why Don't Rural Migrants Want Urban Hukou?' *The China Review* 16 (3): 9-39.

Chevalier, Alexandre, Elena Marinova, and Leonor Peña-Chocarro. 2014. 'Factors and Issues in Plant Choice'. In *Plants and People: Choices and Diversity Through Time*, edited by Alexandre Chevalier, Elena Marinova, and Leonor Peña-Chocarro, 3-14. Oxford: Oxbow.

Chien, Shiuh-Shen. 2015. 'Local Farmland Loss and Preservation in China—A Perspective of Quota Territorialization'. *Land Use Policy* 49: 65-74.

Chu, Julie Y. 2010. *Cosmologies of Credit: Transnational Mobility and the Politics of Destination in China*. Durham: Duke University Press.

Clifford Collard, Niamh Jane. 2016. 'Social Strategies and Material Fixes in Agotime Weaving'. In *Craftwork as Problem Solving: Ethnographic Studies of Design and Making*, edited by Trevor H.J. Marchand, 153-168. Farnham: Ashgate.

Cong, Zhen, and Merril Silverstein. 2011. 'Intergenerational Exchange Between Parents and Migrant and Nonmigrant Sons in Rural China'. *Journal of Marriage and Family* 73 (1): 93-104.

Dalakoglou, Dimitris. 2010. 'Migrating-remitting-"building"-dwelling: House-making as "Proxy" Presence in Postsocialist Albania'. *Journal of the Royal Anthropological Institute* 16 (4): 761-777.

De Brauw, Alan. 2010. 'Seasonal Migration and Agricultural Production in Vietnam'. *The Journal of Development Studies* 46 (1): 114-139.

De La Rupelle, Maëlys, Quheng Deng, Li Shi, and Thomas Vendryes. 2008. 'Land Rights and Rural-Urban Migration in China'. *China Perspectives* 2008 (2): 25-35.

Eyferth, Jacob. 2009. *Eating Rice from Bamboo Roots: The Social History of a Community of Handicraft Papermakers in Rural Sichuan, 1920-2000*. Harvard: Harvard University Press.

Fan, C. Cindy, Mingjie Sun, and Siqi Zheng. 2011. 'Migration and Split Households: A Comparison of Sole, Couple, and Family Migrants in Beijing, China'. *Environment and Planning* 43: 2164-2185.

Fan, C. Cindy, and Wenfei Winnie Wang. 2008. 'The Household as Security: Strategies of Rural-Urban Migrants in China'. In *Migration and Social Protection in China*, edited by Ingrid Nielsen and Russel Smyth, 205-243. New Jersey: World Scientific Publishing.

Fang, Xiangxin 方向新. 1998. 农村变迁论：当代中国农村变革与发展研究 *Nongcun bian qian lun: Dangdai Zhongguo nongcun biange yu fazhan yanjiu (Study on Rural Transformation: Current Chinese Rural Reform and Development)*. Changsha: Hunan People's Press.

Farquhar, Judith. 2006. 'Food, Eating, and the Good Life'. In *Handbook of Material Culture*, edited by Christopher Tilley, Webb Keane, Susanne Küchler, Mike Rowlands, and Patricia Spyer, 145-160. Los Angeles: Sage Publications.

Feder, Gershon, Lawrence J. Lau, Justin Y. Lin, and Luo Xiaopeng. 1992. 'The Determinants of Farm Investment and Residential Construction in Post-Reform China'. *Economic Development and Cultural Change* 41 (1): 1-26.

Gaetano, Arianne M., and Tamara Jacka, eds. 2004. *On the Move: Women and Rural-to-Urban Migration in Contemporary China*. New York: Columbia University Press.

Gao, Qihui. 2011. '2m Hectares of Farmland Abandoned per Year'. *China Daily*. 13 September 2011. Accessed 3 September 2015. http://www.chinadaily.com.cn/china/2011-09/13/content13678049.htm.

Glenn, Elias, and Kevin Yao. 2016. 'China Loosens Land Transfer Rules to Spur Larger, More Efficient Farms'. *Reuters*. 3 November 2016. Accessed 4 November 2016. http://www.reuters.com/article/us-china-economy-landrights-idUSKBN12Y09F?il=0.

Harries, John. 2017. 'A Stone That Feels Right in the Hand: Tactile Memory, the Abduction of Agency and Presence of the Past'. *Journal of Material Culture* 22 (1): 110-130.

He, Congzhi, and Jingzhong Ye. 2014. 'Lonely Sunsets: Impacts of Rural-Urban Migration on the Left-behind Elderly in Rural China'. *Population, Space and Place* 20 (4): 352-369.

He, Shenjing, and Desheng Xue. 2014. 'Identity Building and Communal Resistance against Landgrabs in Wukan Village, China'. *Current Anthropology* 55 (9): S126-S137.

Heerink, Nico, Futian Qu, Marijke Kuiper, Xiaoping Shi, and Shuhao Tan. 2007. 'Policy Reforms, Rice Production and Sustainable Land Use in China: A Macro-Micro Analysis'. *Agricultural Systems* 94: 784-800.

Ho, Peter. 2001. 'Who Owns China's Land? Policies, Property Rights and Deliberate Institutional Ambiguity'. *The China Quarterly* 166: 394-421.

HPBS, Hunan Provincial Bureau of Statistics. 2014. *Hunan Statistical Yearbook 2014*. Beijing: China Statistics Press.

—. 2019. *Hunan Statistical Yearbook 2019*. Beijing: China Statistics Press.

Huang, Philip C.C. 2016. 'China's Hidden Agricultural Revolution, 1980-2010, in Historical and Comparative Perspective'. *Modern China* 42 (4): 339-376.

IRRI, International Rice Research Institute. 2015. 'Direct Seeding'. Accessed 8 March 2016. http://www.knowledgebank.irri.org/step-by-step-production/growth/planting/direct-seeding.

—. 2016. 'How to Plant Rice'. Accessed 12 May 2016. http://www.knowledgebank.irri.org/step-by-step-production/growth/planting.

Jacka, Tamara. 1997. *Women's Work in Rural China: Change and Continuity in an Era of Reform*. Cambridge: Cambridge University Press.

Jin, Songqing, and Klaus Deininger. 2009. 'Land Rental Markets in the Process of Rural Structural Transformation: Productivity and Equity Impacts from China'. *Journal of Comparative Economics* 37 (3): 629-646.

Kamoshita, Akihiko, Sareath Chea, Satoshi Hayashi, Hiroyuki Ikeda, and Boonrat Jongdee. 2009. 'A Case Study on Farmers' Choice of Direct Seeding and Transplanting in Rain-Fed Lowlands in Northeast Thailand and Northwest Cambodia'. *Tropical Agriculture and Development* 53 (2): 43-54.

Labrada, Ricardo. 2003. 'The Need for Improved Weed Management in Rice'. In *Sustainable Rice Production for Food Security*. Rome: Food and Agriculture Organization of the United Nations.

Lave, Jean, and Etienne Wenger. 1991. *Situated Learning: Legitimate Peripheral Participation*. Cambridge: Cambridge University Press.

Levin, Iris, and Ruth Fincher. 2010. 'Tangible Transnational Links in the Houses of Italian Immigrants in Melbourne'. *Global Networks* 10 (3): 401-423.

Li, Jiming, Yeyun Xin, and Longping Yuan. 2009. 'Hybrid Rice Technology Development: Ensuring China's Food Security'. IFPRI Discussion Paper 00918, 2020 Vision Initiative.

Li, Yuyu. 2006. 'The Impact of Rural Migration on Village Development: A Comparative Study in Three Chinese Villages'. PhD diss., Johns Hopkins University.

Liu, Jieyu. 2017. 'Ageing in Rural China: State, Family and Gendered Care Responsibilities'. In *Handbook of Welfare in China*, edited by Beatriz Carrillo, Johanna Hood, and Paul Kadetz, 285-296. Northampton: Edward Elgar Publishing.

Lo, Kuei-Mei, and Hsin-Hsing Chen. 2011. 'Technological Momentum and the Hegemony of the Green Revolution: A Case Study of an Organic Rice Cooperative in Taiwan'. *East Asian Science, Technology and Society* 5 (2): 135-172.

Luhmann, Niklas. 1988. 'Familiarity, Confidence, Trust: Problems and Alternatives'. In *Trust: Making and Breaking Cooperative Relations*, edited by Diego Gambetta, 94-107. New York: Blackwell.

McKay, Deirdre. 2005. 'Reading Remittance Landscapes: Female Migration and Agricultural Transition in the Philippines'. *Danish Journal of Geography* 105 (1): 89-99.

Miller, Daniel. 2008. 'Migration, Material Culture and Tragedy: Four Moments in Caribbean Migration'. *Mobilities* 3 (3): 397-413.

Murphy, Rachel. 2002. *How Migrant Labor Is Changing Rural China*. Cambridge: Cambridge University Press.

—. 2006. 'Citizenship Education in Rural China: The Dispositional and Technical Training of Cadres and Farmers'. In *Chinese Citizenship: Views from the Margins*, edited by Vanessa L. Fong and Rachel Murphy, 9-26. London: Routledge.

NBSC, National Bureau of Statistics of China. 2019. *China Statistical Yearbook 2019*. Beijing: China Statistics Press.

Nguyen, Tuan Anh, Jonathan Rigg, and Annuska Derks. 2015. 'Migration and Agricultural Production in a Vietnamese Village'. Max Planck Institute for Social Anthropology, Working Paper No. 164.

NPC, The National People's Congress of the People's Republic of China. 2002. 'Law of the People's Republic of China on Land Contract in Rural Areas'. Beijing: Database of Laws and Regulations.

Obendiek, Helena. 2016. *'Changing Fate': Education, Poverty and Family Support in Contemporary Chinese Society*. Berlin: Lit.

OECD. 2005. *OECD Review of Agricultural Policies: China*. Paris: OECD Publishing.

—. 2014. *Agricultural Policy Monitoring and Evaluation 2014: OECD Countries*. Paris: OECD Publishing.

Ogura, Chikara, Somsak Sukuchan, and Hajime Narioka. 2011. 'Introduction of Direct Seeding and Selection of Farmland in Rainfed Rice Fields in Small Watersheds in Northeast Thailand'. *The Japan Agricultural Research Quarterly* 45 (3): 337-347.

Oi, Jean, and Andrew Walder, eds. 1999. *Property Rights and Economic Reform in China*. Stanford: Stanford University Press.

Ortner, Sherry B. 2006. *Anthropology and Social Theory: Culture, Power, and the Acting Subject*. Durham: Duke University Press.

Oshiro, Kenji K. 1985. 'Mechanization of Rice Production in Japan'. *Economic Geography* 61 (4): 323-331.

Oxfeld, Ellen. 2017. *Bitter and Sweet: Food, Meaning, and Modernity in Rural China*. Oakland: University of California Press.

Pandey, Sushil, Martin Mortimer, Leonard Wade, To Phuc Tuong, Katherine Lopez, and Bill Hardy, eds. 2002. *Direct Seeding: Research Strategies and Opportunities*. Los Baños: International Rice Research Institute.

Pauli, Julia. 2015. 'Gebauter Lebenssinn. Häuser in transnationalen mexikanischen Familien'. *Sociologus* 65 (2): 153-176.

Perdue, Peter C. 1987. *Exhausting the Earth: State and Peasant in Hunan, 1500-1850*. Cambridge, MA: Harvard University Press.

Pieke, Frank N. 2002. 'The Politics of Rural Land Use Planning in China'. Max Planck Institute for Social Anthropology, Working Paper No. 40.

Qin, Hua. 2010. 'Rural-to-Urban Labor Migration, Household Livelihoods, and the Rural Environment in Chongqing Municipality, Southwest China'. *Human Ecology* 38 (5): 675-690.

Rawski, Evelyn Sakakida. 1972. *Agricultural Change and the Peasant Economy of South China*. Cambridge, MA: Harvard University Press.

Rigg, Jonathan, Albert Salamanca, and Eric C. Thompson. 2016. 'The Puzzle of East and Southeast Asia's Persistent Smallholder'. *Journal of Rural Studies* 43: 118-133.

Roberts, Kenneth D. 1997. 'China's "Tidal Wave" of Migrant Labour: What Can We Learn From Mexican Undocumented Migration to the United States?' *International Migration Review* 31 (2): 249-293.

Santos, Gonçalo. 2017. 'Multiple Mothering and Labor Migration in Rural South China'. In *Transforming Patriarchy: Chinese Families in the Twenty-First Century*, edited by Gonçalo Santos and Stevan Harrell, 91-110. Seattle: University of Washington Press.

Sargeson, Sally. 2002. 'Subduing "The Rural House-Building Craze": Attitudes Towards Housing Construction and Land Use Controls in Four Zhejiang Villages'. *The China Quarterly* 172: 927-955.

Schippers, Thomas K. 1992. 'Le proverbe comme forme de transmission de savoirs: Quelques réflexions à propos d'un corpus de proverbes météorologiques'. In *Les voies de la parole*, edited by Jean-Noël Pelen and Claude Martel, 99-105. Aix-en-Provence: Université de Provence.

—. 2014a. 'Agricultural Practices: Change and Stability'. In *Exploring and Explaining Diversity in Agricultural Technology*, edited by Annelou van Gijn, John C. Whittaker, and Patricia C. Anderson, 339-341. Oxford: Oxbow.

—. 2014b. 'Seasonal Variations in Crop Tending and Folk Knowledge in Southern France'. In *Exploring and Explaining Diversity in Agricultural Technology*, edited by Annelou van Gijn, John C. Whittaker, and Patricia C. Anderson, 79-80. Oxford: Oxbow.

Song, Yiching. 1998. '"New" Seed in "Old" China: Impact of CIMMYT Collaborative Programme on Maize Breeding in South-Western China'. PhD diss., Wageningen University.

Steinmüller, Hans. 2013. *Communities of Complicity: Everyday Ethics in Rural China*. New York: Berghahn.

Tian, Qing, Daniel G. Brown, Lin Zheng, Shuhua Qi, Ying Liu, and Luguang Jiang. 2015. 'The Role of Cross-Scale Social and Environmental Contexts in Household-Level Land-Use Decisions, Poyang Lake Region, China'. *Annals of the Association of American Geographers* 105 (6): 1240-1259.

Tilt, Bryan. 2008. 'Smallholders and the "Household Responsibility System": Adapting to Institutional Change in Chinese Agriculture'. *Human Ecology* 36: 189-199.

Walsh, Katie. 2011. 'Migrant Masculinities and Domestic Space: British Home-making Practices in Dubai'. *Transactions of the Institute of British Geographers* 36 (4): 516-529.

Wang, Hongfei, and Robert J. Hijmans. 2019. 'Climate Change and Geographic Shifts in Rice Production in China'. *Environmental Research Communications* 1 (1): 011008.

Wu, Chunyuan 伍春元, ed. 2010. 益相史话 *Yixiang shihua* (*Historical Narrative of Yixiang*). Beijing: Jiuzhou Press.

Xie, Yong, and Quanbao Jiang. 2016. 'Land Arrangements for Rural-Urban Migrant Workers in China: Findings from Jiangsu Province'. *Land Use Policy* 50: 262-267.

Xinhua. 2020. 'China Focus: Major Rice Producer Promotes Double-Cropping Rice to Ensure Food Security'. Xinhua Net. 28 April 2020. Accessed 5 June 2020. http://www.xinhuanet.com/english/2020-04/28/c_139015447.htm.

Xu, Dingde, Shili Guo, Fangting Xie, Shaoquan Liu, and Sha Cao. 2017. 'The Impact of Rural Laborer Migration and Household Structure on Household Land Use Arrangements in Mountainous Areas of Sichuan Province, China'. *Habitat International* 70: 72-80.

Yan, Yunxiang. 2003. *Private Life under Socialism: Love, Intimacy and Family Change in a Chinese Village, 1949-1999*. Stanford: Stanford University Press.

Yu, Zhonglei, Lei Liu, Hua Zhang, and Jinshe Liang. 2017. 'Exploring the Factors Driving Seasonal Farmland Abandonment: A Case Study at the Regional Level in Hunan Province, Central China'. *Sustainability* 9 (187): 1-18.

Yuan, Juanwen, and Anke Niehof. 2011. 'Agricultural Technology Extension and Adoption in China: A Case from Kaizuo Township, Guizhou Province'. *The China Quarterly* 206: 412-425.

Zhang, Ying, Xiubin Li, and Wei Song. 2014. 'Determinants of Cropland Abandonment at the Parcel, Household Andvillage Levels in Mountain Areas of China: A Multi-Level Analysis'. *Land Use Policy* 41: 186-192.

Zhi, Fujing 致富经. 2010. '灯芯草"肥"了龙市乡 Dengxin cao "fei" le Longshi xiang (Lamp Rush "Fertilized" Longshi Township)'. 湖南科技报 *Hunan Science and Technology Newspaper*. 20 December 2010. Accessed 24 March 2013. http://www.zgxncjs.com/xnc/xnc/showclass.jsp-id=19322.htm.

Zhou, Daming 周大鸣. 2005. *渴望生存: 农民工流动的人类学考察 Kewang shengcun: Nongmingong liudong de renleixue kaocha* (*Longing for Survival: An Anthropological Investigation of the Flow of Migrant Workers*). Guangzhou: Zhongshan University Press.

Conclusion: A Skill Perspective on Migration

Abstract

This concluding chapter discusses four advantages of investigating migration settings from a socio-technical skill perspective. First, it provides an understanding of a particular form of peasant agency that is commonly overlooked. Second, focusing on skill allows us to better understand farmers' decision-making. Third, it provides new insights into technology and Chinese modernity. Finally, it contributes to understanding migration beyond the common dichotomies such as between people and things, or migrants and those left behind. It concludes that even those who move to the cities remain part of their village communities of practice. They maintain their ties to the land through the ongoing management of their paddy fields – whether hands-on in person or at a distance using other household farming strategies.

Keywords: materialities of migration, rural-urban migration, skill perspective, socio-technical peasant agency, farming community of practice, Chinese modernity

Since I first set out to study Chinese rural migrants in 2007, new transformations have occurred. Green Water villagers have begun to adopt transplanting machines, which the local government had failed to popularize since their introduction to the area in the 1950s. Moreover, smartphones have become widespread, easing not only the continued conversation between my interlocutors and myself but, importantly, also flows of and access to knowledge, as well as video conversations between migrated and staying household members. In addition, the restrictive birth control policy has officially been relaxed, the *hukou* system is gradually being abolished, and the changing climate is presenting new challenges for sustaining rice yields (Muehe et al. 2019). Meanwhile, Chinese policy makers are already

Kaufmann, Lena, *Rural-Urban Migration and Agro-Technological Change in Post-Reform China*. Taylor & Francis Group 2021
DOI: 10.5117/9789463729734_CONCL

taking their first steps towards the next transformation. Much of future rural policy is targeted at scaling-up agriculture; reforming land-use and transfer rights; increasing urbanization, including in the countryside; and augmenting agro-industrialization. At some point, it might also comprise rice-seeding and pesticide-spraying drones, remote-controlled weeding machines; smart farming methods supported by Artificial Intelligence; and the commercial distribution of genetically modified rice. All of these are already being tested and might eventually result in farmers 'wearing leather shoes to farm', instead of getting themselves dirty in the muddy fields (Tang, Jiang, and Xin 2020).[1]

At the same time, Chinese agriculture is becoming more global. In view of China's quest for national grain sovereignty, this may not be so apparent at first sight. Still, as Bray et al. (2015) have shown for rice, China, too, has a global history of agriculture. Today this history continues to be written, whether on the African continent (Brautigam 2015), in the framework of China's 'going out' strategy and securing offshore food supplies (McMichael 2020), or in the dispute over agricultural products in the ongoing Sino-American trade war. Today, rice is the staple food of more than half of the global population (CGIAR 2020). Changing Chinese land-use and land-arrangement strategies are thus also a global question, the impacts of which will only become clear in future. In all of this, paddy fields are material interfaces which sensitively display these transformations – whether political, technological, social or global.

Since my fieldwork period, the actors in this book have moved on, and so have I. The left-behind school children I met have now grown up. Most of them have become migrants themselves, some for educational reasons, others to take over the small enterprises of their migrant parents. Some of these parents have returned home. New challenges are constantly arising, such as when migrant quarters in Beijing were brutally being torn down towards the end of 2017 to make way for Beijing's new urban planning policy, stripping thousands of inhabitants suddenly and ruthlessly of their new city homes and workplaces.

Meanwhile, I moved to Switzerland to become what is commonly called a highly-skilled migrant. In Switzerland, whenever I open my mouth and people hear my standard German accent instead of Swiss German, it is immediately revealed that I am not a local. Especially at the beginning, I often felt an invisible barrier and distance in the way people on the street reacted to me. Although the context is very different, this always reminds

1 In Chinese: '穿着皮鞋能种田' *chuanzhe pixie neng zhongtian* (Tang, Jiang, and Xin 2020).

me of the Chinese migrants' experience. Their rural origin is commonly exposed to their urban counterparts through their accents – along with their often suntanned skin and what are perceived as distinctively 'rural' ways of dressing and behaving. For example, when Yuemei began working in Beijing, one of the first things her urban boss told her to do was to buy new clothes and dress more appropriately. She tried her best to do so and to fit in.

Yet, in general, even as we try to adapt to new socio-cultural environments, our backgrounds are hard to shake off. Obviously, personal feelings of belonging tie us to specific places and communities. More unconsciously, we also incorporate different layers of practical experiences that we have made previously and continue to make. Moving on, we often retain what Marcel Mauss has called 'techniques of the body' (*les techniques du corps*) (Mauss 1934). These techniques manifest themselves in mundane everyday activities such as washing clothes by hand, using a knife for cooking, eating with chopsticks or forks, using the toilet, walking along the street, playing sports or – as my interlocutors from Hunan say – 'doing paddy rice'.

In this book, I have examined precisely such ways of 'doing paddy rice' and the related everyday life strategies. In the light of China's massive rural-urban migration and rapid agro-technological transformation, I have focused on one exemplary resource – paddy fields – arguing that we need to pay more attention to socio-material resources in migration. I have explored this resource in view of a particular predicament that Chinese rice farmers are confronted with: the pressure to migrate to the cities, and the simultaneous need to continuously cultivate their paddy fields in order to preserve them as a safety net resource in a context of uncertainty. I have suggested that studies which deal with the material side of migration rarely take skill and knowledge into consideration, even though this knowledge is central for preserving paddy fields – for both those who migrate out of rural villages and those who stay at home. I have therefore identified the need to rethink the notion of 'migrant worlds' (Basu and Coleman 2008) as, instead, a 'community of practice worlds'.

Moreover, I have demonstrated that the rice farmers' predicament under study is, first of all, a socio-material one. It is a situation that is virtually 'objectified' in various material resources, so it requires socio-material solutions. Hence, migrants and left-behind people form a 'community of practice' (Lave and Wenger 1991) that centres on the central question of how the paddy fields' soil quality can be preserved, which requires knowledge, techniques, and skilful cultivation. Taking a socio-material lens to scrutinize their actions is indispensable for properly understanding this community of rice famers' strategic responses to their predicament. Such a lens, focusing on the actual practices of people engaging with their material world, has several implications with regard to our understanding of: (1) agricultural practices, (2) farmers' decision-making, (3) technology and modernity, and (4) migration.

Agency beyond resistance

I have suggested that focusing on the intimate, often tacit interaction of rice farmers with their material world allows us to sense the specific agency of rice farming households between the countryside and city. This type of agency lies beyond overt resistance. It manifests itself in farmers' concrete everyday practices of engaging with their fields. This may be directly by farming them. Alternatively, it may be indirectly, by delegating tasks from afar, and by using skills newly acquired in the city to earn an income with which to 'feed' the fields. This includes, for example, an elderly left-behind woman choosing to farm a particular crop such as lamp rush, a farmer using a specific technique such as direct seeding, or a migrant paying for the use of technology such as a combine harvester. Precisely because this agency 'resides in the fields' (van der Ploeg 2007), being deeply immersed in the material and the everyday, it is easily overlooked.

As I have shown, the skills that migrating and remaining farmers apply to their fields are, indeed, much more than merely technical abilities. They have an underlying intention, entailing multi-dimensional reasoning for potentially manifold purposes, whether short, medium or long-term, social or economic goals. These range from ensuring one's own old-age care and health, the family's status within village society, and the continuity of the patriline, to preserving the fields as a safety net for future generations. Not acknowledging this means denying farmers a large share of their own agency.

Yet, while some of the farmers' strategies I have described, such as abandoning fields, may readily be read as resistance in James Scott's sense (1985), in choosing the term 'agency' I advocate a more differentiated perception of farmers' actions (see also Ortner 2006). As I have shown, in pursuing their own projects, farmers are actually often acting in line with state objectives – whether deliberately or not. A good example of this is the strategy of renting out fields. As mentioned in Chapter 5, this practice implicitly facilitates the transition towards big household (*dahu*) farming. This is, in fact, an important state policy that aims to increase agricultural productivity through large-scale commercial farming, and ensure that deserted farmland is used (Li 2006, 401; OECD 2014). Thus, if we do not pay attention to the more subtle agency of farmers, we could easily misunderstand farmers' actions as simple responses to market forces and government policies when they are, in fact, much more complex than this.

Decision making beyond economic reasoning

Considering farmers as agents in this way also reveals a great deal about the decisions they make between home and migration. Previous studies have tended to highlight the economic factors behind farmers' choices (see, e.g., Barlett 1980; Chibnik 2011). While these are doubtlessly important to Chinese villagers, skill is also a factor. Such skills become visible, in the Chinese case, by focusing on the particular predicament moment of decision making and the challenges faced by the community of practice worlds. As I have shown previously for Chinese migrants in the city, and here with regard to the migration-affected countryside, skill has important implications for migration decisions and processes. It not only structures where and in which constellations people migrate, what they do and how they organize themselves in their places of arrival, but it also has an impact on how people deal with their home resources, and whether they stay or migrate from their home villages.

As the strategies depicted in Chapter 5 suggest, skill plays a role as a decision-making factor, for example, in the case of left-behind Granny Li, who takes certain production decisions (planting lamp rush) in view of her own skills and the lacking availability of skills in her migration-affected household. Another example is the case of Yuemei and her siblings. It is through the particular skills they have acquired that they can migrate and earn sufficient money in the city to finance machine harvesting, which ensures the cultivation of the household's fields during their absence.

Additionally, I suggest that it is no coincidence that migrants prefer to leave their own family members behind to take care of the household's fields. This is not only due to the insecure situation of land tenure in China. It is also related to the fact that, in the community of practice world, people trust their family members to have the necessary intimate knowledge of the allocated plots, and to be more willing and prepared to put their efforts and skill into preserving the fields.

With regard to skill as a decision-making factor, it would certainly be rewarding to look more deeply below the household level to discern answers to several questions: How are the various factors in complex decisions weighed against each other? Which skills do the individual household members possess and how do these enter into the actual practice of negotiating decisions within the household? What do shifts in individual skill repertoires imply, for example with regard to gender roles, and the values ascribed to these skills?

Technology beyond linear progress

A skill perspective also has important implications for how we understand technology and, connected to this, modernity. As Francesca Bray notes:

> We live in a world where most people, from technocrats to schoolchildren to academics, believe that technology means iPhones and biotech but not refrigerators, string or latrines; that technology is culture-free and politics-free; that it is a force irresistibly propelling us into the future; and that its history confirms that West is Best (while nervously recognizing that the East is Catching Up Fast). (Bray 2017, 95.)

Looking at farmers' strategies of technology adoption, where technology is understood beyond tractors and combine harvesters as 'ways of *making* and *doing*' (ibid., 97; original italics), shows that Chinese farmers are neither 'backward', as commonly portrayed in public discourse, where '[k]nowledge and technology were (and remain) understood to be by definition the antithesis of the traditional and the peasant' (Schmalzer 2016, 108). Nor are they simply passive adopters of new technologies, blindly reproducing the state narrative of technological innovations and progress. Instead, they strategically draw on a whole repertoire of solutions to deal with their paddy field predicament.

These findings resonate with David Edgerton's (2007) 'history of technology-in-use', i.e. a history of technology that focuses on actual practices, rather than mere inventions. This challenges our perceptions of technological time as being innovation-based, of the importance of certain technologies, and ultimately also of modernity (Edgerton 2007, xi). Applying a skill perspective makes it possible to grasp the agency of rural Chinese people who actually choose and use technologies. Among other things, this provides valuable insights into farmer-state relations. More generally, it enables us to grasp and describe a Chinese modernity that is grounded in practices of 'non-synchronicities', where stone mills, oxen-pulled ploughs, combine harvesters and hybrid rice coexist. This modernity is clearly distinct from a notion of modernity as merely technological advancement.

Related to this, a skill perspective refutes the assumption that economies of scale, which are commonly seen as the necessary precondition for industrializing farming, are automatically a marker of modernity. In fact, the case of contemporary Chinese farming demonstrates clearly that, despite the issue of renting out fields that encourages bigger farms, smallholdings generally continue to prevail and to make sense to farmers today, being an integral part of the Chinese modernity (van der Ploeg and Ye 2016). This is, in my view, related to Sigaut's findings (1994) about the necessary size of a 'skill-producing group': if the socio-technical system of farming is to work well, the skill-producing group must not be too small but, crucially, not too big either. In this regard, compared to farming at scale using industrial methods, small-scale family farming fosters knowledge transmission within the skill-producing group. Focusing on technology as entailing skilful ways of making and doing is, therefore, highly effective for describing and understanding the past and current transformations in and beyond the Chinese countryside.

Migration beyond dichotomies

Applying this skill perspective to migration studies negates many of the dichotomies that have been prevalent in migration studies for a long time, such as differentiations between skilled and unskilled migrants, internal and international migration, migrants and non-migrants, and people and things. This book has shown that these alleged divisions are not, however, clear-cut. In fact, my community of practice worlds approach disproves the very notion of an either/or situation in the reality of people's lived experience.

Regarding the differentiation between skilled and unskilled, my findings imply that it is not appropriate to call some migrants unskilled and others highly skilled. All of those I researched are indeed skilled – some more in one area, and some more in another. As long as the 'skill-producing group' has not completely disintegrated, all of these people are part of their community of practice worlds. Within these worlds, migrants can be seen as members who have incorporated, and retain, crucial 'tactile memory' (Harries 2017).

In view of this, the difference lies much more at the discursive level, i.e. in how migrants' skills are valued. To appreciate their skills thus requires rethinking our own values and the ways we perceive migrants. This applies as much to spontaneous rural-urban migration in China as it does, for example, to the current wave of people fleeing to Europe from war and hardship, often facing considerable hostility in their places of arrival. On a very practical level, perceiving these people as part of a community of practice worlds means reconsidering the political efforts being employed by European governments and societies to try to integrate migrants into their places of arrival. It also requires reconsidering initiatives that exist to prevent migration altogether, by barricading borders or imposing top-down development projects. It requires, instead, acknowledging the potential of migrants and the skills they possess – not only for their own survival strategies, but also the contribution they could make to their host societies, as well as the beneficial impact they could have on the wellbeing of their places of origin. For the study of migration, this means that we should focus on investigating what people are actually capable of, rather than what they lack.

Reflecting on the Chinese case to review the professed distinction between migrants and non-migrants, my book clearly demonstrates that migrant work and farm work, and migrant and left-behind farmers are closely interlinked. There is no simple urban/rural dichotomy when it comes to Chinese migration patterns – even those who move to the cities remain part of their village communities, sustaining relationships with their families and friends through visits and interactions, on top of maintaining their ties to the land through the ongoing management of their paddy fields – whether hands-on in person or at a distance using other farming strategies. Farmers not only circulate between the two spheres, but their actions always take both sides into consideration – the countryside and the city, in a highly productive way. By focusing on both the migrants' places of origin and places of arrival, taking a skill perspective considerably extends our understanding of migration processes and migrant-home relations.

Regarding the interlinkage of the places of origin and those of arrival, the material aspect plays a crucial role. In objectifying the situation, the material provides valuable insights into more implicit aspects of migration, including how migrants make their material world and how this material world makes them (Basu and Coleman 2008). This not only comprises tangible material items, however, but also the skills connected to them. As I have shown, the skills needed to preserve a crucial home resource feature prominently in farmers' connections, not just to the countryside and the city, but also to the people and their material world through their communities of practice. Since migrants embody the migration process, we should not perceive the individuals separately from their material world that connects them with different places. On the contrary, perceiving them as one entity opens up an entirely new perspective on the socio-material transitions that occur in China and elsewhere. It shows how migrants in various places are dynamic actors dealing with specific socio-material challenges and predicaments.

References

Barlett, Peggy F., ed. 1980b. *Agricultural Decision Making*. New York: Academic Press.

Basu, Paul, and Simon Coleman. 2008. 'Introduction: Migrant Worlds, Material Cultures'. *Mobilities* 3 (3): 313-330.

Brautigam, Deborah. 2015. *Will Africa Feed China?* New York: Oxford University Press.

Bray, Francesca. 2017. 'Science, Technology and Late Imperial History'. *The Chinese Historical Review* 24 (1): 93-104.

Bray, Francesca, Peter A. Coclanis, Edda L. Fields-Black, and Dagmar Schäfer, eds. 2015. *Rice: Global Networks and New Histories*. New York: Cambridge University Press.

CGIAR, Research Program on Rice. 2020. 'Ricepedia: The Global Staple'. 2020. Accessed 9 January 2020. http://ricepedia.org/rice-as-food/the-global-staple-rice-consumers.

Chibnik, Michael. 2011. *Anthropology, Economics, and Choice*. Austin: University of Texas Press.

Edgerton, David. 2007. *The Shock of the Old: Technology and Global History since 1900*. Oxford: Oxford University Press.

Harries, John. 2017. 'A Stone That Feels Right in the Hand: Tactile Memory, the Abduction of Agency and Presence of the Past'. *Journal of Material Culture* 22 (1): 110-130.

Lave, Jean, and Etienne Wenger. 1991. *Situated Learning: Legitimate Peripheral Participation*. Cambridge: Cambridge University Press.

Li, Yuyu. 2006. 'The Impact of Rural Migration on Village Development: A Comparative Study in Three Chinese Villages'. PhD diss., Johns Hopkins University.

Mauss, Marcel. 1934. 'Les techniques du corps'. *Journal de psychologie* 32 (3-4).

McMichael, Philip. 2020. 'Does China's "Going out" Strategy Prefigure a New Food Regime?' *The Journal of Peasant Studies* 47 (1): 116-154.

Muehe, E. Marie, Tianmei Wang, Carolin F. Kerl, Britta Planer-Friedrich, and Scott Fendorf. 2019. 'Rice Production Threatened by Coupled Stresses of Climate and Soil Arsenic'. *Nature Communications* 10 (4985): 1-10.

OECD. 2014. *Agricultural Policy Monitoring and Evaluation 2014: OECD Countries*. Paris: OECD Publishing.

Ortner, Sherry B. 2006. *Anthropology and Social Theory: Culture, Power, and the Acting Subject*. Durham: Duke University Press.

van der Ploeg, Jan Douwe. 2007. 'Resistance of the Third Kind and the Construction of Sustainability'. Paper presented to the ESRS Conference, August 23, 2007, Wageningen.

van der Ploeg, Jan Douwe, and Jingzhong Ye, eds. 2016. *China's Peasant Agriculture and Rural Society: Changing Paradigms of Farming*. London: Routledge.

Schmalzer, Sigrid. 2016. *Red Revolution, Green Revolution: Scientific Farming in Socialist China*. Chicago: University of Chicago Press.

Scott, James C. 1985. *Weapons of the Weak: Everyday Forms of Peasant Resistance*. New Haven: Yale University Press.

Sigaut, François. 1994. 'Technology'. In *Companion Encyclopedia of Anthropology*, edited by Tim Ingold, 420-459. London: Routledge.

Tang, Xiaoqing 唐小晴, Junjun Jiang 蒋军君, and Xuan Xin 新轩. 2020. '中国农民春耕新景象："共享农机"开进田间 穿着皮鞋能种田 Zhongguo nongmin chungeng xin jingxiang: "Gongxiang nongji" kai jin tianjian chuanzhe pixie neng zhongtian (A New Scene of Spring Farming for Chinese Farmers: "Shared Agricultural Machinery" Enters the Field – Wearing Leather Shoes to Farm)'. *China News Hunan*. 18 March 2020. Accessed 4 June 2020. http://www.hn.chinanews.com/news/2020/0318/381860.html.

Appendix

I Glossary

ba, pa	耙	harrow, rake
bao, pao	刨	plane, digger
baihe	百合	lily (*Lilium brownii* F. E. Brown var. *viridulum* Baker)
banchu	板锄	board hoe
bantong	扮桶	wooden threshing barrel
biandan	扁担	shoulder pole
bieren	别人	other people
bingfei	饼肥	cake fertilizer
bu fada de jiqi	不发达的机器	underdeveloped machine
bu ru hu xue, yan de hu zi	不入虎穴，焉得虎子	how can you catch tiger cubs without venturing into the tiger's den (saying)
chayangji	插秧机	transplanting machine
changgui shuidao	常规水稻	conventional rice
chang xin	尝新	tasting the new [i.e. freshly harvested rice]
chengbao	承包	to contract (land out to other people)
chengyu	成语	set phrase, idiom (written language)
chu	锄	hoe
chuantong	传统	traditional
cunmin weiyuanhui	村民委员会	villagers' committee
dadaoji	打稻机	threshing machine
dagong	打工	working under contract/for a boss (used to denote migrant work)
dahu	大户	big household
dazhong chuangye, wanzhong chuangxin	大众创业、万众创新	mass innovation and entrepreneurship (slogan)
da zhucao	打猪草	collect pig weed
dan	担	load, unit of weight, equal to 50 kg
dan shou dan zhong fangfa	单种单收方法	single harvest, single sow method

dangfei	凼肥	pit fertilizer, wet compost
dao	刀	knife
dengxin cao	灯芯草	lamp rush (*Juncus effusus*)
dibao	低保	Minimum Livelihood Guarantee (abbr.)
dichan tian	低产田	low production fields
difangzhi	地方志	local gazetteer
dianli	电犁	ploughing machine
dianxing shifan, wenbu tuiguang	典型示范，稳步推广	setting up an example with a model, popularizing steadily
du kouzi	堵口子	blocking the water inlets
dulun che	独轮车	wheelbarrow
dui	碓	treadle-operated tilt hammer for hulling rice
erchi chu	二齿锄	'two teeth' hoe
Ershisi jieqi ge	二十四节气歌	Song of the 24 Solar Terms
fei tian	肥田	fertile field
fen	分	unit of area, equal to 0.1 *mu* (about 66.67 m^2)
fengche	风车	winnowing machine
fu	斧	axe
gailiang li	改良犁	improved plough
gaochan tian	高产田	high production fields
geheqi	割禾器	grain cutting machine
genzhe bieren zuo	跟着别人做	follow others (here: copy others)
gengdao, jingdao	粳稻	round-grained *japonica* rice
gengtianji	耕田机	ploughing machine
gongmin	工民	artisans
gongmin	公民	citizens
guanli de hao	管理得好	good to manage
hu	户	household
hukou	户口	household registration
huaxingqi	划行器	paddle instruments
huang	荒	waste, desolate
huangdi nü'er bu chou jia	皇帝的女儿不愁嫁	you don't have to worry about marrying off the emperor's daughter (saying)
hui lai zai xuexi	回来再学习	learn again upon return
ji	箕	winnowing basket, dustpan

ji gunchuan	机滚船	motorized 'roll boats', a tilling machine
jia	家	family, home
jiapu	家谱	family record, family tree
jiefang laodong li	解放劳动力	set free the labour force
jieqi	节气	solar term
jin	斤	unit of weight, equal to 500 grams
ju	锯	saw
kai yangtian men	开秧田门	opening the door of the seedbed
ku	苦	bitter
ku de budeliao	苦得不得了	terribly bitter
laojia	老家	ancestral home
li	里	500 m (Chinese measurement)
li	犁	plough
li ling zhi hun	利令智昏	to lose one's head through material greed
litianji	犁田机	ploughing machine
liandao	镰刀	sickle
liangzhong	良种	improved variety
liu	留	remain, stay
liushou	留守	stay behind, stay to take care of
liushou ertong	留守儿童	left-behind children
liusuan pa	硫酸怕	sulphuric acid fear
long	砻	rice huller
longxian	龙线	dragon line
luo	箩	square-bottomed bamboo basket
luohou	落后	backward
mapo doufu	麻婆豆腐	tofu dish from Sichuan Province
mijiu	米酒	rice wine, fermented rice
mixin	迷信	superstitious
mianzi	面子	social face, reputation
minjian wenxue	民间文学	folk literature
ming	命	life, fate
mo	磨	grinder
mu	亩	unit of area, equal to 1/15 hectare
neibu	内部	internal (publication)
nian	碾	roller
niangjia	娘家	home of a married woman's parents
nong	农	agriculture; also: agriculture, the countryside, and peasants (*san nong*)

Nongda	农达	Roundup
nongli	农历	farmers' calendar
nongmin	农民	farmers, peasants
nongmin gong	农民工	peasant workers, labour migrants
nongyao	农药	farm chemicals, pesticides
nuomi	糯米	glutinous rice
penwuqi	喷雾器	back-carried atomizer
pugun	蒲滚	a type of tilling tool or machine
putong hua	普通话	standard Chinese, Mandarin
qichun	起春	the beginning of spring
qin gong jian xue	勤工俭学	part-work and part-study system
qingsong duo le	轻松多了	much more relaxed
qunzhong	群众	the masses
ren ding sheng tian	人定胜天	man must conquer nature (slogan attributed to Mao Zedong)
sanchi chu	三齿锄	'three teeth' hoe
san fen caotian yang die niang	三分草田养爹娘	three *fen* of grassfield [a field planted with lamp rush] support father and mother (saying)
Santao Jicheng	三套集成	Three Comprehensive Collections (of Chinese Folk Literature)
shaizi	筛子	sieve
shangmin	商民	merchants
shao yangtian zhi	烧秧田纸	burning of seedbed paper
shi ba yi mu gengdi	十八亿亩耕地	1.8 billion *mu* farmland preservation
hongxian	红线	redline (policy)
shimin	士民	gentry, scholars
shimin	市民	city residents, (urban) citizens
shou	守	guard, conserve, protect
shoufu tuolaji	手扶拖拉机	walking tractor
shou tian	瘦田	unfertile field
shuangqiang	双枪	the 'double rush' of simultaneously harvesting and transplanting
sichi chu	四齿锄	'four teeth' hoe
simin	四民	four classes of people
suzhi	素质	quality (discourse)
tili huo	体力活	physical labour
tianjiu	甜酒	rice wine, fermented rice
tuopen	拖盆	roller for levelling the ploughed field
wandao	晚稻	late-season rice

weixing shuilunji	微型水轮机	mini water turbine
wenhua	文化	culture, education
wugu fengdeng	五谷丰登	an abundant harvest of all food crops
xiandao	籼稻	long-grained, non-glutinous indica rice
xiang	乡	township
xinku	辛苦	hard, laborious
yahe	丫禾	yahe-technique of intercropping rice
yanyu	谚语	proverb, saying
yi nongjiafei weizhu, yi huafei weifu	以农家肥为主，以化肥为辅	relying primarily on farmers' fertilizers, and secondarily on chemical fertilizers (Mao-era policy)
yi tiao niu de fen, san miao tian de fei	一条牛的粪，三苗田的肥	the dung of an ox fertilizes three fields of seedlings (saying)
yizi chu	一字锄	'one line' hoe
yong jiqi	用机器	to use machines
yuankou chu	圆口锄	'round mouth' hoe
zajiao shuidao	杂交水稻	hybrid paddy rice
zaliang	杂粮	miscellaneous grain crop
zaodao	早稻	early-season rice
zeng fei buru huan zhong	增肥不如换种	it is better to change seeds than to increase fertilizer (saying)
zhen	镇	town
zhongdao	中稻	middle-season rice
zhonggengqi	中耕器	intertilling machines
Zhongguo minjian wenxue san tao jicheng	中国民间文学三套集成	Three Comprehensive Collections of Chinese Folk Literature
zhushai	竹筛	bamboo sieve (for winnowing)
ziku mi	spoken language	*ziku* rice (name of a local rice variety)
ziliu di	自留地	private plot
zuo daogu	做稻谷	to cultivate rice
zuo hantu	做旱土	to cultivate dry fields
zuo tian mao qiao, san nian liang tiao	作田冒巧，三年两斢	when preparing the field you need to put in skill, you need to change [seeds] twice in three years (saying)

II Solar terms

The following table lists the solar terms (*jieqi*) and their equivalent dates in the Gregorian calendar. As explained in Chapter 2, the solar terms subdivide the agricultural year of the luni-solar Chinese farmers' calendar (*nongli*) into 24 sections, indicating specific agricultural and ritual activities.

Table 3 The 24 solar terms and equivalent dates in the Gregorian calendar (based on Qi 1986, 141-142)

Solar term	Date in the Gregorian calendar
*lichun*立春 (Beginning of Spring)	February 4/5
yushui 雨水 (Rain Water)	February 19/20
jingzhe 惊蛰 (Awakening of Insects)	March 5/6
chunfen 春分 (Spring Equinox)	March 20/21
qingming 清明 (Clear and Bright)	April 4/5
guyu 谷雨 (Grain Rain)	April 20/21
lixia 立夏 (Beginning of Summer)	May 5/6
xiaoman 小满 (Lesser Fullness of Grain)	May 21/22
mangzhong 芒种 (Grain in Ear)	June 5/6
xiazhi 夏至 (Summer Solstice)	June 21/22
xiaoshu 小暑 (Lesser Heat)	July 7/8
dashu 大暑 (Greater Heat)	July 23/24
liqiu 立秋 (Beginning of Autumn)	August 7/8
chushu 处暑 (End of Heat)	August 23/24
bailu 白露 (White Dew)	September 7/8
qiufen 秋分 (Autumn Equinox)	September 23/24
hanlu 寒露 (Cold Dew)	October 8/9
shuangjiang 霜降 (Frost Descent)	October 23/24
lidong 立冬 (Beginning of Winter)	November 7/8
xiaoxue 小雪 (Lesser Snow)	November 22/23
daxue 大雪 (Greater Snow)	December 7/8
dongzhi 冬至 (Winter Solstice)	December 21/22
xiaohan 小寒 (Lesser Cold)	January 5/6
dahan 大寒 (Greater Cold)	January 20/21

III Song of the 24 Solar Terms

This well-known Chinese four-verse song, which I copied from a Green Water villager's farmers' calendar, lists the abbreviated names of the 24 solar terms in a rhythmic, rhyming form. It is a mnemonic aid for the six relevant calendric dates in each of the four seasons, i.e. the basic structure of the Chinese agricultural year. Today it can easily be accessed on the Internet, accompanied by pop music and cartoon images:

春雨惊春清谷天，	*Chun yu jing chun qing gu tian,*
夏满芒夏暑相连	*xia man mang xia shu xianglian.*
秋处露秋寒霜降，	*Qiu chu lu qiu han shuang jiang,*
冬雪雪冬小大寒	*dong xue xue dong xiao da han.*
每月两节不变更，	*Mei yue liang jie bu biangeng,*
最多相差一两天	*zuiduo xiangcha yi liang tian.*
上半年来六廿一，	*Shangbannian lai liu nian yi,*
下半年是八廿三	*xiabannian shi ba nian san.*

The song can be roughly translated as follows (free translation by the author):

> Spring begins, rain water, the insects awake, the spring equinox comes, clear and bright grain rain,
> Summer begins, the grain awns are full and in ear, the summer solstice comes, lesser and greater heat are connected.
> Autumn begins, the heat ends, there is white dew, the autumn equinox comes, cold dew, and frost descends,
> Winter begins, there is lesser and greater snow, the winter solstice comes, then lesser and greater cold.
> The two solar terms per month do not change,
> They differ at most by one or two days.
> Six twenty-one [i.e. the sixth or twenty-first day of the month] comes in the first half of the year,
> Eight twenty-three [i.e. the eighth or twenty-third day of the month] comes in the second half of the year.

IV Examples of proverbs and encoded knowledge

The following are examples of proverbs and the rice farming knowledge encoded within them. I list almost 150 sayings, including several proverbs for each cultivation step, in order to give an idea of the richness and complexity of this special, everyday communicative medium. This selection still only constitutes about one seventh of the agricultural proverbs that have made it into the two anthologies, and not even 0.4 percent of the sayings originally collected for the Chenzhou anthology (CZ 1988; XT 1988).

The proverbs are divided into three groups: first, in line with the knowledge needed about the individual steps of rice cultivation (A.1-11), second, relating to embodied knowledge, the working body, and tool use (A.12-14) and, third, around the intensification of rice farming (A.15). While most of the proverbs contain codified knowledge about how to properly cultivate the paddy field resource, the few proverbs that have to do with intensification may provide some clues about how to manage the paddy field resource when confronted with various situations, such as lacking labour.

Proverb category 1: rice cultivation

A.1 Proverbs about the seasons

The proverbs which centre on the topics of time and seasons are especially numerous. On the one hand, some proverbs remind farmers how important it is to know the seasons and the solar terms (see Chapter 3): the perfect moment for each cultivation task is critical because, if you miss the right moment, especially during springtime, this will have a negative influence on the entire rice cultivation cycle:

人误地一时，地误人一年。	*Ren wu di yi shi, di wu ren yi nian.*	If people miss the fields for a little while, the fields will miss the people for one year. (XT 1988, 207; see also CZ 1988, 158)

On the other hand, some proverbs are used as an aide-memoire for the specific tasks that need to be completed in particular seasons. There are proverbs that relate to all four seasons, most being those that remind farmers to begin certain jobs in springtime. Other proverbs are even more concrete, linking certain tasks to specific solar terms. The following proverbs focus on spring, and the first two mention which precise solar term the tasks should be accomplished in:

惊蛰忙送粪，春分犁不空。	*Jingzhe mang song fen, chunfen li bu kong.*	Being busy applying night soil on *jingzhe* (Awaking of Insects), ploughing without free time on *chunfen* (Spring Equinox). (XT 1988, 209)
早禾不吃清明水，二禾不吃谷雨水。	*Zao he bu chi qingming shui, er he bu chi guyu shui.*	The early rice does not eat *qingming*-(Clear and Bright) water, the late rice does not eat *guyu*-(Grain Rain) water. (XT 1988, 209)

四月采茶茶叶黄，田中插秧人人忙。	*Si yue cai cha chaye huang, tian zhong cha yang ren ren mang.*	Plucking the tea in the fourth month, when the tea leaves are yellow, the people in the field are busy transplanting seedlings. (XT 1988, 209)

The following sayings concentrate on summer:

六月六，早禾熟；七月七，吃新米。	*Liu yue liu, zao he shu, qi yue qi, chi xin mi.*	On the sixth day of the sixth month, the early rice is ripe,[1] on the seventh day of the seventh month, the new rice is eaten. (XT 1988, 209)
芒种忙忙种。	*Mangzhong mang mang zhong.*	Planting busily on *mangzhong* (Grain in Ear). (XT 1988, 209)

The autumn is mentioned in these three proverbs:

晚稻不要粪，只要秋风秋雨喷。	*Wandao bu yao fen, zhiyao qiufeng qiuyu pen.*	The late rice does not need manure, it only needs the autumn wind and the sprinkling of the autumn rain. (XT 1988, 210)
秋后不插晚秧，禾遭霜打难灌浆。	*Qiu hou bu cha wan yang, he zao shuang da nan guanjiang.*	Don't transplant late rice seedlings after the autumn, if the grain encounters frost, the milk stage will be difficult. (XT 1988, 210)[2]
白露不秀，寒露不收。	*Bailu bu xiu, hanlu bu shou.*	If [the late rice crop] did not produce ears on *bailu* (White Dew), you won't harvest on *hanlu* (Cold Dew). (XT 1988, 210)

1 In these proverbs, 禾 *he* ('standing grain'), 粮 *liang*, 庄稼 *zhuangjia*, and 谷 *gu* ('grain', 'crop') all refer to rice.

2 灌浆 *guanjiang* relates to the fruit development or milk stage of grain development, the seventh of ten (0-9) principal growth stages of the rice plant (Meier 1997, 20-23).

There are also some proverbs about winter, advising farmers not to relax during wintertime, and claiming that providence is important for preparing for the following year.

要想来年收成好，要在今年赶冬搞。	*Yao xiang lainian shoucheng hao, yao zai jinnian gan donggao.*	For the harvest in the coming year to be good, you need to work this winter. (XT 1988, 211)
一年之计在于冬，修塘修坝莫放松。	*Yi nian zhi ji zaiyu dong, xiu tang xiu ba mo fangsong.*	Planning for next year occurs in the winter, [when you have to] repair the ponds and dykes and should not relax. (XT 1988, 211)
犁田过冬，草死泥松。	*Li tian guo dong, cao si ni song.*	If you get through the winter with ploughed fields, the weeds will die and the mud will be soft. (CZ 1988, 161; XT 1988, 215)

A whole range of sayings also state during which natural phenomenon (e.g. the flowering of the *tong* tree) which agricultural task has to be accomplished. This category clearly shows that local knowledge is largely context-related:

燕子来，齐插秧；燕子去，稻花香。	*Yanzi lai, qi cha yang; yanzi qu, daohua xiang.*	When the swallow comes, transplant the seedlings, when the swallow leaves, the rice flowers will be fragrant. (XT 1988, 212)

A.2 Proverbs about fields

Knowledge about the fields in general, and about the construction and structuring of wet fields in particular, is also reflected in these proverbs:

田塍田塍，只要过得根绳。	*Tiancheng tiancheng, zhi yao guo de gen xian.*	Field ridge, field ridge, only a rope should fit on it. (XT 1988, 213)[3]

3 田塍 *tiancheng* or 田埂*tiangeng* refer to earth ridges inside a field, while 田墈 *tiankàn*, 田坎 *tiankăn*, or 田堪 *tiankān* relate to the higher and elevated steep banks, i.e. the walls around a compound of fields.

田塂田塂，只要过得一只鸭。	*Tiankan tiankan, zhiyao guo de yi zhi ya.*	Field ridge, field ridge, only one duck should [be able to] pass by. (XT 1988, 213)
田边不栽树，地头不种花。	*Tianbian bu zai shu, ditou bu zhong hua.*	Don't plant trees on the side of the field, don't plant flowers on the edge of the field. (XT 1988, 213)
人要好看靠整面，土要好看靠整边。	*Ren yao hao kan kao zheng mian, tu yao hao kan kao zheng bian.*	If people are to look good, their face has to be even, if the soil is to look good, the side [of the field] has to be neat and tidy. (CZ 1988, 163)
旱田靠沟，涝田靠埂。	*Han tian kao gou, lao tian kao geng.*	The dry field depends on the ditch, the flooded field depends on the ridge. (XT 1988, 225)

The distinction mentioned by my interlocutors, between fertile and unfertile fields, as well as the way to deal with these fields is also brought up:

人病要吃药，地瘦要施肥。	*Ren bing yao chi yao, di shou yao shi fei.*	When people are sick, they have to take medicine, if the soil is unfertile you have to apply fertilizer. (XT 1988, 227)

The proverbs also explain which type of field or soil is suitable for rice farming, e.g.:

不是肥田不种秧，不是肥土不栽姜。	*Bu shi feitian bu zhong yang, bu shi feitu bu zai jiang.*	Don't plant [rice] seedlings if there is no fertile [paddy] field, don't plant ginger if there is no fertile soil. (CZ 1988, 165)

A.3 Proverbs about rice varieties and seed selection

Knowledge about varieties and seeds is also expressed in proverbs. Many of this type are rather general in content and advise that good fruits depend on good seeds (see CZ 1988, 158-159; XT 1988, 217). Others suggest that good seeds are as important, if not more important, than a field with good soil:

十分田，八分种。	*Shi fen tian, ba fen zhong.*	Ten parts are the field, eight parts are the seeds [i.e. the seeds make up 80 percent of a paddy field]. (CZ 1988, 158; XT 1988, 217)
肥田不如肥种。	*Fei tian buru fei zhong.*	A fertile field is not as good as fertile seeds. (CZ 1988, 158; XT 1988, 217)

The following saying cautions what will happen if the seeds are not chosen carefully:

种子不选好，满田长稗草。	*Zhongzi bu xuan hao, man tian zhang picao.*	If the seeds are not well-selected, barnyard grass will grow on the whole field. (CZ 1988, 158; XT 1988, 217)

Moreover, the practices of exchanging seeds and keeping seeds are addressed:

好花十里香，好种千里传。	*Hao hua shi li xiang, hao zhong qian li zhuan.*	The scent of good flowers lasts for ten *li* [1 *li* = 500 m], good seeds are passed on for one thousand *li.* (XT 1988, 218; see also CZ 1988, 159)
好花要人栽，好种要人留。	*Hao hua yao ren zai, hao zhong yao ren liu.*	Good flowers need a person to plant them, good seeds need a person to keep them. (XT 1988, 218)
宁可饿肚子，不能吃种子。	*Ningke e duzi, bu neng chi zhongzi.*	It is better to have a hungry stomach than to eat the seeds. (CZ 1988, 159; XT 1988, 218)

There are also instructions on keeping seeds:

留种要晒干，藏种要常翻。	*Liu zhong yao shai gan, cang zhong yao chang fan.*	When you keep seeds, you have to dry them in the sun, when you store seeds, you have to turn them over often. (CZ 1988, 159; XT 1988, 218)

Some adages add that farmers should change their seeds for new ones every year (CZ 1988, 159). Knowledge about simultaneously cultivating different varieties and minimizing risk is also expressed:

家有十样种，不怕老天哄。	*Jia you shi yang zhong, bu pa laotian hong.*	If you have ten varieties of seeds at home, you don't need to be afraid of Heaven roaring. (XT 1988, 218)

The following proverb dwells on the link between the ripening time and crop of high-yielding varieties:

早熟品种不高产，高产品种不早熟。	*Zaoshu pinzhong bu gaochan, gaochan pinzhong bu zaoshu.*	Early-ripening varieties don't yield much, high-yielding varieties don't ripen early. (XT 1988, 218)

A.4 Proverbs about sowing and cultivating seedlings

How to treat various seeds prior to germination is also covered. Proverbs list techniques to prevent illnesses or pests and to accelerate germination. Some sayings also give instructions regarding timings:

播前把种晒，播后发芽快。	*Bo qian ba zhong shai, bo hou fayang kuai.*	If you dry the seeds before sowing, they will sprout quicker after sowing. (XT 1988, 218)
好种晒干湿水浸，长出秧苗不得病。	*Hao zhong shai gan shi shui jin, zhangchu yangmiao bu de bing.*	If good seeds are dried in the sun [and then] immersed in wet water, the seedlings won't fall sick. (XT 1988, 218)
水浸三天，正合时宜。	*Shui jin san tian, zheng he shiyi.*	Immersing [the early rice seeds] in water for three days, is just enough time. (XT 1988, 219)
种浸一天，不长不短。	*Zhong jin yi tian, bu chang bu duan.*	Immersing the [late rice] seeds for one day, is not too long and not too short. (XT 1988, 219)

Further sayings apply to the organization of seedbeds and sowing. The following proverb implicitly mentions the right order the tasks should be completed in:

宁可田等秧，不可秧等田。	*Ningke tian deng yang, bu ke yang deng tian.*	It is better if the field waits for the seedlings, than the seedlings for the field. (CZ 1988, 163; XT 1988, 220)
秧田整得平，还要三日晴。	*Yangtian zheng de ping, hai yao san tian ri qing.*	When the seedbed has been levelled flat, it still needs three days of sunshine. (XT 1988, 220)

Other sayings that are listed together with this proverb emphasize that seeds in the seedbed will not tolerate rain (see XT 1988, 221). They also specify the correct density to sow seeds in the seedbed:

要想收成强，定要稀种秧。	*Yao xiang shoucheng qiang, ding yao xi zhong yang.*	If you want a good harvest, you must plant the seedlings sparsely. (XT 1988, 220)

As further proverbs on the same page of the collection explain, the seedlings will then grow horizontally, rather than vertically (XT 1988, 220). In contrast, the following proverb warns what will happen if a farmer sows carelessly and mixes up their seeds:

种好出苗好，种杂收成少。	*Zhong hao chu miao hao, zhong za shoucheng shao.*	If you sow well, the seedlings come out well, if you sow mixed, the harvest will be smaller. (CZ 1988, 159)

A.5 Proverbs about transplanting

The proverbs refer to the necessity of transplanting and the fatigue of this task:

种子要选，秧苗要移。	*Zhongzi yao xuan, yangmiao yao yi.*	The seeds have to be selected, the seedlings have to be transplanted. (XT 1988, 218)

宁可过坏一个年，不可插坏一丘田。	*Ningke guo huai yi ge nian, buke cha huai yi qiu tian.*	Rather spend one bad year, than to transplant one field badly. (XT 1988, 232)

插田如上阵。	*Cha tian ru shang zhen.*	Transplanting is like going into battle. (XT 1988, 232)

The latter proverb also informs that the transplanting period is highly labour-intensive. One saying details how and why transplanting should be accomplished:

插得正，等于上次粪；插得匀，抽穗一齐平；插得浅，有利分蘖与生长。	*Cha de zheng, dengyu shangci fen; cha de yun, chousui yi qi ping; cha de qian, you li fennie yu shengzhang.*	Transplanting [early rice] straight is equivalent to applying manure once; transplanting evenly the ears will be produced simultaneously and evenly; transplanting flat is better for tillering and growing. (XT 1988, 232)[4]

The proverbs note that early and late rice need to be transplanted differently:

头禾水上漂，晚禾插齐腰。	*Tou he shui shang piao, wan he cha qi yao.*	Early rice floats on the water, late rice has to be neatly transplanted up to its waist. (XT 1988, 232)

Transplanting also differs in fertile and unfertile fields:

肥田插稀，瘦田插密。	*Fei tian cha xi, shou tian cha mi.*	Fertile fields have to be planted sparsely, unfertile fields have to be planted densely. (XT 1988, 233)

4 See also CZ (1988, 163). 抽穗 *choushui* ('producing ears') and 分蘖 *fennie* ('tillering') are technical terms that refer to particular growth stages of the rice plant.

Moreover, the planting density is indicated:

稀三箩，密三箩，不稀不密收九箩。	*Xi san luo, mi san luo, bu xi bu mi shou jiu luo.*	Three baskets sparsely, three baskets densely, if it is not too sparse and not too dense, you will harvest nine baskets. (XT 1988, 233)

A.6 Proverbs about general plant protection

Some sayings define how much time farmers should spend on plant protection, as compared to the time spent on planting or harvesting:

三分种，七分管。	*San fen zhong, qi fen guan.*	Thirty percent is planting, seventy percent is taking care [of the plants]. (XT 1988, 242; see also CZ 1988, 163)

The following proverbs convey the importance of plant protection:

有收无收在于种，多收少收在于管。	*You shou wu shou zaiyu zhong, duo shou shao shou zaiyu guan.*	Whether you harvest depends on planting, how much you harvest depends on taking care. (XT 1988, 242; see also CZ 1988, 161)
禾要好，除虫草。	*He yao hao, chu chong cao.*	If the grain is to be well, you have to get rid of pests and weeds. (CZ 1988, 164)
田间管理好，没有病虫草。	*Tianjian guanli hao, meiyou bing chong cao.*	If you take good care of the fields, there won't be illnesses, pests, and weeds. (XT 1988, 242)

A.7 Proverbs about fertilizing

As mentioned in Chapter 3, there are large numbers of diverse proverbs about fertilizing. On the one hand, they emphasize the importance of fertilizing to attain high yields. This is often done in the form of analogies, for instance, when fertilizer for the field is compared to food or milk for humans, to wood for a fire, grass for a horse, water for a fish, a walking cane for a blind person, or oil for a lamp (e.g. CZ 1988, 159, 161; XT 1988, 226-227). On the other hand, the proverbs contain practical instructions regarding suitable substances, correct timing, what type of fertilizer is appropriate for which type of cultivation and field, how fertilizer should be applied, and how much:

水过田肥。	*Shui guo tian fei.*	When the water flows, the field is fertile. (XT 1988, 222)
冬季积堆草，春天就是宝。	*Dongji ji dui cao, chuntian jiu shi bao.*	If you collect a pile of grass in the winter, it really becomes a treasure in spring. (XT 1988, 228)
要有大粪堆，天天保存灰。	*Yao you da fendui, tian tian baocun hui.*	If you want to have a big dunghill, you have to preserve ash every day. (XT 1988, 229)
泥干百日成粪，粪干百日成泥。	*Ni gan bai ri cheng fen, fen gan bai ri cheng ni.*	When mud dries for one hundred days it becomes fertilizer, when fertilizer dries for one hundred days it becomes mud. (XT 1988, 229)
养猪养牛，肥料不愁。	*Yang zhu yang niu, feiliao bu chou.*	If you keep pigs and cows, you don't have to worry about fertilizer. (XT 1988, 230)
尿发煤渣面，大粪也不换。	*Niao fa meizha mian, da fen ye bu huan.*	If you mix urine with coal dust, you wouldn't [even want] to exchange it with a lot of manure from excrement. (XT 1988, 230)
尿是银子粪是金，它是农家命根根。	*Niao shi yinzi fen shi jin, ta shi nongjia ming gen gen.*	Urine is silver, excrement is gold, these are the roots of life for the farming family. (XT 1988, 230)
石灰田中宝，丘丘少不了。	*Shihui tian zhong bao, qiu qiu shao bu liao.*	Lime is the treasure of the field, no field should have too little of it. (XT 1988, 230)
人吃五谷粮，地吃多样肥。	*Ren chi wu gu liang, di chi duoyang fei.*	People eat five types of grain, the soil eats many types of fertilizer. (XT 1988, 230)

大粪一季， 枯饼一年。	*Dafen yi ji, kubing yi nian.*	Excrement [fertilizes] for one season, oil cake [fertilizes] for one year. (XT 1988, 231)[5]

As already mentioned in Chapter 3, there are also sayings that give instructions for producing 'pit fertilizer':

烂泥沤青草， 就是作田宝。	*Lan ni ou qing cao, jiu shi zuo tian bao.*	Soaking green grass in the mud, produces a great treasure. (XT 1988, 229)
凼肥沤得烂， 一旦顶两旦。	*Dangfei ou de lan, yi dan ding liang dan.*	If you soak pit fertilizer until it rots, one day becomes two days. (XT 1988, 229)[6]

Below, information about different phases of fertilizing is provided:

春肥保一季， 冬肥保一年。	*Chunfei bao yi ji, dongfei bao yi nian.*	Spring fertilizer keeps for one season, winter fertilizer keeps for one year. (XT 1988, 231)

Further proverbs add that fertilizer should be applied if the seedlings turn green, while ash should be applied to a rice field in which the plants already bear panicles. Furthermore, lukewarm fertilizer should be applied to a cold field, and lime is appropriate if the grain gets cold (see XT 1988, 231). The following saying depicts in detail what should be applied to which crop:

牛粪下冷田， 猪粪下藕田， 青菜惟有人粪 好，人尿淋蒜 苗。	*Niufen xia leng tian, zhufen xia outian, qingcai weiyou renfen hao, renniao lin suanmiao.*	Cow dung on the cold field, pig dung on the field of lotus roots, human excrement is only good for green vegetables, human urine dapples the garlic sprouts. (XT 1988, 231)

5 Oil cake is the residue produced from pressing oil.

6 It is also possible, and would make just as much sense, if the proverb referred to *dan* 担, a measurement of two baskets (of fertilizer), instead of *dan* 旦 ('day').

苗黄上粪，苗黑撒灰。	*Miao huang shang fen, miao hei sa hui.*	Apply human excrement to yellow sprouts, apply ash to black sprouts. (XT 1988, 231)

Other proverbs note that too much fertilizer is harmful, because it causes the grain to lodge, likewise, insufficient fertilizer is also harmful, because it causes the grain to turn yellow (XT 1988, 231). The depth of the fertilizer applied is also given:

底肥扎根，追肥提苗。	*Difei zha gen, zhuifei ti miao.*	The base fertilizer [which is applied to the soil before the planting of crops] is for taking root, the top fertilizer is for raising the sprouts. (XT 1988, 231)

Another proverb mentions that river mud should be used as fertilizer for the deeper layer of the soil, while pig dung should be applied to the roots (ibid.). The sayings also explain how fertilizer should be applied:

施肥一大片，不如一条线。	*Shi fei yi da pian, bu ru yi tiao xian.*	To apply the fertilizer in a big slice is not as good as in one line. (XT 1988, 232)

A.8 Proverbs about irrigation

As well as the topic of fertilizing, irrigation is a key concern. The large number of proverbs about water reflects both the centrality of water for wet rice farming, and of fertilization, which is often mentioned alongside irrigation. On the one hand, some sayings emphasize the importance of water for rice cultivation in general. For instance, water for the field is compared to blood, milk, to a mother, or oil for a lamp (XT 1988, 221-222). The vitality of water for rice plants is stressed, and it is stated that water is even more important than fertilizer:

多收少收在肥，有收无收在水。	*Duo shou shao shou zai fei, you shou wu shou zai shui.*	How much you harvest depends on the fertilizer, whether you harvest at all depends on the water. (XT 1988, 222)
有水无肥一半谷，有肥无水望天哭。	*You shui wu fei yi ban gu, you fei wu shui wang tian ku.*	If you have water and no fertilizer, you will only have half a harvest, if you have fertilizer and no water, you will watch the sky and cry. (CZ 1988, 160)

人要吃饱， 地要灌好。	*Ren yao chi bao, di yao guan hao.*	People have to be fully fed, the soil has to be well irrigated. (XT 1988, 222)

On the other hand, certain adages convey detailed technical knowledge about irrigation, for example regarding the water level in the fields during particular seasons or tasks:

深水泡田， 浅水耙田。	*Shen shui pao tian, qian shui ba tian.*	Soak the field in deep water, harrow the field in flat water. (XT 1988, 215)

In turn, the following saying gives instructions on irrigating the seedbed according to the weather:

晴天满沟水， 阴天半沟水， 雨天排干水。	*Qingtian man gou shui, yintian ban gou shui, yutian pai gan shui.*	A full ditch of water [in the seedbed] when it's sunny, half a ditch of water when it's cloudy, drain the water when it rains. (XT 1988, 221)

Other proverbs contain advice about the water level required at different growth stages of the rice plant:

插田水平掌， 踩田水平腰。	*Cha tian shuiping zhang, cai tian shuiping yao.*	When transplanting, the water level should be one palm/foot high, when 'stepping into the field', the water level should be waist-high. (CZ 1988, 163)[7]
浅水好发蔸， 深水好打苞。	*Qian shui hao fa dou, shen shui hao da bao.*	Flat water is good for taking root, deep water is good for budding. (XT 1988, 222)
谷含苞， 水齐腰。	*Gu han bao, shui qi yao.*	When the grain contains buds, the water has to reach its waist. (XT 1988, 223)

7 See below for the term 踩田 *cai tian* ('stepping on the field'). 'Waist-high' refers to the 'waist' of the rice plant.

深水回青， 浅水分蘖。 （指晚稻插后）	*Shen shui huiqing, qian shui fennie.*	Turning green in deep water, tillering in flat water [late rice after transplanting]. (XT 1988, 223)

Some proverbs talk about the varying water needs of early and late rice, and the dangers of too much water, while others provide guidance about the technologies of irrigation and water storage, for example by reminding farmers that pools and dykes always have to be maintained (XT 1988, 222-224):

金库银库， 不如水库； 金山银山， 不如肥山。	*Jinku yinku, buru shuiku; jinshan yinshan, buru feishan.*	A reservoir of gold and a reservoir of silver, are not as good as a reservoir of water; a mountain of gold and a mountain of silver, are not as good as a mountain of fertilizer. (XT 1988, 224; CZ 1988)
修塘如修仓， 蓄水如蓄粮。	*Xiu tang ru xiu cang, xu shui ru xu liang.*	Repairing the ponds is like repairing the storehouse, storing water is like storing grain. (XT 1988, 224)

The ridges between the fields also need to be maintained, as well as the water inlets and outlets:

晴天不开沟， 落雨遍地流。	*Qingtian bu kai gou, luo yu biandi liu.*	When it's sunny, don't open the ditch, when rain falls, let the water flow everywhere. (XT 1988, 226)
晴天开水道， 莫待雨淋头。	*Qingtian kai shuidao, mo dai yu lin tou.*	When it's sunny, open the waterway [i.e. inlet], don't wait for the rain to shower your head. (XT 1988, 226)
田边开条流水沟，荒年也有七成收。	*Tian bian kai tiao liushuigou, huangnian ye you qi cheng shou.*	By opening a ditch with flowing water on the side of the field, you [will still] have 70 percent of your harvest in famine years. (XT 1988, 226)

The importance of saving water is emphasized, especially the spring rain, as well as maintaining and repairing ponds and dykes, specifically during the winter. In this way, farmers will not need to fear a summer drought. They can rely on their own work, instead of the moods of the sky (CZ 1988, 160; XT 1988, 224-225), therefore reducing risks:

修好塘和坝，旱涝都不怕。	*Xiu hao tang he ba, han lao dou bu pa.*	If you repair the ponds and dykes well, you don't need to fear droughts and floods. (CZ 1988, 160)

Finally, the twofold benefit of the water for fishery and rice farming is stated:

塘修一口，鱼米都有。	*Tang xiu yi kou, yu mi dou you.*	If you repair the pond once, you have both fish and rice. (XT 1988, 224)

A.9 Proverbs about weed and pest management

There are not many proverbs about weed and pest management, especially in comparison to the topics of fertilizing and irrigation. As stated above, this might be related to the complexity of fertilization knowledge, as well as the possibilities for taking action. Besides, traditional forms of weed and pest management require body techniques, which are not easily elucidated in the form of sayings.

The contents of many proverbs on weed management are rather general, warning that there will be losses in yields if weeds are not eliminated (XT 1988, 246). Only a few sayings are more specific, singling out particular pest management techniques:

若要来年虫子少，冬天烧去田边草。	*Yao xiang lainian chongzi shao, dongtian shaoqu tianbian cao.*	If you want few pests in the coming year, you have to burn down the weeds at the side of the field in winter. (XT 1988, 245)
田塍三面光，害虫无处藏。	*Tiancheng san mian guang, haichong wuchu cang.*	If the ridges of the field are bare on three sides, the harmful insects have no place to hide. (CZ 1988, 164)

In addition, some proverbs also name the tools needed for pest and weed management. Methods of pest management are given as fire and the manual catching of insects (XT 1988, 244-246; also see the proverbs on farm implements below). The tools for weed removal are hoes, feet, ploughs, harrows, shovels and scissors, and the importance of removing the whole weed, including the roots is noted. Further sayings also remind farmers to start weeding early. From the second proverb below, we may infer that hands are used to pull out the weeds:

斩草不除根，来春又发青。	*Zhan cao bu chu gen, lai chun you faqing.*	If you don't eliminate the roots when chopping the weeds, in the coming spring they will become green again. (XT 1988, 246)
杂草不扯完，丢掉半年粮。	*Zacao bu che wan, diudiao bannian liang.*	If you don't pull out the weeds completely, you will lose six months of grain. (CZ 1988, 164)

Pests have to be removed early on:

要想虫少，除虫要早。	*Yao xiang chong shao, chu chong yao zao.*	If you want few insects, you have to begin eliminating pests early. (XT 1988, 244)
种前防虫，种后治虫。	*Zhong qian fang chong, zhong hou zhi chong.*	Prevent pests before planting, exterminate pests after planting. (XT 1988, 244)

A.10 Proverbs about harvesting

As for harvesting, apart from the sickle, which is mentioned in the section on farm implements below, the threshing tub is often referred to. The following proverb draws an analogy between transplanting and threshing, since both tasks entail a race against time and weather:

插秧如赶考，扮禾如抢宝。	*Cha yang ru gan kao, ban he ru qiang bao.*	Transplanting seedlings is like taking the imperial examinations, threshing is like seizing a treasure. (XT 1988, 247)

A range of harvesting proverbs reminds farmers not to miss the right time for harvesting:

收谷如收金。季节不等人。	*Shou gu ru shou jin. Jijie bu deng ren.*	Harvesting grain is like harvesting gold. The seasons don't wait for the people. (XT 1988, 248)
早割是谷，迟割是芽。	*Zao ge shi gu, chi ge shi ya.*	If you harvest [with a sickle] early, it is grain, if you harvest late [and the kernels fall on the ground], it is sprouts. (XT 1988, 249)
插田不躲雨，扮禾不歇凉。	*Cha tian bu duo yu, ban he bu xie liang.*	When transplanting, don't hide from the rain, when threshing don't relax in a cool place. (CZ 1988, 163; XT 1988, 248)

The last proverb also implies knowledge about the seasons and the right time for particular cultivation steps. Moreover, the busy seasons are mentioned. Households have to know about these, because they require a lot of labour, which has to be taken into account when making strategic decisions about cultivating plants with different ripening times, rural emigration or the seasonal return of migrants. The reason the harvesting period is so busy is apparent in the following saying, which outlines a sequence of harvesting tasks:

秋收有五忙：杀、扮、挑、晒、藏。	*Qiu shou you wu mang: sha, ban, tiao, shai, cang.*	During the autumn harvest there are five busy tasks: killing [i.e. cutting], threshing, carrying on the shoulder, drying in the sun, and storing. (XT 1988, 248)

The sayings below advise farmers to pay attention during the harvesting and post-harvesting tasks:

地不丢穗，场不丢粒。	*Di bu diu sui, chang bu diu li.*	Don't lose ears [of grain] on the field, don't lose grains in the [drying] place. (XT 1988, 249)
打谷要抢先，晒谷要抢天。	*Da gu yao qiang xian, shai gu yao qiang tian.*	When threshing grain you have to be first, when drying grain you have to catch the right weather. (XT 1988, 249)

Farmers are also reminded about the right time for harvesting and threshing:

生扮全收，熟扮错失田租。	*Sheng ban quan shou, shu ban cuoshi tianzu.*	If you thresh unripe, you harvest everything, if you thresh ripe, you miss [paying] the land tax. (XT 1988, 248)

A.11 Proverbs about cropping patterns

Occasionally, proverbs mention the need to plant diverse crops:

要得富，庄稼开个杂货铺。	*Yao de fu, zhuangjia kai ge zahuopu.*	If you want to become rich, make your crops into a variety store.[8] (XT 1988, 216)

Moreover, a whole range of proverbs cover crops other than rice, which indicates the practice of mixed cropping. These are about buckwheat, maize, millet, pulses, sweet potatoes, wheat, cotton, sesame, groundnuts, rape, tobacco, hemp, tea, and tubers (CZ 1988, 160-161, 165; XT 1988, 233-242). The following is an example of a proverb about rape seed. Implicitly it also refers to knowledge about the seasons, crop rotation, and agricultural tools:

夏锄冬挖，油菜大发。	*Xia chu dong wa, youcai da fa.*	If you hoe in the summer and dig in the winter, the rape will grow quickly. (CZ 1988, 164)

The saying below is an example about cultivating different plants depending on the weather:

雨种豆子晴种棉，种菜最好在阴天。	*Yu zhong douzi qing zhong mian, zhong cai zui hao zai yintian.*	Plant beans when it's rainy, plant cotton when it's sunny, best to plant vegetables when it's cloudy. (CZ 1988, 165)

8 The same proverb, also found in a Guangzhou collection, is quoted in Schmalzer, describing how this proverb was politically and scientifically reinterpreted in the framework of promoting inter-cropping during the Cultural Revolution (Schmalzer 2016, 106). I use her translation here.

Proverb category 2: embodied knowledge

A.12 Proverbs about the craft of farming

Famers' knowledge and skills are sometimes explicitly referred to. Some sayings explain what people have to learn in order to become a farmer:

犁田不挨边，还要学三天。	*Li tian bu aibian, hai yao xue san tian.*	Not to be near the edge while ploughing still needs three days of learning. (XT 1988, 215)
当兵学打枪，作田学育秧。	*Dangbing xue daqiang, zuo tian xue yu yang.*	A soldier learns how to shoot, a farmer learns how to raise rice seedlings. (XT 1988, 219)

Sometimes the skills of a knowledgeable farmer are compared to the skills needed in other professions. The next proverb clearly has a political undertone, referring to the categories of 'peasant' and 'worker':

农民一双手，瘦地出清油；工人一双手，平地起高楼。	*Nongmin yi shuang shou, shou di chu qingyou; gongren yi shuang shou, pingdi qi gaolou.*	With the two hands of a peasant, clear oil comes out of the unfertile soil; with the two hands of a worker, a high building rises from the flat ground. (XT 1988, 206)
十年难称作田汉，三年学个手艺人。	*Shi nian nan cheng zuo tian han, san nian xue ge shouyi ren.*	It is hard to become a farmer in ten years, but in three years you learn to be a craftsman. (XT 1988, 207)
农民看粪堆，商人看货堆。	*Nongmin kan fendui, shangren kan huodui.*	The farmer watches over the dunghill, the merchant watches over the pile of goods. (XT 1988, 228)

In another category of proverbs, farming skills are named in relation to practical cultivation tasks:

会作田的作一丘，不会作田的作一洲。	*Hui zuo tian de zuo yi qiu, bu hui zuo tian de zuo yi zhou.*	Those who know how to farm cultivate a field, those who don't know how to farm cultivate an island. (XT 1988, 214)
若要田里功夫搞得好，必须冬天犁得早，一来烂泥，二来死草。	*Ruo yao tian li gongfu gao de hao, bixu dongtian li de zao, yi lai lanni, er lai si cao.*	If you want to practice your abilities and skills in the field well, you have to plough early in winter, first comes the mud, second dies the grass. (XT 1988, 215)
会种田土土也光，不会种田到处脏。	*Hui zhong tiantu tu ye guang, bu hui zhong tian daochu zang.*	If you know how to farm, the soil is also polished, if you don't know how to farm, it is dirty everywhere. (XT 1988, 216)
有钱易买种，千金难买苗。	*You qian yi mai zhong, qian jin nan mai miao.*	With money it is easy to buy seeds, but with one thousand pieces of gold it is difficult to buy seedlings. (XT 1988, 219)
管秧如绣花，一针不能差。	*Guan yang ru xiuhua, yi zhen bu neng cha.*	Taking care of seedlings is like embroidery, you shouldn't miss even one stitch. (XT 1988, 221)
多上粪，庄稼好，还看巧不巧。	*Duo shang fen, zhuangjia hao, hai kan qiao bu qiao.*	If you apply a lot of fertilizer, the crops do well, but you still have to see if you are skilled at it. (XT 1988, 231)
作田不用问，一半功夫一半粪。	*Zuo tian bu yong wen, yi ban gongfu yi ban fen.*	In farming there's no question, one half is skill, one half is fertilizing. (XT 1988, 242)
不懂庄稼脾气，枉费一年力气。	*Bu dong zhuangjia piqi, wang fei yi nian liqi.*	If you don't understand the temper of the crop, you waste one year of effort. (XT 1988, 242)

只要功夫深，田土出黄金。	*Zhi yao gongfu shen, tian tu chu huangjin.*	You only need deep skill, for gold to come out of the soil. (CZ 1988, 158)
不懂季节，不算农人。	*Bu dong jijie, bu suan nongren.*	If you don't understand the seasons, you are not a farmer. (XT 1988, 207)

A.13 Proverbs about the working body

Implicitly, the physical side of farm work, which contains a reference to embodied knowledge, is also found in the proverbs, for example through mentions of using the hands, feet and legs, physical hardship, hard work, and diligence:

人勤地献宝，人懒地长草。	*Ren qin di xian bao, ren lan di zhang cao.*	If people are hard-working, the soil gives treasures, if people are lazy, grass grows on the soil. (XT 1988, 213)
十丘田，九块地，不犁挖，吃个屁。	*Shi qiu tian, jiu kuai di, bu li wa, chi ge pi.*	Ten pieces of field, nine pieces of soil, if you don't plough and dig, you eat a fart. (XT 1988, 212)

Muscles and joints are explicitly brought up in the next saying:

钢要安在刀口上，肥要施在筋节上。	*Gang yao an zai daokou shang, fei yao shi zai jin jie shang.*	Steel has to be stabilized on the edge of the knife, fertilizer has to be applied with the muscles and joints. (XT 1988, 230)

Legs and hands are touched upon in relation to accumulating fertilizer:

积肥没巧，退勤就好。	*Ji fei mei qiao, tui qin jiu hao.*	If you have no skill at collecting manure, then hard-working legs are just fine. (XT 1988, 228)
常垫猪栏掏鸡窝，退勤手快积肥多。	*Chang dian zhulan tao jiwo, tui qin shou kuai ji fei duo.*	Cushioning the pigsty and taking out the chicken nest often, with hard-working legs and quick hands you collect a lot of manure. (XT 1988, 230)

A whole series of proverbs centres on practices of treading with the feet, particularly mentioning the soles (see especially XT 1988, 246-247):

田要过脚板，地要过铁板。	*Tian yao guo jiaoban, di yao guo tieban.*	The field has to pass under the soles of the feet, the soil has to pass under the iron plate [i.e. plough or hoe]. (XT 1988, 213)
脚板到底，稗尽草死。	*Jiaoban dao di, pi jin cao si.*	When the sole reaches the bottom, the barnyard grass is exhausted and the grass [i.e. weeds] dies. (XT 1988, 247)

This group of proverbs on the feet are often connected to the weeding technique of 'treading green' or 'treading on the field' (*cai tian*),[9] as in the following:

禾踩三道，谷会得报。	*He cai san dao, gu hui de bao.*	If you tread on the [growing] crop three times, the [harvested] grain will respond. (XT 1988, 246; see also CZ 1988, 161)
踩田要圆蔸，不留宝塔洲。	*Cai tian yao yuan dou, bu liu baota zhou.*	If you tread on the field you have to circle the roots, you shouldn't leave behind pagoda islands [i.e. you need to flatten the field thoroughly]. (XT 1988, 247)

In the proverb collection from Xiangtan, sayings about treading on the field are listed together with proverbs that explain how to thin out or add transplanted seedlings (XT 1988, 247). Hands are mentioned particularly frequently, both metaphorically and practically:

秧田要整一掌平，下种手要撒得匀。	*Yangtian yao zheng yi zhang ping, xia zhong shou yao sa de yun.*	The seedbed has to be tidied as level as a palm, when sowing the hands have to sow evenly. (XT 1988, 220)

9 This practice also has mythical origins, being seen in connection with the Divine Farmer Shennong (see Zhuzhou Network of Social Sciences 2012).

若要五谷香，不离两手脏。	*Ruo yao wu gu xiang, bu li liang shou zang.*	If you want the five grains to be fragrant, you can't do it without getting both hands dirty. (XT 1988, 226)
只要勤动手，肥料到处有。	*Zhi yao qin dong shou, feiliao daochu you.*	You only need to move your hands diligently, to get fertilizer everywhere. (XT 1988, 228)

According to the proverb below, farmers also need visual skills to evaluate the weather, the soil and the plants:

作田施肥冇得巧，看天看地又看苗。	*Zuo tian shi fei mao de qiao, kan tian kan di you kan miao.*	In farming and fertilizing there is no skill [i.e. this is the only way to do it, there is no other skilful solution]: watch the sky, watch the soil, and also watch the seedlings. (XT 1988, 230)

A.14 Proverbs about tools

Furthermore, knowledge about the handling of farm implements also plays a role in the proverbs. Here, too, on the one hand embodied knowledge is needed to handle the tools. On the other hand, the conscious choice of particular technologies also offers possibilities for intensifying agriculture, saving labour, and economic diversification, as in the cases of mixing crops and labour migration (see Chapter 5). The proverbs mention many of the implements used in Green Water, especially ploughs, harrows, hoes, and sickles. The plough is the most frequently mentioned technology in the sayings, often in connection with the harrow:

犁好耙好，光长苗来不长草。	*Li hao ba hao, guang zhang miao lai bu zhang cao.*	Ploughing well, harrowing well, only sprouts will grow, grass won't grow. (XT 1988, 214)
犁上结谷，耙上死草。	*Li shang jie gu, ba shang si cao.*	Ploughing bears grain, harrowing kills weeds. (XT 1988, 214)
犁田要深，耙田要平。	*Li tian yao shen, ba tian yao ping.*	You have to plough deeply, you have to harrow shallowly. (XT 1988, 214; see also CZ 1988, 161)

犁多死草，耙多烂泥。	*Li duo si cao, ba duo lan ni.*	When ploughing much, the weeds die, when harrowing much, the mud softens. (CZ 1988, 161)
土地深耕有三好，保水灭虫又除草。	*Tudi shen geng you san hao, bao shui hai chong you chu cao.*	Ploughing the soil deeply has three advantages: to retain water, to get rid of pests and weeds. (XT 1988, 214; see also CZ 1988, 163)
随收深耕有三好，肥田除虫又除草。	*Sui shou shen geng you san hao, fei tian chu chong you chu cao.*	Ploughing deeply after the harvest has three advantages: it fertilizes the field, it gets rid of pests and weeds. (XT 1988, 215)
犁三遍，耙三遍，不怕老天晒半年。	*Li san bian, ba san bian, bu pa laotian shai ban nian.*	By ploughing three times, harrowing three times, you don't have to fear that the sky is sunny for six months of the year. (CZ 1988, 161)
冬耕要深，春耕要平。	*Dong geng yao shen, chun geng yao ping.*	The winter ploughing has to be deep, the spring ploughing has to be shallow. (CZ 1988, 161; XT 1988, 215)

The hoe is multifunctional; it is used to loosen the soil, to regulate the water inlets and outlets, to apply fertilizer, and for weeding:

一把锄头放水。	*Yi ba chutou fang shui.*	Hoe once to release water. (XT 1988, 224)
种在犁上，收在锄上。	*Zhong zai li shang, shou zai chu shang.*	Planting depends on ploughing, harvesting depends on hoeing. (XT 1988, 245)
锄头响，粪堆长。	*Chutou xiang, fendui zhang.*	When the hoe sounds, the dunghill grows. (XT 1988, 228)

Brooms, shovels, knives, scissors, and carrying poles are only mentioned few times:

扫把响， 肥堆长。	*Saoba xiang, feidui zhang.*	When the brush sounds, the dunghill grows. (XT 1988, 228)
虫口一把刀， 草口一把剪。	*Chongkou yi ba dao, caokou yi ba jian.*	A knife for the mouth of the insect, a pair of scissors for the mouth of the weed. (XT 1988, 244)
冬天铲去草， 春天虫子少。	*Dongtian chanqu cao, chuntian chongzi shao.*	If you shovel away the weeds in the winter, you will have fewer pests in the spring. (XT 1988, 245)
作田老倌不知闲，放下锄头拿扁担。	*Zuo tian lao guan bu zhi xian, fang xia chutou na biandan.*	In farming, the old herdsman knows no rest, as soon as he puts down his hoe, he picks up his carrying pole. (XT 1988, 243)

The sickle appears, among others, in the following harvesting proverb:

寒露到， 割晚稻。	*Hanlu dao, ge wandao.*	Cutting the late rice on *hanlu* [Cold Dew]. (XT 1988, 210)

The next proverb focuses on the visual skills around the harvesting task. Farmers need to observe when the plants are ripe and ready to be cut:

见黄就割， 不割就落。	*Jian huang jiu ge, bu ge jiu luo.*	If you see yellow [grain], just cut, if you don't cut, it will fall off. (XT 1988, 248)

The three proverbs below remind farmers that they need to harvest quickly and carefully. Moreover, the second proverb advises how to thresh:

镰刀不快不割禾。	*Liandao bu kuai bu ge he.*	If the sickle is not quick, it won't cut the grain. (XT 1988, 247)
割禾不轻， 粒粒落空。	*Ge he bu qing, li li luo kong.*	If you don't cut gently, all grains will be fruitless. (XT 1988, 249)

Finally, the maintenance of the implements is also part of their skilful handling:

不怕墙上加土，只怕铁上加泥。	*Bu pa qiang shang jia tu, zhi pa tie shang jia ni.*	Don't be afraid to get soil on the wall, fear only getting mud on the iron. (XT 1988, 247)

Proverb category 3: farming intensification

A.15 Proverbs about the intensification of rice farming

In addition to the proverbs mentioned in Chapter 3, these two proverbs contain references about the possibilities of intensifying rice farming. This gains increased importance in view of migration decisions around household labour:

庄稼不管爹和娘，精耕细作多打粮。	*Zhuangjia bu guan die he niang, jinggeng xizuo duo da liang.*	The crop doesn't care about father and mother, therefore you have to cultivate intensively and carefully to thresh more grain. (XT 1988, 214)[10]

The following proverb reminds farmers that fields and soil alone are not enough:

好田不如好耕，好土不如好种。	*Hao tian buru hao geng, hao tu buru hao zhong.*	A good field is not as good as ploughing well, good soil is not as good as planting well. (CZ 1988, 160)

Moreover, the proverbs above that state which times of the year require the most labour are also significant for decision making about labour migration or seasonal returns.

10 See also XT (1988, 242) and CZ (1988, 158). *Jinggeng xizuo* or 'intensive cultivation' was actively promoted in Mao Zedong's 1957 essay 'Be Activists in Promoting the Revolution' (see Schmalzer 2016, 106).

References

Abranches, Maria. 2013. 'When People Stay and Things Make Their Way: Airports, Mobilities and Materialities of a Transnational Landscape'. *Mobilities* 8 (4): 506-527.

ACBS, Anren County Bureau of Statistics. 2020. '安仁县2019 年国民经济和社会发展统计公报 Anren xian 2019 nian guomin jingji he shehui fazhan tongji gonggao (Anren County 2019 Statistical Communiqué of the Development of the People's Economy and Society)'. 13 March 2020. Accessed 3 July 2020. http://www.anrenzf.gov.cn/15/51015/51019/content_3126960.html.

ACGCC, 安仁县志编纂委员会 Anren Xianzhi Bianji Weiyuanhui (Anren County Gazetteer Compilation Committee). 1996. 安仁县志 *Anren Xianzhi* (*Anren County Gazetteer*). Beijing: Zhongguo Shehui Publishing.

—. 2011. 安仁县志 *Anren Xianzhi, 1989-2003* (*Anren County Gazetteer, 1989-2003*). Changsha: Hunan Renmin Publishing.

ACIGCC, 安仁县工业志编纂委员会 Anren Xian Gongyezhi Bianji Weiyuanhui (Anren County Industrial Gazetteer Compilation Committee). 1993. 安仁县工业志 *Anren Xian Gongye Zhi* (*Anren County Industrial Gazetteer*). Hefei: Huangshan Publishing.

ACWF, All China Women's Federation. 2013. '我国农村留守儿童、城乡流动儿童状况研究报告 Woguo nongcun liushou ertong, chengxiang liudong ertong zhuangkuang yanjiu baogao (Research Report on the Situation of Left-behind Children and Migrant Children in the Chinese Countryside)'. *Zhongguo Funü Xinwen*. May 10, 2013. Accessed 9 December 2017. http://acwf.people.com.cn/n/2013/0510/c99013-21437965.html.

Ahearn, Laura M. 2001. *Invitations to Love: Literacy, Love Letters, and Social Change in Nepal*. Ann Arbor: University of Michigan Press.

Alber, Erdmute, and Heike Drotbohm. 2015. 'Introduction'. In *Anthropological Perspectives on Care: Work, Kinship, and the Life-Course*, edited by Erdmute Alber and Heike Drotbohm, 1-20. New York: Palgrave Macmillan.

Altieri, Miguel A. 1987. *Agroecology: The Scientific Bases of Alternative Agriculture*. Boulder: Westview Press.

Anagnost, Ann. 1997. *National Past-Times: Narrative, Representation, and Power in Modern China*. Durham: Duke University Press.

Anderson, Patricia C. 2014. 'Trampling the Crops with Animals'. In *Exploring and Explaining Diversity in Agricultural Technology*, edited by Annelou van Gijn, John C. Whittaker, and Patricia C. Anderson, 138-140. Oxford: Oxbow.

Anderson, Patricia C., Annelou van Gijn, John C. Whittaker, and François Sigaut. 2014. 'The Dimension of Tools, Skills and Processes: Exploring Diversity'. In

Exploring and Explaining Diversity in Agricultural Technology, edited by Annelou van Gijn, John C. Whittaker, and Patricia C. Anderson, 3-15. Oxford: Oxbow.

Anderson, Patricia C, and François Sigaut. 2014. 'Introduction: Reasons for Variability in Harvesting Techniques and Tools'. In *Exploring and Explaining Diversity in Agricultural Technology*, edited by Annelou van Gijn, John C. Whittaker, and Patricia C. Anderson, 85-92. Oxford: Oxbow.

Andreas, Joel, and Shaohua Zhan. 2016. 'Hukou and Land: Market Reform and Rural Displacement in China'. *The Journal of Peasant Studies* 43 (4): 798-827.

Appadurai, Arjun. 1986a. 'Introduction: Commodities and the Politics of Value'. In *The Social Life of Things: Commodities in Cultural Perspective*, edited by Arjun Appadurai, 3-63. Cambridge: Cambridge University Press.

—. 1986b. *The Social Life of Things: Commodities in Cultural Perspective*. Cambridge: Cambridge University Press.

—. 1999. *Modernity at Large: Cultural Dimensions of Globalization*. Minneapolis: University of Minnesota Press.

Aubert, Claude. 2003. 'Landwirtschaftspolitik'. In *Das grosse China-Lexikon*, edited by Brunhild Staiger, Stefan Friedrich, and Hans-Wilm Schütte, 424-427. Darmstadt: Wissenschaftliche Buchgesellschaft.

Barlett, Peggy F. 1980a. 'Adaptive Strategies in Peasant Agricultural Production'. *Annual Review of Anthropology* 9 (1): 545-573.

—., ed. 1980b. *Agricultural Decision Making*. New York: Academic Press.

Barth, Fredrik. 2002a. 'An Anthropology of Knowledge'. *Current Anthropology* 43 (1): 1-18.

—. 2002b. 'Toward a Richer Description and Analysis of Cultural Phenomena'. In *Anthropology Beyond Culture*, edited by Richard Fox and Barbara King, 23-36. Oxford: Berg.

Basu, Paul, and Simon Coleman. 2008. 'Introduction: Migrant Worlds, Material Cultures'. *Mobilities* 3 (3): 313-330.

van den Berg, M. Marrit, Huib Hengsdijk, Joost Wolf, Martin K. van Ittersum, Wang Guanghuo, and Reimund P. Rötter. 2007. 'The Impact of Increasing Farm Size and Mechanization on Rural Income and Rice Production in Zhejiang Province, China'. *Agricultural Systems* 94 (3): 841-850.

Bergstedt, Cecilia. 2016. *Cultivating Gender: Meanings of Place and Work in Rural Vietnam*. Copenhagen: NIAS Press.

Boserup, Ester. 1965. *The Conditions of Agricultural Growth: The Economics of Agrarian Change under Population Pressure*. London: Allen & Unwin.

Bossen, Laurel. 1994. '中国农村妇女什么原因使她们留在农田里 Zhongguo noncun funü: Shenme yuanyin shi tamen liu zai nongtian li? (Chinese Rural Women:

What Keeps Them Down on the Farm?)'. In 性别与中国 *Xingbie Yu Zhongguo* (*Gender and China*), edited by Xiaojiang Li, Hong Zhu, and Xiuyu Dong, 128-154. Beijing: Sanlian Shudian.

—. 2011. 'Reproduction and Real Property in Rural China: Three Decades of Development and Discrimination'. In *Women, Gender and Rural Development in China*, edited by Tamara Jacka and Sally Sargeson, 97-123. Cheltenham: Edward Elgar Publishing.

Bourdieu, Pierre. 1977. *Outline of a Theory of Practice*. Cambridge: Cambridge University Press.

Brandt, Loren, Susan H. Whiting, Linxiu Zhang, and Tonglong Zhang. 2017. 'Changing Property-Rights Regimes: A Study of Rural Land Tenure in China'. *The China Quarterly* 232: 1026-1049.

Brautigam, Deborah. 2015. *Will Africa Feed China?* New York: Oxford University Press.

Braverman, Harry. 1974. *Labour and Monopoly Capital: The Degradation of Work in the Twentieth Century*. New York: Monthly Review Press.

Bray, Francesca. 1984. *Agriculture*. In *Science and Civilisation in China*, Vol. 6, Part 2, edited by Joseph Needham. Cambridge: Cambridge University Press.

—. 1994. *The Rice Economies: Technology and Development in Asian Societies*. Reprint. Berkeley: University of California Press.

—. 2004. 'Rice, Technology, and History, the Case of China'. *Education About Asia* 9 (3): 14-20.

—. 2008. 'Science, Technique, Technology: Passages between Matter and Knowledge in Imperial Chinese Agriculture'. *The British Journal for the History of Science* 41 (3): 319-344.

—. 2013. *Technology, Gender and History in Imperial China: Great Transformations Reconsidered*. New York: Routledge.

—. 2017. 'Science, Technology and Late Imperial History'. *The Chinese Historical Review* 24 (1): 93-104.

Bray, Francesca, Peter A. Coclanis, Edda L. Fields-Black, and Dagmar Schäfer, eds. 2015. *Rice: Global Networks and New Histories*. New York: Cambridge University Press.

Brettell, Caroline B. 2008. 'Theorizing Migration in Anthropology: The Social Construction of Networks, Identities, Communities, and Globalscapes'. In *Migration Theory: Talking Across Disciplines*, edited by Caroline B. Brettell and James F. Hollifield, 113-160. New York: Routledge.

Buch, Elana D. 2015. 'Anthropology of Aging and Care'. *Annual Review of Anthropology* 44 (1): 277-293.

Burrell, Kathy. 2011. 'Going Steerage on Ryanair: Cultures of Air Travel for Migration from Poland to the UK'. *Journal of Transport Geography* 19 (5): 1023-1030.

Carter, Michael R., and Yang Yao. 2002. 'Local versus Global Separability in Agricultural Household Models: The Factor Price Equalization Effect of Land Transfer Rights'. *American Journal of Agricultural Economics* 84 (3): 702-715.

Castles, Stephen, Heine de Haas, and Mark J. Miller. 2014. *The Age of Migration: International Movements in the Modern World*. 5th ed. New York: Guilford.

Central Government of the People's Republic of China. 2016. 'The 13th Five-Year Plan for Economic and Social Development of the People's Republic of China (2016-2020)'. Accessed 9 January 2020. https://en.ndrc.gov.cn/newsrelease_8232/201612/P020191101481868235378.pdf.

—. 2017. '李克强谈"中国制造2025": 从制造大国迈向制造强国 Li Keqiang tan "Zhongguo zhizao 2025": Cong zhizao daguo maixiang zhizao qiangguo (Li Keqiang on "Made in China 2025": From a Manufacturing Country towards a Manufacturing Power)'. Accessed 9 January 2020. http://www.gov.cn/guowuyuan/2017-08/10/content_5216727.htm.

CGIAR, Research Program on Rice. 2020. 'Ricepedia: The Global Staple'. 2020. Accessed 9 January 2020. http://ricepedia.org/rice-as-food/the-global-staple-rice-consumers.

Chan, Kam Wing. 2010. 'The Global Financial Crisis and Migrant Workers in China: "There Is No Future as a Labourer; Returning to the Village Has No Meaning"'. *International Journal of Urban and Regional Research* 34 (3): 659-677.

—. 2019. 'China's Hukou System at 60: Continuity and Reform'. In *Handbook on Urban Development in China*, edited by Ray Yep, June Wang, and Thomas Johnson, 59-79. Cheltenham: Edward Elgar Publishing.

Chang, Te-Tzu. 2000. 'Rice'. In *The Cambridge World History of Food*, edited by Kenneth F. Kiple and Kriemhild Coneè Ornelas, 132-149. Cambridge: Cambridge University Press.

Chard, Robert L. 1990. 'Folktales of the God of the Stove'. *Hanxue Yanjiu* (*Chinese Studies*) 8 (1): 149-182.

Chayanov, Aleksandr Vasilevich. 1966. *The Theory of Peasant Economy*, edited by Daniel Thorner, Basile Kerblay, and R.E.F. Smith. Homewood: Irvin.

Chen, An. 2020. 'The Impact of Land Requisition on Peasant Life in China'. *Modern China* 46 (1): 79-110.

Chen, Chen. 2015. 'Rural-Urban Migration and Circularity in China: Evidence from Anhui Province'. PhD diss., University of California Los Angeles.

Chen, Chen, and C. Cindy Fan. 2018. 'Rural-Urban Circularity in China: Analysis of Longitudinal Surveys in Anhui, 1980-2009'. *Geoforum* 93 (July): 97-104.

Chen, Chuanbo, and C. Cindy Fan. 2016. 'China's Hukou Puzzle: Why Don't Rural Migrants Want Urban Hukou?' *The China Review* 16 (3): 9-39.

Chen, Chunhong 陈春红. 2016. '晾晒难 收割难 水稻收获两痛点如何破解 Liangshai nan shouge nan shuidao shouhuo liang tongdian ruhe pojie (Drying

Is Difficult, Harvesting Is Difficult: How to Break the Two Sorrows of Harvesting Paddy Rice)'. 中国农机化导报 *China Agricultural Mechanization Herald*. 30 August 2016. Accessed 9 September 2016. http://www.camn.agri.gov.cn/Html/2016_08_30/2_1842_2016_08_30_30001.html.

Chen, Nancy N. 2001. 'Health, Wealth, and the Good Life'. In *China Urban: Ethnographies of Contemporary Culture*, edited by Nancy N. Chen, Constance D. Clark, Suzanne Z. Gottschang, and Lyn Jeffery, 165-182. Durham: Duke University Press.

Chen, Ruishan, Chao Ye, Yunlong Cai, Xiaoshi Xing, and Qiong Chen. 2014. 'The Impact of Rural Out-Migration on Land Use Transition in China: Past, Present and Trend'. *Land Use Policy* 40: 101-110.

Cheng, Tiejun, and Mark Selden. 1994. 'The Origins and Social Consequences of China's Hukou System'. *The China Quarterly* 139: 644-668.

Chevalier, Alexandre, Elena Marinova, and Leonor Peña-Chocarro. 2014. 'Factors and Issues in Plant Choice'. In *Plants and People: Choices and Diversity Through Time*, edited by Alexandre Chevalier, Elena Marinova, and Leonor Peña-Chocarro, 3-14. Oxford: Oxbow.

Chi, Ch'ao-ting. 1936. *Key Economic Areas in Chinese History*. 2nd ed. New York: Paragon Reprint Corp.

Chibnik, Michael. 2011. *Anthropology, Economics, and Choice*. Austin: University of Texas Press.

Chien, Shiuh-Shen. 2015. 'Local Farmland Loss and Preservation in China—A Perspective of Quota Territorialization'. *Land Use Policy* 49: 65-74.

Chu, Julie Y. 2010. *Cosmologies of Credit: Transnational Mobility and the Politics of Destination in China*. Durham: Duke University Press.

Clifford Collard, Niamh Jane. 2016. 'Social Strategies and Material Fixes in Agotime Weaving'. In *Craftwork as Problem Solving: Ethnographic Studies of Design and Making*, edited by Trevor H.J. Marchand, 153-168. Farnham: Ashgate.

Cohen, Jeffrey H., and Ibrahim Sirkeci. 2011. *Cultures of Migration: The Global Nature of Contemporary Mobility*. Austin: University of Texas Press.

Cong, Zhen, and Merril Silverstein. 2011. 'Intergenerational Exchange Between Parents and Migrant and Nonmigrant Sons in Rural China'. *Journal of Marriage and Family* 73 (1): 93-104.

Cooke, Fang Lee. 2011. 'Labour Market Disparities and Inequalities'. In *China's Changing Workplace: Dynamism, Diversity and Disparity*, edited by Peter Sheldon, Sunghoon Kim, Yiqiong Li, and Malcolm Warner, 259-276. New York: Routledge.

Cortada, James W. 2012. *The Digital Flood: The Diffusion of Information Technology across the U.S., Europe and Asia*. Oxford: Oxford University Press.

Coy, Michael W., ed. 1989. *Apprenticeship: From Theory to Method and Back Again*. Albany: SUNY Press.

Croll, Elisabeth J., and Huang Ping. 1997. 'Migration For and Against Agriculture in Eight Chinese Villages'. *The China Quarterly* 149: 128-146.

Cuhls, Kerstin, Cheng Fan, Bruno Gransche, Erduana Shala, and Benjamin Teufel. 2016. *Zusammenfassender Endbericht schrumpfende Gesellschaften im Vergleich*. Karlsruhe: Fraunhofer ISI.

CZ, 郴州地区民间文学集成编委会 Chenzhou Diqu Minjian Wenxue Jicheng Bianweihui (Editorial Board of the Comprehensive Collection of the Folk Literature of Chenzhou District), ed. 1988. 中国民间谚语集成湖南卷: 郴州地区分卷 *Zhongguo minjian yanyu jicheng Hunan juan: Chenzhou diqu fen juan* (*Hunan Volume of the Comprehensive Collection of Chinese Folk Proverbs: Section Volume of Chenzhou District*). Chenzhou: N.p.

Dalakoglou, Dimitris. 2010. 'Migrating-remitting-"building"-dwelling: House-making as "Proxy" Presence in Postsocialist Albania'. *Journal of the Royal Anthropological Institute* 16 (4): 761-777.

Davis, Benjamin, Gero Carletto, and Paul C. Winters. 2010. 'Migration, Transfers and Economic Decision Making among Agricultural Households: An Introduction'. *The Journal of Development Studies* 46 (1): 1-13.

Day, Alexander F. 2013. *The Peasant in Postsocialist China: History, Politics, and Capitalism*. Cambridge: Cambridge University Press.

De Brauw, Alan. 2010. 'Seasonal Migration and Agricultural Production in Vietnam'. *The Journal of Development Studies* 46 (1): 114-139.

De La Rupelle, Maëlys, Quheng Deng, Li Shi, and Thomas Vendryes. 2008. 'Land Rights and Rural-Urban Migration in China'. *China Perspectives* 2008 (2): 25-35.

Dixon, John. 1981. *The Chinese Welfare System, 1949-1979*. New York: Praeger.

Douglass, Mike. 2006. 'Global Householding in Pacific Asia'. *International Development Planning Review* 28 (4): 421-446.

Duckett, Jane, and Beatriz Carrillo. 2011. 'China's Changing Welfare Mix: Introducing the Local Perspective'. In *China's Changing Welfare Mix: Local Perspectives*, edited by Beatriz Carrillo and Jane Duckett, 1-19. New York: Routledge.

Edgerton, David. 2007. *The Shock of the Old: Technology and Global History since 1900*. Oxford: Oxford University Press.

Eisenman, Joshua. 2018. *Red China's Green Revolution: Technological Innovation, Institutional Change, and Economic Development Under the Commune*. New York: Columbia University Press.

Eyferth, Jacob. 2009. *Eating Rice from Bamboo Roots: The Social History of a Community of Handicraft Papermakers in Rural Sichuan, 1920-2000*. Harvard: Harvard University Press.

Fan, C. Cindy. 2008. *China on the Move: Migration, the State, and the Household.* London: Routledge.

—. 2016. 'Household-Splitting of Rural Migrants in Beijing, China'. *Trialog* 1/2 (116/117): 19-24.

Fan, C. Cindy, and Tianjiao Li. 2019. 'Familization of Rural-Urban Migration in China: Evidence from the 2011 and 2015 National Floating Population Surveys'. *Area Development and Policy* 4 (2): 134-156.

Fan, C. Cindy, Mingjie Sun, and Siqi Zheng. 2011. 'Migration and Split Households: A Comparison of Sole, Couple, and Family Migrants in Beijing, China'. *Environment and Planning* 43: 2164-2185.

Fan, C. Cindy, and Wenfei Winnie Wang. 2008. 'The Household as Security: Strategies of Rural-Urban Migrants in China'. In *Migration and Social Protection in China*, edited by Ingrid Nielsen and Russel Smyth, 205-243. New Jersey: World Scientific Publishing.

Fang, Xiangxin 方向新. 1998. 农村变迁论：当代中国农村变革与发展研究 *Nongcun bian qian lun: Dangdai Zhongguo nongcun biange yu fazhan yanjiu (Study on Rural Transformation: Current Chinese Rural Reform and Development*). Changsha: Hunan People's Press.

FAO, Food and Agriculture Organization. 2019. 'Cereal Yield (Kg per Hectare): China'. The World Bank Data. 2019. Accessed 9 January 2020. https://data.worldbank.org/indicator/AG.YLD.CREL.KG?locations=CN&view=chart.

—. 2020. 'FAO in China: China at a Glance'. 2020. Accessed 9 January 2020. http://www.fao.org/china/fao-in-china/china-at-a-glance.

Farquhar, Judith. 2006. 'Food, Eating, and the Good Life'. In *Handbook of Material Culture*, edited by Christopher Tilley, Webb Keane, Susanne Küchler, Mike Rowlands, and Patricia Spyer, 145-160. Los Angeles: Sage Publications.

Feder, Gershon, Lawrence J. Lau, Justin Y. Lin, and Luo Xiaopeng. 1992. 'The Determinants of Farm Investment and Residential Construction in Post-Reform China'. *Economic Development and Cultural Change* 41 (1): 1-26.

Feuchtwang, Stephan. 2002. *An Anthropological Analysis of Chinese Geomancy.* 2nd ed. Bangkok: White Lotus.

Finnegan, Ruth. 1992. *Oral Traditions and the Verbal Arts: A Guide to Research Practices*. London: Routledge.

Flitsch, Mareile. 1994. *Der Ginsengkomplex in den Han-chinesischen Erzähltraditionen des Jiliner Changbai-Gebietes*. Frankfurt am Main: Peter Lang.

—. 2002. 'Mündliche Überlieferungen und die Problematik ihrer Aufzeichnung in der VR China'. *Oriens Extremus* 43: 221-236.

—. 2004. *Der Kang: Eine Studie zur materiellen Alltagskultur bäuerlicher Gehöfte in der Manjurei*. Wiesbaden: Harrassowitz.

—. 2008. 'Knowledge, Embodiment, Skill and Risk: Anthropological Perspectives on Women's Everyday Technologies in Rural North China'. *East Asian Science, Technology and Society* 2 (2): 265-288.

Frykman, Maja Povrzanović. 2009. 'Material Aspects of Transnational Social Fields: An Introduction'. *Dve Domovini / Two Homelands* 29: 105-114.

Gaetano, Arianne M., and Tamara Jacka, eds. 2004. *On the Move: Women and Rural-to-Urban Migration in Contemporary China*. New York: Columbia University Press.

Gale, Fred. 2013. *Growth and Evolution in China's Agricultural Support Policies*. Washington DC: USDA Economic Research Service.

Gaibazzi, Paolo. 2015. *Bush Bound: Young Men and Rural Permanence in Migrant West Africa*. New York: Berghahn.

Gao, Qi-jie, and Chuan-hong Zhang. 2010. 'Agricultural Technology Extension System in China: Current Situation and Reform Direction'. *Management Science and Engineering* 2 (4): 47-58.

Gao, Qihui. 2011. '2m Hectares of Farmland Abandoned per Year'. *China Daily*. 13 September 2011. Accessed 3 September 2015. http://www.chinadaily.com.cn/china/2011-09/13/content13678049.htm.

Gao, Yan, and Shailaja Fennell. 2018. *China's Rural-Urban Inequality in the Countryside*. Singapore: Springer.

Geertz, Clifford. 1963. *Agricultural Involution: The Processes of Ecological Change in Indonesia*. Berkeley: University of California Press.

Gell, Alfred. 1998. *Art and Agency: An Anthropological Theory*. Oxford: Clarendon.

Gianessi, Leonard, and Ashley Williams. 2011. 'China's Agricultural Success: The Role of Herbicides'. Washington DC: Crop Protection Research Institute.

van Gijn, Annelou, John Whittaker, and Patricia C. Anderson, eds. 2014. *Exploring and Explaining Diversity in Agricultural Technology*. Oxford: Oxbow.

Gilbert, Erik. 2015. 'Asian Rice in Africa: Plant Genetics and Crop History'. In *Rice: Global Networks and New Histories*, edited by Francesca Bray, Peter A. Coclanis, Edda L. Fields-Black, and Dagmar Schäfer, 212-228. New York: Cambridge University Press.

Glenn, Elias, and Kevin Yao. 2016. 'China Loosens Land Transfer Rules to Spur Larger, More Efficient Farms'. *Reuters*. 3 November 2016. Accessed 4 November 2016. http://www.reuters.com/article/us-china-economy-landrights-idUSKBN12Y09F?il=0.

Glick Schiller, Nina, Linda Basch, and Cristina Szanton Blanc, eds. 1992. *Towards a Transnational Perspective on Migration: Race, Class, Ethnicity, and Nationalism Reconsidered*. New York: New York Academy of Sciences.

Goh, Esther C. L. 2013. *China's One-Child Policy and Multiple Caregiving: Raising Little Suns in Xiamen*. London: Routledge.

Granet, Marcel. 1934. *La pensée chinoise*. Paris: Albin Michel.

Grasseni, Christina. 2009. *Developing Skill, Developing Vision: Practices of Locality at the Foot of the Alps*. Oxford: Berghahn.

Gray, Clark L. 2009. 'Rural Out-Migration and Smallholder Agriculture in the Southern Ecuadorian Andes'. *Population and Environment* 30 (4-5): 193-217.

Greenpeace China. 2010. 'The Real Cost of Nitrogen Fertilizer'. Accessed 31 July 2013. http://www.greenpeace.org/canada/Global/canada/report/2010/8/Summary-En.pdf.

GRiSP, Global Rice Science Partnership. 2013. *Rice Almanac*. Los Baños: International Rice Research Institute.

Hanks, Lucien M. 1972. *Rice and Man: Agricultural Ecology in Southeast Asia*. Chicago: Aldine.

Hannerz, Ulf. 1996. *Transnational Connections: Cultures, People, Places*. London: Routledge.

Harries, John. 2017. 'A Stone That Feels Right in the Hand: Tactile Memory, the Abduction of Agency and Presence of the Past'. *Journal of Material Culture* 22 (1): 110-130.

Haudricourt, André-Georges, and Mariel Jean-Brunhes Delamarre. 1955. *L'homme et la charrue à travers le monde*. Paris: Galimard.

He, Congzhi, and Jingzhong Ye. 2014. 'Lonely Sunsets: Impacts of Rural-Urban Migration on the Left-behind Elderly in Rural China'. *Population, Space and Place* 20 (4): 352-369.

He, Shenjing, and Desheng Xue. 2014. 'Identity Building and Communal Resistance against Landgrabs in Wukan Village, China'. *Current Anthropology* 55 (9): S126-S137.

Heerink, Nico, Futian Qu, Marijke Kuiper, Xiaoping Shi, and Shuhao Tan. 2007. 'Policy Reforms, Rice Production and Sustainable Land Use in China: A Macro-Micro Analysis'. *Agricultural Systems* 94: 784-800.

Heger, Isabel. 2020. 'More Than "Peasants Without Land": Individualisation and Identity Formation of Landless Peasants in the Process of China's State-Led Rural Urbanisation'. *Journal of Current Chinese Affairs* 49 (1): 1-25.

Heiss, Jan Patrick. 2003. *Zur Komplexität bäuerlicher Feldarbeit in Afrika: Eine Fallstudie in einem Manga-Dorf (Niger)*. Münster: Lit.

Hexun. 2020. '农用化肥施用量 Nongyong huafei shiyong liang, 1984-2018 (Agricultural Chemical Fertilizer Consumption, 1984-2018)'. Accessed 13 July 2020. http://calendar.hexun.com/area/dqzb_430000_D0610000.shtml.

Hickey, Maureen. 2016. 'Modernisation, Migration, and Mobilisation: Relinking Internal and International Migrations in the "Migration and Development Nexus"'. *Population, Space and Place* 22 (7): 681-692.

Ho, Peter. 2001. 'Who Owns China's Land? Policies, Property Rights and Deliberate Institutional Ambiguity'. *The China Quarterly* 166: 394-421.

Ho, Peter, Jennifer H. Zhao, and Dayuan Xue. 2009. 'Access and Control of Agro-biotechnology: Bt Cotton, Ecological Change and Risk in China'. *The Journal of Peasant Studies* 36 (2): 345-364.

Ho, Ping-ti. 1959. *Studies on the Population of China, 1368-1853*. Cambridge, MA: Harvard University Press.

Hoang, Lan Anh, and Brenda Yeoh, eds. 2015. *Transnational Labour Migration, Remittances, and the Changing Family in Asia*. Basingstoke: Palgrave Macmillan.

HPBS, Hunan Provincial Bureau of Statistics. 1984. *1982*湖南省统计年鉴 *Hunan sheng tongji nianjian* (*Hunan Province Statistical Yearbook 1982*). Changsha: Hunan Renmin Publishing.

—. 2012a. '郴州·安仁县2011 年国民经济和社会发展统计公报 Chenzhou, Anren xian 2011 nian guomin jingji he shehui fazhan tongji gongbao (Chenzhou, Anren County 2011 Statistical Communiqué of the Development of the People's Economy and Society)'. Accessed 9 March 2016. http://www.hntj.gov.cn/tjgb/xqgb/czgb/201204/t20120401_92218.htm.

—. 2012b. 'Statistical Communiqué of Hunan Province on the 2011 National Economic and Social Development'. Accessed 19 August 2015. http://hntj.gov.cn/sjfb/tjnj/12tjnj/html/gbe.htm.

—. 2013. '郴州·安仁县2012 年国民经济和社会发展统计公报 Chenzhou, Anren xian 2012 nian guomin jingji he shehui fazhan tongji gongbao (Chenzhou, Anren County 2012 Statistical Communiqué of the Development of the People's Economy and Society)'. Accessed 9 March 2016. http://www.hntj.gov.cn/tjgb/xqgb/czgb/201304/t20130417_99892.htm.

—. 2014. *Hunan Statistical Yearbook 2014*. Beijing: China Statistics Press.

—. 2018. *Hunan Statistical Yearbook 2018*. Beijing: China Statistics Press.

—. 2019. *Hunan Statistical Yearbook 2019*. Beijing: China Statistics Press.

Huang, Jikun, and Scott Rozelle. 2009. 'Agricultural Development and Nutrition: The Policies behind China's Success'. *Asian Journal of Agriculture and Development* 7 (1): 93-126.

Huang, Jikun, Shukun Wang, and Zhihua Xiao. 2017. 'Rising Herbicide Use and Its Driving Forces in China'. *The European Journal of Development Research* 29 (3): 614-627.

Huang, Jikun, Xiaobing Wang, and Huanguang Qiu. 2012. *Small-Scale Farmers in China in the Face of Modernisation and Globalisation*. London: International Institute for Environment and Development.

Huang, Jikun, Xiaobing Wang, and Scott Rozelle. 2013. 'The Subsidization of Farming Households in China's Agriculture'. *Food Policy* 41: 124-132.

Huang, Martin W. 1995. *Literati and Self-Re-Presentation: Autobiographical Sensibility in the Eighteenth-Century Chinese Novel*. Stanford: Stanford University Press.

Huang, Philip C.C. 2016. 'China's Hidden Agricultural Revolution, 1980-2010, in Historical and Comparative Perspective'. *Modern China* 42 (4): 339-376.

Hull, James R. 2007. 'Migration, Remittances, and Monetization of Farm Labor in Subsistence Sending Areas'. *Asian and Pacific Migration Journal* 16 (4): 451-484.

Hunan Government. 2015. 'About Hunan'. Accessed 31 August 2015. http://www.enghunan.gov.cn/AboutHunan/HunanFacts/.

Ingold, Tim. 2000. *The Perception of the Environment: Essays on Livelihood, Dwelling and Skill*. London: Routledge.

—. 2006. 'Walking the Plank: Meditations on a Process of Skill'. In *Defining Technological Literacy: Towards an Epistemological Framework*, edited by John R. Dakers, 65-80. New York: Palgrave Macmillan.

IRRI, International Rice Research Institute. 2015. 'Direct Seeding'. Accessed 8 March 2016. http://www.knowledgebank.irri.org/step-by-step-production/growth/planting/direct-seeding.

—. 2016. 'How to Plant Rice'. Accessed 12 May 2016. http://www.knowledgebank.irri.org/step-by-step-production/growth/planting.

Jacka, Tamara. 1997. *Women's Work in Rural China: Change and Continuity in an Era of Reform*. Cambridge: Cambridge University Press.

—. 2006. *Rural Women in Urban China: Gender, Migration, and Social Change*. London: Routledge.

—. 2012. 'Migration, Householding and the Well-Being of Left-behind Women in Rural Ningxia'. *China Journal* 67 (2012): 1-22.

—. 2014. 'Left-behind and Vulnerable? Conceptualising Development and Older Women's Agency in Rural China'. *Asian Studies Review* 38 (2): 186-204.

Jepson, Anne. 2014. 'Gardening and Wellbeing: A View from the Ground'. In *Making and Growing: Anthropological Studies of Organisms and Artefacts*, edited by Elizabeth Hallam and Tim Ingold, 147-162. Farnham: Ashgate.

Jin, Songqing, and Klaus Deininger. 2009. 'Land Rental Markets in the Process of Rural Structural Transformation: Productivity and Equity Impacts from China'. *Journal of Comparative Economics* 37 (3): 629-646.

Kamoshita, Akihiko, Sareath Chea, Satoshi Hayashi, Hiroyuki Ikeda, and Boonrat Jongdee. 2009. 'A Case Study on Farmers' Choice of Direct Seeding and Transplanting in Rain-Fed Lowlands in Northeast Thailand and Northwest Cambodia'. *Tropical Agriculture and Development* 53 (2): 43-54.

Kaufmann, Lena. 2011. *Mala Tang – Alltagsstrategien ländlicher Migranten in Shanghai*. Wiesbaden: Harrassowitz.

—. 2016. 'Everyday Strategies of Rural Migrants: Assembling Skills in Mala Tang Production'. *Journal of Chinese Dietary Culture* 12 (2): 63-112.

—. 2019. 'Staying Behind, Moving On: Women, Technology and Migration in Contemporary Rural China'. *ZANTHRO*, Zurich Anthropology Working Paper No. 8: 1-20.

Kelly, Philip F., ed. 2013. *Migration, Agrarian Transition, and Rural Change in Southeast Asia*. London: Routledge.

Kleinhenz, Volker, Wilfried H. Schnitzler, and David J. Midmore. 1996. 'Diversification and Transformation of Asian Paddy Rice Fields to Upland Vegetable Production'. *Plant Research and Development* 43: 81-94.

Knutsson, Helena. 2014a. 'Blades as Messengers of Agriculture: A Case Study from Scandinavia'. In *Exploring and Explaining Diversity in Agricultural Technology*, edited by Annelou van Gijn, John C. Whittaker, and Patricia C. Anderson, 278-285. Oxford: Oxbow.

—. 2014b. 'The Complex Art of Changing Lifestyles on the Verge of the Neolithic'. In *Exploring and Explaining Diversity in Agricultural Technology*, edited by Annelou van Gijn, John C. Whittaker, and Patricia C. Anderson, 295-310. Oxford: Oxbow.

Kolb, Raimund Th. 2003. 'Reis'. In *Das grosse China-Lexikon*, edited by Brunhild Staiger, Stefan Friedrich, and Hans-Wilm Schütte, 618-622. Darmstadt: Wissenschaftliche Buchgesellschaft.

Kong, Sherry Tao, Xin Meng, and Dandan Zhang. 2010. 'Impact of Economic Slowdown on Migrant Workers'. In *China's New Place in a World in Crisis*, edited by Ross Garnaut, Ligang Song, and Wing Thye Woo, 233-260. Canberra: ANU E Press.

Kong, Sherry Tao, and Jonathan Unger. 2013. 'Egalitarian Redistributions of Agricultural Land in China through Community Consensus: Findings from Two Surveys'. *The China Journal* 69: 1-19.

Kovach, Michael J., Megan T. Sweeney, and Susan R. McCouch. 2007. 'New Insights into the History of Rice Domestication'. *Trends in Genetics* 23 (11): 578-587.

Labrada, Ricardo. 2003. 'The Need for Improved Weed Management in Rice'. In *Sustainable Rice Production for Food Security*. Rome: Food and Agriculture Organization of the United Nations.

LaFleur, Robert André. 2020. 'Round and Square: China's Lunar Calendar 2020'. 2020. Accessed 10 January 2020. http://robert-lafleur.blogspot.com/.

Latour, Bruno. 1988. *The Pasteurization of France*. Cambridge, MA: Harvard University Press.

—. 1999. *Pandora's Hope: Essays on the Reality of Science Studies*. Cambridge, MA: Harvard University Press.

Lave, Jean, and Etienne Wenger. 1991. *Situated Learning: Legitimate Peripheral Participation*. Cambridge: Cambridge University Press.

Lawson, Victoria A. 1998. 'Hierarchical Households and Gendered Migration in Latin America: Feminist Extensions to Migration Research'. *Progress in Human Geography* 22: 39-53.

Lemonnier, Pierre. 1993a. 'Introduction'. In *Technological Choices: Transformation in Material Cultures Since the Neolithic*, edited by Pierre Lemonnier, 1-34. Oxford: Oxford University Press.

—. , ed. 1993b. *Technological Choices: Transformation in Material Cultures Since the Neolithic*. London: Routledge.

—. 2012. 'Technology'. In *The Oxford Handbook of Linguistic Fieldwork*, edited by Nick Thieberger, 298-316. Oxford: Oxford University Press.

Leroi-Gourhan, André. 1964. *Le geste et la parole I: Techniques et langage*. Paris: Albin Michel.

Levin, Iris, and Ruth Fincher. 2010. 'Tangible Transnational Links in the Houses of Italian Immigrants in Melbourne'. *Global Networks* 10 (3): 401-423.

Li, Jiming, Yeyun Xin, and Longping Yuan. 2009. 'Hybrid Rice Technology Development: Ensuring China's Food Security'. IFPRI Discussion Paper 00918, 2020 Vision Initiative.

Li, Kwok-sing. 1995. *A Glossary of Political Terms of the People's Republic of China*. Translated by Mary Lok. Hong Kong: The Chinese University Press.

Li, Yuyu. 2006. 'The Impact of Rural Migration on Village Development: A Comparative Study in Three Chinese Villages'. PhD diss., Johns Hopkins University.

Liang, Shumin. 2010. 'Agricultural Pollution Zoning and Comprehensive Control Measures'. In *The China Environment Yearbook, Volume 4: Tragedy and Hope – From the Sichuan Earthquake to the Olympics*, edited by Dongping Yang and Friends of Nature, 145-160. Leiden: Brill.

Lin, Justin Yifu. 1991. 'The Household Responsibility System Reform and the Adoption of Hybrid Rice in China'. *Journal of Development Economics* 36 (2): 353-372.

Lin, Justin Yifu, Ran Tao, and Mingxing Liu. 2007. 'Rural Taxation and Local Governance Reform in China's Economic Transition: Origins, Policy Responses and Remaining Challenges'. Stanford Center for International Development, Working Paper No. 317.

Linares, Olga F. 2003. 'Going to the City … and Coming Back? Turnaround Migration Among the Jola of Senegal'. *Africa* 73 (1): 113-132.

Liu, Jieyu. 2017. 'Ageing in Rural China: State, Family and Gendered Care Responsibilities'. In *Handbook of Welfare in China*, edited by Beatriz Carrillo, Johanna Hood, and Paul Kadetz, 285-296. Northampton: Edward Elgar Publishing.

Lo, Kuei-Mei, and Hsin-Hsing Chen. 2011. 'Technological Momentum and the Hegemony of the Green Revolution: A Case Study of an Organic Rice Cooperative in Taiwan'. *East Asian Science, Technology and Society* 5 (2): 135-172.

Long, Hualou. 2014. 'Land Use Policy in China: Introduction'. *Land Use Policy* 40: 1-5.

Looney, Kristen E. 2008. 'Village Gazetteers: A New Source in the China Field'. *The China Journal* 60: 135-147.

Lora-Wainwright, Anna. 2011. '"If You Can Walk and Eat, You Don't Go to Hospital": The Quest for Healthcare in Rural Sichuan'. In *China's Changing Welfare Mix: Local Perspectives*, edited by Beatriz Carrillo and Jane Duckett, 104-125. New York: Routledge.

Lu, Ming, and Yiran Xia. 2016. 'Migration in the People's Republic of China'. ADBI Working Paper No. 593. Tokyo: Asian Development Bank Institute, 1-35.

Luhmann, Niklas. 1988. 'Familiarity, Confidence, Trust: Problems and Alternatives'. In *Trust: Making and Breaking Cooperative Relations*, edited by Diego Gambetta, 94-107. New York: Blackwell.

Madsen, Richard. 1991. 'The Countryside under Communism'. In *The Cambridge History of China*, Vol. 15, Part 2, edited by Roderick MacFarquhar and John K. Fairbank, 619-681. Cambridge: Cambridge University Press.

Malthus, Thomas. 1798. *An Essay on the Principle of Population*. Reprinted in 1986. Harmondsworth: Penguin Classics.

Manivong, Vongpaphane, Rob Cramb, and Jonathan Newby. 2014. 'Rice and Remittances: Crop Intensification Versus Labour Migration in Southern Laos'. *Human Ecology* 42 (3): 367-379.

Mao, Zedong. 1965. *On Practice*. Beijing: Foreign Languages Press.

Marchand, Trevor H.J., ed. 2010. 'Making Knowledge'. *Special Issue, Journal of the Royal Anthropological Institute* 16 (May).

Marchand, Trevor H.J. 2014. 'Skill and Aging: Perspectives from Three Generations of English Woodworkers'. In *Making and Growing Anthropological Studies of Organisms and Artefacts*, edited by Elizabeth Hallam and Tim Ingold, 183-201. Farnham: Ashgate.

Mauss, Marcel. 1934. 'Les techniques du corps'. *Journal de psychologie* 32 (3-4).

McKay, Deirdre. 2005. 'Reading Remittance Landscapes: Female Migration and Agricultural Transition in the Philippines'. *Danish Journal of Geography* 105 (1): 89-99.

McMichael, Philip. 2020. 'Does China's "Going out" Strategy Prefigure a New Food Regime?' *The Journal of Peasant Studies* 47 (1): 116-154.

Meier, Uwe, ed. 1997. *Growth Stages of Mono- and Dicotyledonous Plants: BBCH-Monograph*. Berlin: Blackwell.

Meng, Xiangdan. 2014. 'The Feminization of China's Agriculture: A Sociological Analysis'. PhD diss., Wageningen University.

Mieder, Wolfgang. 2004. *Proverbs: A Handbook*. Westport: Greenwood Press.

Miles, Douglas. 1979. 'The Finger Knife and Ockham's Razor: A Problem in Asian Culture History and Economic Anthropology'. *American Ethnologist* 6 (2): 223-243.

Miller, Daniel. 1987. *Material Culture and Mass Consumption*. Oxford: Basil Blackwell.

—. 2008. 'Migration, Material Culture and Tragedy: Four Moments in Caribbean Migration'. *Mobilities* 3 (3): 397-413.

Mintz, Sidney W., and Christine M. Du Bois. 2002. 'The Anthropology of Food and Eating'. *Annual Review of Anthropology* 31 (1): 99-119.

MoA, Ministry of Agriculture of the People's Republic of China. 1989. 中国农村经济统计大全 *Zhongguo nongcun jingji tongji daquan, 1949-1986* (*Comprehensive Collection of the Statistics of China's Rural Economy, 1949-1986*). Beijing: Agricultural Press.

—. 2012. 'Excerpt of China's 12th Five-Year Plan: Agriculture Part'. Accessed 21 May 2015. http://english.agri.gov.cn/hottopics/five/201301/t20130115_9545.htm.

—. 2015. 'SCIO Briefing on Agricultural Modernization'. 3 February 2015. Accessed 6 June 2016. http://english.agri.gov.cn/hottopics/cpc/201502/t20150204_24960.htm.

—. 2016. 'China's No.1 Central Document Focuses on Agriculture for 13th Consecutive Year'. Accessed 28 January 2016. http://english.agri.gov.cn/news/dqnf/201601/t20160128_164966.htm.

Muehe, E. Marie, Tianmei Wang, Carolin F. Kerl, Britta Planer-Friedrich, and Scott Fendorf. 2019. 'Rice Production Threatened by Coupled Stresses of Climate and Soil Arsenic'. *Nature Communications* 10 (4985): 1-10.

Müller, Daniel, and Thomas Sikor. 2006. 'Effects of Postsocialist Reforms on Land Cover and Land Use in South-Eastern Albania'. *Applied Geography* 26: 175-191.

Murphy, Rachel. 2002. *How Migrant Labor Is Changing Rural China*. Cambridge: Cambridge University Press.

—. 2004. 'Turning Peasants into Modern Chinese Citizens: "Population Quality" Discourse, Demographic Transition and Primary Education'. *The China Quarterly* 177: 1-20.

—. 2006. 'Citizenship Education in Rural China: The Dispositional and Technical Training of Cadres and Farmers'. In *Chinese Citizenship: Views from the Margins*, edited by Vanessa L. Fong and Rachel Murphy, 9-26. London: Routledge.

Naughton, Barry. 2007. *The Chinese Economy: Transitions and Growth*. Cambridge, MA: MIT Press.

NBSC, National Bureau of Statistics of China. 2015. 'National Data: Annual by Province'. Accessed 20 July 2015. http://data.stats.gov.cn/english/mapdata.htm?cn=E0103&zb=A0301.

—. 2018. *China Statistical Yearbook 2018*. Beijing: China Statistics Press.

—. 2019a. *China Statistical Yearbook 2019*. Beijing: China Statistics Press.

—. 2019b. 'National Data: Annual by Province'. Accessed 6 January 2020. http://data.stats.gov.cn/english/easyquery.htm?cn=E0103.

NBSC Rural Social and Economic Investigation Division. 2010. *2010*中国农村统计年鉴 *Zhongguo nongcun tongji nianjian* (*2010 Statistical Yearbook of the Chinese Countryside*). Beijing: China Statistics Press.

Needham, Joseph. 1974. 'The Nature of Chinese Society: A Technical Interpretation'. *University of Hong Kong Gazette* XXIII (5): part 2.

Netting, Robert McC. 1993. *Smallholders, Householders: Farm Families and the Ecology of Intensive, Sustainable Agriculture*. Stanford: Stanford University Press.

Nguyen, Tuan Anh, Jonathan Rigg, and Annuska Derks. 2015. 'Migration and Agricultural Production in a Vietnamese Village'. Max Planck Institute for Social Anthropology, Working Paper No. 164.

NPC, The National People's Congress of the People's Republic of China. 2002. 'Law of the People's Republic of China on Land Contract in Rural Areas'. Beijing: Database of Laws and Regulations.

—. 2004. 'Land Administration Law of the People's Republic of China'. Beijing: Database of Laws and Regulations.

—. 2007. 'Property Law of the People's Republic of China'. Beijing: Database of Laws and Regulations.

—. 2009. 'Law of the People's Republic of China on Mediation and Arbitration of Disputes over Rural Land Contract and Management'. Beijing: Database of Laws and Regulations.

Nyíri, Pál. 2010. *Mobility and Cultural Authority in Contemporary China*. Seattle: University of Washington Press.

Oakes, Tim, and Louisa Schein. 2005. *Translocal China: Linkages, Identities and the Reimagining of Space*. London: Taylor & Francis.

Obendiek, Helena. 2016. *'Changing Fate': Education, Poverty and Family Support in Contemporary Chinese Society*. Berlin: Lit.

OECD. 2003. 'OECD Glossary of Statistical Terms: Market Price Support (MPS)'. 7 March 2003. Accessed 6 May 2020. https://stats.oecd.org/glossary/detail.asp?ID=1600.

—. 2005. *OECD Review of Agricultural Policies: China*. Paris: OECD Publishing.

—. 2013. *Agricultural Policy Monitoring and Evaluation 2011: OECD Countries and Emerging Economies*. Paris: OECD Publishing.

—. 2014. *Agricultural Policy Monitoring and Evaluation 2014: OECD Countries*. Paris: OECD Publishing.

—. 2018. *Innovation, Agricultural Productivity and Sustainability in China*. Paris: OECD Publishing.

—. 2019. *Agricultural Policy Monitoring and Evaluation 2019*. Paris: OECD Publishing.

Ogura, Chikara, Somsak Sukuchan, and Hajime Narioka. 2011. 'Introduction of Direct Seeding and Selection of Farmland in Rainfed Rice Fields in Small Watersheds in Northeast Thailand'. *The Japan Agricultural Research Quarterly* 45 (3): 337-347.

Ohnuki-Tierney, Emiko. 1995. *Rice as Self: Japanese Identities through Time*. Princeton: Princeton University Press.

Oi, Jean. 1989. *State and Peasant in Contemporary China: The Political Economy of Village Government*. Berkeley: University of California Press.

Oi, Jean, and Andrew Walder, eds. 1999. *Property Rights and Economic Reform in China*. Stanford: Stanford University Press.

Ong, Aihwa. 1999. *Flexible Citizenship: The Cultural Logics of Globalization*. Durham: Duke University Press.

Ong, Walter J. 2002. *Orality and Literacy: The Technologizing of the Word*. 2nd ed. London: Routledge.

Ortner, Sherry B. 2006. *Anthropology and Social Theory: Culture, Power, and the Acting Subject*. Durham: Duke University Press.

Oshiro, Kenji K. 1985. 'Mechanization of Rice Production in Japan'. *Economic Geography* 61 (4): 323-331.

Oxfeld, Ellen. 2017. *Bitter and Sweet: Food, Meaning, and Modernity in Rural China*. Oakland: University of California Press.

Pandey, Sushil, Martin Mortimer, Leonard Wade, To Phuc Tuong, Katherine Lopez, and Bill Hardy, eds. 2002. *Direct Seeding: Research Strategies and Opportunities*. Los Baños: International Rice Research Institute.

Patton, Dominique. 2014. 'China Looking to Curb Fertilizer, Pesticide Use'. *Reuters*. 5 December 2014. Accessed 6 December 2014. https://www.reuters.com/article/us-china-grain-fertilizers/china-looking-to-curb-fertilizer-pesticide-use-idUSKCN0JJ0SO20141205.

Pauli, Julia. 2015. 'Gebauter Lebenssinn. Häuser in transnationalen mexikanischen Familien'. *Sociologus* 65 (2): 153-176.

Peng, Mu. 2008. *Shared Practice, Esoteric Knowledge, and Bai: Envisioning the Yin World in Rural China*. PhD diss., University of Pennsylvania.

Perdue, Peter C. 1987. *Exhausting the Earth: State and Peasant in Hunan, 1500-1850*. Cambridge, MA: Harvard University Press.

Perkins, Dwight H. 1991. 'China's Economic Policy and Performance'. In *The Cambridge History of China*, Vol. 15, Part 2, edited by Roderick MacFarquhar and John K. Fairbank, 475-539. Cambridge: Cambridge University Press.

Pfaffenberger, Bryan. 1992. 'Social Anthropology of Technology'. *Annual Review of Anthropology* 21: 491-516.

Pieke, Frank N. 2002. 'The Politics of Rural Land Use Planning in China'. Max Planck Institute for Social Anthropology, Working Paper No. 40.

Pieke, Frank N., Pál Nyíri, Mette Thunø, and Antonella Ceccagno. 2004. *Transnational Chinese: Fujianese Migrants in Europe*. Stanford: Stanford University Press.

van der Ploeg, Jan Douwe. 2007. 'Resistance of the Third Kind and the Construction of Sustainability'. Paper presented to the ESRS Conference, August 23, 2007, Wageningen.

van der Ploeg, Jan Douwe, and Jingzhong Ye. 2010. 'Multiple Job Holding in Rural Villages and the Chinese Road to Development'. *The Journal of Peasant Studies* 37 (3): 513-530.

—., eds. 2016. *China's Peasant Agriculture and Rural Society: Changing Paradigms of Farming*. London: Routledge.

Postill, John. 2010. 'Introduction: Theorising Media and Practice'. In *Theorising Media and Practice*, edited by Birgit Bräuchler and John Postill, 1-34. New York: Berghahn.

Potter, Sulamith Heins, and Jack M. Potter. 1990. *China's Peasants: The Anthropology of a Revolution*. Cambridge: Cambridge University Press.

Qi, Xing. 1986. *Traditionelle Feste und Gebräuche in China*. Beijing: Foreign Language Press.

Qin, Hua. 2010. 'Rural-to-Urban Labor Migration, Household Livelihoods, and the Rural Environment in Chongqing Municipality, Southwest China'. *Human Ecology* 38 (5): 675-690.

Qin, Hua, and Tim F. Liao. 2016. 'Labor Out-Migration and Agricultural Change in Rural China: A Systematic Review and Meta-Analysis'. *Journal of Rural Studies* 47: 533-541.

Rawski, Evelyn Sakakida. 1972. *Agricultural Change and the Peasant Economy of South China*. Cambridge, MA: Harvard University Press.

Resurreccion, Bernadette P., and Ha Thi Van Khanh. 2007. 'Able to Come and Go: Reproducing Gender in Female Rural-Urban Migration in the Red River Delta'. *Population, Space and Place* 13 (3): 211-224.

Rice, Tom. 2010. 'Learning to Listen: Auscultation and the Transmission of Auditory Knowledge'. *Journal of the Royal Anthropological Institute* 16: S41-S61.

Richards, Paul. 1985. *Indigenous Agricultural Revolution: Ecology and Food Production in West Africa*. London: Hutchinson.

Rigg, Jonathan. 2019. *More than Rural: Textures of Thailand's Agrarian Transformation*. Honolulu: University of Hawai'i Press.

Rigg, Jonathan, Albert Salamanca, and Eric C. Thompson. 2016. 'The Puzzle of East and Southeast Asia's Persistent Smallholder'. *Journal of Rural Studies* 43: 118-133.

Roberts, Kenneth D. 1997. 'China's "Tidal Wave" of Migrant Labour: What Can We Learn From Mexican Undocumented Migration to the United States?' *International Migration Review* 31 (2): 249-293.

Salazar, Noel B. 2017. 'Key Figures of Mobility: An Introduction'. *Social Anthropology* 25 (1): 5-12.

Santos, Gonçalo. 2011. 'Rethinking the Green Revolution in South China: Technological Materialities and Human-Environment Relations'. *East Asian Science, Technology and Society* 5 (4): 479-504.

—. 2017. 'Multiple Mothering and Labor Migration in Rural South China'. In *Transforming Patriarchy: Chinese Families in the Twenty-First Century*, edited by Gonçalo Santos and Stevan Harrell, 91-110. Seattle: University of Washington Press.

Santos, Gonçalo, and Stevan Harrell, eds. 2017. *Transforming Patriarchy: Chinese Families in the Twenty-First Century*. Seattle: University of Washington Press.

Sargeson, Sally. 2002. 'Subduing "The Rural House-Building Craze": Attitudes Towards Housing Construction and Land Use Controls in Four Zhejiang Villages'. *The China Quarterly* 172: 927-955.

—. 2012. 'Villains, Victims and Aspiring Proprietors: Framing "Land-Losing Villagers" in China's Strategies of Accumulation'. *Journal of Contemporary China* 21 (77): 757-777.

Schäfer, Ingo. 1983. *Populäre Sprachformen und politische Argumentation: Zur Funktion der Idiomatik in den Schriften Mao Zedongs*. Frankfurt am Main: Haag + Herchen.

Schippers, Thomas K. 1992. 'Le proverbe comme forme de transmission de savoirs: Quelques réflexions à propos d'un corpus de proverbes météorologiques'. In *Les voies de la parole*, edited by Jean-Noël Pelen and Claude Martel, 99-105. Aix-en-Provence: Université de Provence.

—. 2014a. 'Agricultural Practices: Change and Stability'. In *Exploring and Explaining Diversity in Agricultural Technology*, edited by Annelou van Gijn, John C. Whittaker, and Patricia C. Anderson, 339-341. Oxford: Oxbow.

—. 2014b. 'Seasonal Variations in Crop Tending and Folk Knowledge in Southern France'. In *Exploring and Explaining Diversity in Agricultural Technology*, edited by Annelou van Gijn, John C. Whittaker, and Patricia C. Anderson, 79-80. Oxford: Oxbow.

—. 2014c. 'Skills as Identity Markers'. In *Exploring and Explaining Diversity in Agricultural Technology*, edited by Annelou van Gijn, John C. Whittaker, and Patricia C. Anderson, 276-277. Oxford: Oxbow.

—. 2014d. 'The Contemporary Use of Iberian Threshing Sledges: Some Ethnographic Observations about an Obsolete Choice'. In *Exploring and Explaining Diversity in Agricultural Technology*, edited by Annelou van Gijn, John Whittaker, and Patricia C. Anderson, 152-154. Oxford: Oxbow.

Schmalzer, Sigrid. 2016. *Red Revolution, Green Revolution: Scientific Farming in Socialist China*. Chicago: University of Chicago Press.

Schneider, Mindi. 2015. 'What, Then, Is a Chinese Peasant? Nongmin Discourses and Agroindustrialization in Contemporary China'. *Agriculture and Human Values* 32 (2): 331-346.

Schweizer, Thomas. 1989. *Reisanbau in einem javanischen Dorf.* Köln: Böhlau.

Scoones, Ian. 2009. 'Livelihoods Perspectives and Rural Development'. *The Journal of Peasant Studies* 36 (1): 171-196.

Scott, James C. 1977. *The Moral Economy of the Peasant: Rebellion and Subsistence in Southeast Asia*. New Haven: Yale University Press.

Scott, James C. 1985. *Weapons of the Weak: Everyday Forms of Peasant Resistance.* New Haven: Yale University Press.

Shakespeare, Tom. 2006. *Disability Rights and Wrongs*. London: Routledge.

Shapiro, Judith. 2001. *Mao's War against Nature: Politics and the Environment in Revolutionary China*. Cambridge: Cambridge University Press.

Sheller, Mimi, and John Urry. 2006. 'The New Mobilities' Paradigm'. *Environment and Planning* 38: 207-226.

Shen, Xiaobai. 2010. 'Understanding the Evolution of Rice Technology in China: From Traditional Agriculture to GM Rice Today'. *The Journal of Development Studies* 46 (6): 1026-1046.

Sigaut, François. 1991. 'Les techniques de récolte des grains: Identification, localisation, problèmes d'interprétation'. In *Rites et rythmes agraires: Séminaire de recherche*, edited by Marie-Claire Cauvin, 31-43. Lyon: GDR-Maison de l'Orient.

—. 1994. 'Technology'. In *Companion Encyclopedia of Anthropology*, edited by Tim Ingold, 420-459. London: Routledge.

Skeldon, Ronald. 1996. 'Migration from China'. *Journal of International Affairs* 49 (2): 434-455.

Smerdel, Inja. 2014a. 'Concluding Remarks'. In *Exploring and Explaining Diversity in Agricultural Technology*, edited by Annelou van Gijn, John C. Whittaker, and Patricia C. Anderson, 286. Oxford: Oxbow.

—. 2014b. '"Training Oxen Meant Training for the Children"'. In *Exploring and Explaining Diversity in Agricultural Technology*, edited by Annelou van Gijn, John C. Whittaker, and Patricia C. Anderson, 258-268. Oxford: Oxbow.

Smil, Václav. 2004. *China's Past, China's Future: Energy, Food, Environment*. New York: RoutledgeCurzon.

Song, Yiching. 1998. '"New" Seed in "Old" China: Impact of CIMMYT Collaborative Programme on Maize Breeding in South-Western China'. PhD diss., Wageningen University.

Stark, Oded. 1991. *The Migration of Labor*. Oxford: Blackwell.

Stark, Oded, and David E. Bloom. 1985. 'The New Economics of Labour Migration'. *American Economic Review* 75: 173-178.

State Council. 1984. '国务院关于农民进入集镇落户问题的通知 Guowuyuan guanyu nongmin jinru jizhen luohu wenti de tongzhi (Notice of the State Council on the Issue of Peasants Entering Towns and Settling Down)'. Notice No. 141, 13 October 1984.

Statista. 2018. 'China: Pesticide Use by Region 2016'. Accessed 12 December 2019. https://www.statista.com/statistics/300970/china-pesticide-use-by-region/.

Stavis, Benedict. 1974. *Making Green Revolution: The Politics of Agricultural Development in China*. Ithaca, NY: Rural Development Committee, Cornell University.

—. 1978. *The Politics of Agricultural Mechanization in China*. Ithaca, NY: Cornell University Press.

Steinmüller, Hans. 2013. *Communities of Complicity: Everyday Ethics in Rural China*. New York: Berghahn.

Sternfeld, Eva. 2009. 'Biologischer Anbau "Made in China."' *EU-China Civil Society Forum Hintergrundinformationen* 12: 1-12.

Stone, Glenn Davis. 2001. 'Theory of the Square Chicken: Advances in Agricultural Intensification Theory'. *Asia Pacific Viewpoint* 42 (2-3): 163-180.

—. 2007. 'Agricultural Deskilling and the Spread of Genetically Modified Cotton in Warangal'. *Current Anthropology* 48 (1): 67-103.

Stross, Randal E. 1986. *The Stubborn Earth: American Agriculturalists on Chinese Soil, 1898-1937*. Berkeley: University of California Press.

Sweeney, Megan, and Susan R. McCouch. 2007. 'The Complex History of the Domestication of Rice'. *Annals of Botany* 100 (5): 951-957.

Takeuchi, Hiroki. 2014. *Tax Reform in Rural China: Revenue, Resistance, and Authoritarian Rule*. New York: Cambridge University Press.

Tang, Xiaoqing 唐小晴, Junjun Jiang 蒋军君, and Xuan Xin 新轩. 2020. '中国农民春耕新景象："共享农机"开进田间 穿着皮鞋能种田 Zhongguo nongmin chungeng xin jingxiang: "Gongxiang nongji" kai jin tianjian chuanzhe pixie neng zhongtian (A New Scene of Spring Farming for Chinese Farmers: "Shared Agricultural Machinery" Enters the Field – Wearing Leather Shoes to Farm)'. *China News Hunan*. 18 March 2020. Accessed 4 June 2020. http://www.hn.chinanews.com/news/2020/0318/381860.html.

Thøgersen, Stig, and Søren Clausen. 1992. 'New Reflections in the Mirror: Local Chinese Gazetteers (Difangzhi) in the 1980s'. *The Australian Journal of Chinese Affairs* 27: 161-184.

Tian, Qing, Daniel G. Brown, Lin Zheng, Shuhua Qi, Ying Liu, and Luguang Jiang. 2015. 'The Role of Cross-Scale Social and Environmental Contexts in Household-Level Land-Use Decisions, Poyang Lake Region, China'. *Annals of the Association of American Geographers* 105 (6): 1240-1259.

Tian, Wei-Ming, and Zhang-Yue Zhou. 2018. 'Developments in China's Grain Policies'. In *Grains in China: Foodgrain, Feedgrain and World Trade*, edited by Zhang-Yue Zhou and Wei-Ming Tian, 11-23. Oxon: Routledge.

Tilley, Christopher. 2006a. 'Introduction'. In *Handbook of Material Culture*, edited by Christopher Tilley, Webb Keane, Susanne Küchler, Michael Rowlands, and Patricia Spyer, 2-8. Los Angeles: Sage Publications.

—. 2006b. 'Objectification'. In *Handbook of Material Culture*, edited by Christopher Tilley, Webb Keane, Susanne Küchler, Michael Rowlands, and Patricia Spyer, 60-73. Los Angeles: Sage Publications.

Tilt, Bryan. 2008. 'Smallholders and the "Household Responsibility System": Adapting to Institutional Change in Chinese Agriculture'. *Human Ecology* 36: 189-199.

Tolia-Kelly, Divya P. 2004. 'Materializing Post-Colonial Geographies: Examining the Textural Landscapes of Migration in the South Asian Home'. *Geoforum* 35 (6): 675-688.

Toyota, Mika, Brenda S.A. Yeoh, and Liem Nguyen. 2007. 'Bringing the "Left behind" Back into View in Asia: A Framework for Understanding the "migration-Left behind Nexus." *Population, Space and Place* 13 (3): 157-161.

Tregear, Thomas R., and Victor C. Falkenheim. 2015. 'Hunan'. In *Encyclopædia Britannica*. Accessed 20 July 2015. http://academic.eb.com/EBchecked/topic/276448/Hunan.

TU Berlin, Center for Cultural Studies on Science and Technology in China. 2017. 'DFG-Projekt "Making Technology Appropriate: Technology Transfer from Germany to China. The Case of Steam and Ordnance Technologies, 1860-1980"'. Accessed 7 February 2018. http://www.china.tu-berlin.de/menue/forschung/projekte/making_technology_appropriate/.

UNDP, United Nations Development Programme. 2014. *Human Development Report 2014 – Sustaining Human Progress: Reducing Vulnerability and Building Resilience.* New York: UNDP.

Vermeer, Eduard B. 1992. 'New County Histories: A Research Note on Their Compilation and Value'. *Modern China* 18 (4): 438-467.

Vertovec, Steven. 2009. *Transnationalism*. London: Routledge.

VMZ, Ethnographic Museum of the University of Zurich. 2017. 'Skills and Aging: New Trajectories in Anthropology of Technology'. 9 May 2016. Accessed 16 February 2018. https://www.musethno.uzh.ch/de/museum/newsarchiv/Seed-Money-Grant.html.

Walsh, Katie. 2011. 'Migrant Masculinities and Domestic Space: British Home-making Practices in Dubai'. *Transactions of the Institute of British Geographers* 36 (4): 516-529.

Wang, Cangbai. 2016. 'Introduction: The "Material Turn" in Migration Studies'. *Modern Languages Open* 26: 1-11.

Wang, Dehua 王德华. 2014. '湖南省失地农民社会保障问题调查研究 Hunan sheng shidi nongmin shehui baozhang wenti diaocha yanjiu (A Study of the Social Security of Landless Peasants in Hunan)'. 怀化学院学报 *Journal of Huaihua University* 33 (2): 75-77.

Wang, Hongfei, and Robert J. Hijmans. 2019. 'Climate Change and Geographic Shifts in Rice Production in China'. *Environmental Research Communications* 1 (1): 011008.

Wang, Nora. 2003. 'Hunan'. In *Das grosse China-Lexikon*, edited by Brunhild Staiger, Stefan Friedrich, and Hans-Wilm Schütte, 319-320. Darmstadt: Wissenschaftliche Buchgesellschaft.

Wang, Qianxi, and Xiaoling Zhang. 2017. 'Three Rights Separation: China's Proposed Rural Land Rights Reform and Four Types of Local Trials'. *Land Use Policy* 63: 111-121.

Wang, Zhiling, and Lu Chen. 2019. 'Destination Choices of Chinese Rural-Urban Migrant Workers: Jobs, Amenities, and Local Spillovers'. *Journal of Regional Science* 59 (3): 586-609.

Watson, Andrew. 2001. 'Agriculture (Policies, Organizations, Units)'. In *Dictionary of the Politics of the People's Republic of China*, edited by Colin Mackerras, Donald H. McMillen, and Andrew Watson, 57-61. London: Routledge.

Wenger, Etienne. 1998. *Communities of Practice: Learning, Meaning, and Identity*. Cambridge: Cambridge University Press.

White, Ben. 2012. 'Agriculture and the Generation Problem: Rural Youth, Employment and the Future of Farming'. *IDS Bulletin* 43 (6): 9-19.

Whittaker, John C. 2014a. 'Some Principles of Technological Decline: The Case of the Tribulum'. In *Exploring and Explaining Diversity in Agricultural Technology*, edited by Annelou van Gijn, John C. Whittaker, and Patricia C. Anderson, 355-356. Oxford: Oxbow.

—. 2014b. 'Threshing Processes and Tools – Exploring Diversity in the Past: An Introduction'. In *Exploring and Explaining Diversity in Agricultural Technology*, edited by Annelou van Gijn, John C. Whittaker, and Patricia C. Anderson, 133-135. Oxford: Oxbow.

Whyte, Susan Reynolds. 2009. 'Epilogue: Insecurity, Contingency and Uncertainty'. In *Situating Uncertainty in Contemporary Africa*, edited by Liv Haram and Bawa Yamba, 213-216. Uppsala: Nordic Africa Institute.

Wilken, Gene C. 1987. *Good Farmers: Traditional Agricultural Resource Management in Mexico and Central America*. Berkeley: University of California Press.

Wittfogel, Karl August. 1957. *Oriental Despotism: A Comparative Study of Total Power*. New Haven: Yale University Press.

Wolf, Diane Lauren. 1992. *Factory Daughters: Gender, Household Dynamics, and Rural Industrialization in Java*. Berkeley: University of California Press.

Wolf, Eric. 1966. *Peasants*. Englewood Cliffs: Prentice Hall.

Wu, Chunyuan 伍春元, ed. 2010. 益相史话 *Yixiang shihua* (*Historical Narrative of Yixiang*). Beijing: Jiuzhou Press.

Wu, Yi. 2016. *Negotiating Rural Land Ownership in Southwest China: State, Village, Family*. Honolulu: University of Hawai'i Press.

Wu, Yiyun, Xican Xi, Xin Tang, Deming Luo, Baojing Gu, Shu Kee Lam, Peter M. Vitousek, and Deli Chen. 2018. 'Policy Distortions, Farm Size, and the Overuse of Agricultural Chemicals in China'. *PNAS* 115 (27): 7010-7015.

Xiang, Biao. 2007. 'How Far Are the Left-behind Left behind? A Preliminary Study in Rural China'. *Population, Space and Place* 13 (3): 179-191.

—. 2016. 'Beyond Methodological Nationalism and Epistemological Behaviouralism: Drawing Illustrations from Migrations within and from China'. *Population, Space and Place* 22 (7): 669-680.

Xie, Yong, and Quanbao Jiang. 2016. 'Land Arrangements for Rural-Urban Migrant Workers in China: Findings from Jiangsu Province'. *Land Use Policy* 50: 262-267.

Xin, Liangjie, Yuzhi Fan, Minghong Tan, and Luguang Jiang. 2009. 'Review of Arable Land-Use Problems Inpresent-Day China'. *Ambio* 38 (2): 112-115.

Xinhua. 2017. 'China Considers Revising Rural Land Contract Law'. *China Daily*. 31 October 2017. Accessed 13 December 2017. https://www.chinadaily.com.cn/china/2017-10/31/content_33937185.htm.

—. 2020. 'China Focus: Major Rice Producer Promotes Double-Cropping Rice to Ensure Food Security'. Xinhua Net. 28 April 2020. Accessed 5 June 2020. http://www.xinhuanet.com/english/2020-04/28/c_139015447.htm.

Xinhua News Agency. 2015. 'China's Focus on Agricultural Reform for 12th Year in a Row'. 3 February 2015. Accessed 31 October 2017. http://english.agri.gov.cn/news/dqnf/201502/t20150203_24957.htm.

Xinhua News Network. 2018. '中共中央 国务院 印发"乡村振兴战略规划（2018—2022年）" Zhonggong zhongyang guowuyuan yinfa "Xiangcun zhenxing zhanlüe guihua (2018-2022 nian)" (The Central Committee of the Communist Party of China and the State Council Issue the "Strategic Plan for Rural Revitalization (2018-2022)"'. 26 September 2018. Accessed 7 August 2020. http://www.xinhuanet.com/politics/2018-09/26/c_1123487123.htm.

—. 2019. '全国人民代表大会常务委员会关于修改"中华人民共和国土地管理法"、"中华人民共和国城市房地产管理法"的决定 Quanguo renmin daibiao dahui changwu weiyuanhui guanyu xiugai "Zhonghua renmin gongheguo tudi guanli fa", "Zhonghua renmin gongheguo chengshi fangdichan guanli fa" de jueding (Decision of the Standing Committee of the National People's Congress on Amending the Land Management Law and the Law on Urban Real Estate Management of the People's Republic of China)'. 26 August 2019. Accessed 6 January 2020. http://www.xinhuanet.com/politics/2019-08/26/c_1124923935.htm.

XT, 湘潭市民间文学集成编委会 Xiangtan Shi Minjian Wenxue Jicheng Bianweihui (Editorial Board of the Comprehensive Collection of the Folk Literature of Xiangtan City), ed. 1988. 中国谚语集成湖南卷: 湘潭市分卷 *Zhongguo yanyu jicheng Hunan juan: Xiangtan shi fen juan* (*Hunan Volume of the Comprehensive Collection of Chinese Proverbs: Section Volume of Xiangtan City*). Xiangtan: N.p.

Xu, Dingde, Shili Guo, Fangting Xie, Shaoquan Liu, and Sha Cao. 2017. 'The Impact of Rural Laborer Migration and Household Structure on Household Land Use Arrangements in Mountainous Areas of Sichuan Province, China'. *Habitat International* 70: 72-80.

Yan, Yunxiang. 2003. *Private Life under Socialism: Love, Intimacy and Family Change in a Chinese Village, 1949-1999*. Stanford: Stanford University Press.

Yang, Lihui, and Deming An. 2008. *Handbook of Chinese Mythology*. New York: Oxford University Press.

Ye, Jingzhong. 2015. 'Land Transfer and the Pursuit of Agricultural Modernization in China'. *Journal of Agrarian Change* 15 (3): 314-337.

—. 2018. 'Stayers in China's "Hollowed-out" Villages: A Counter Narrative on Massive Rural-Urban Migration'. *Population, Space and Place* 24: 1-10.

Ye, Jingzhong 叶敬忠. 2019. '农村留守人口研究: 基本立场、认识误区与理论转向 Nongcun liushou renkou yanjiu: Jiben lichang, renshi wuqu yu lilun zhuanxiang (Research on the Rural Left-behind Population: Basic Positions, Mistakes, and Theoretical Turn)'. 人口研究 *Population Research* 43 (2): 21-31.

Yu, Zhonglei, Lei Liu, Hua Zhang, and Jinshe Liang. 2017. 'Exploring the Factors Driving Seasonal Farmland Abandonment: A Case Study at the Regional Level in Hunan Province, Central China'. *Sustainability* 9 (187): 1-18.

Yuan, Juanwen, and Anke Niehof. 2011. 'Agricultural Technology Extension and Adoption in China: A Case from Kaizuo Township, Guizhou Province'. *The China Quarterly* 206: 412-425.

Zapata, Lydia. 2014. 'Exploring Diversity in the Past: An Introduction'. In *Plants and People: Choices and Diversity Through Time*, edited by Alexandre Chevalier, Elena Marinova, and Leonor Peña-Chocarro, 16-19. Oxford: Oxbow.

Zavoretti, Roberta. 2017. *Rural Origins, City Lives: Class and Place in Contemporary China*. Seattle: University of Washington Press.

Zhang, Jiayan. 2014. *Coping with Calamity: Environmental Change and Peasant Response in Central China, 1736-1949*. Vancouver: UBC Press.

Zhang, Ying, Xiubin Li, and Wei Song. 2014. 'Determinants of Cropland Abandonment at the Parcel, Household Andvillage Levels in Mountain Areas of China: A Multi-Level Analysis'. *Land Use Policy* 41: 186-192.

Zhang, Ze Pu. 2003. 'Development of Chemical Weed Control and Integrated Weed Management in China'. *Weed Biology and Management* 3 (4): 197-203.

Zhi, Fujing 致富经. 2010. '灯芯草"肥"了龙市乡 Dengxin cao "fei" le Longshi xiang (Lamp Rush "Fertilized" Longshi Township)'. 湖南科技报 *Hunan Science and Technology Newspaper*. 20 December 2010. Accessed 24 March 2013. http://www.zgxncjs.com/xnc/xnc/showclass.jsp-id=19322.htm.

Zhou, Cissy. 2020. 'Coronavirus Has Hit China's Migrant Workers Harder than Sars and the Financial Crisis, but Worst yet to Come'. *South China Morning Post*, 25 May 2020. Accessed 29 May 2020. https://www.scmp.com/economy/china-economy/article/3085904/coronavirus-has-hit-chinas-migrant-workers-harder-sars-and.

Zhou, Daming 周大鸣. 2005. 渴望生存*: 农民工流动的人类学考察 Kewang shengcun: Nongmingong liudong de renleixue kaocha* (*Longing for Survival: An Anthropological Investigation of the Flow of Migrant Workers*). Guangzhou: Zhongshan University Press.

Zhou, Xuesong, and Ying Liu. 2012. 'Peasants' Income Structure Evolution and Its Enlightenment'. *Chinese Agricultural Science Bulletin* 28 (14): 210-213.

Zhuzhou Network of Social Sciences 株洲社科网. 2012. '踩田的来历 Cai tian de laili (The Origin of Stepping on the Field)'. 10 April 2012. Accessed 16 November 2016. http://www.zzsskl.gov.cn/?thread-1485-1.html.

Index

Page numbers in *italics* refer to figures and tables.

Africa 127, 158, 173, 230
agency
  concept of 35, 37-38
  farmers' agency 21, 28, 33, 49, 158, 208, 220, 232-34
agricultural practices, concept of 36-37, 114, 189, 206
agriculture
  Chinese Ministry of Agriculture 75, 206
  extension system 66-67, 72-73, 79, 109-10, 112, 160, 180
  global 26, 63, 65, 75-76, 91, 230
  manual techniques 66, 69, 76, 78-80, 86, 107-09, 124-27, 137-38, 169, 174, 176-77, 179, 181, 198-99, 201, 217
  see also calendar, agricultural; Green Revolution; mechanization; rice cultivation process; technologies, agricultural
agro-system 35-37
America 66, 76, 230, 96n31
ancestors 20, 44, 135
Anhui Province 16, 21, 39, *40*, 40-41, 45, 76, 80, 85-87, 116n4, 117, 119, 123, 130, 132, 136n12, 189, 191-93, 195, 208, 213-14
animals 37, 43, 68, 78, 87n22, 83n17, 119, 125, 131, 134, 157, 189n3, 193
  draught animals 64, 78, 82, 116-17, 119, 125, 134, 171-72, 174, 178-79, 182, 193, 234
  insects 78-79, 131, 158
  pigs 43, 68, 78, 125, 189n3, 210
Anren County *42*, 43, 62, 67-74, 76-84, 86-87, 89-90, 93-95, 116, 135, 145, 175, 204, 209, 211-13, 216
  Anren County Department of Agriculture 72, 74, 79, 175
  Anren County Gazetteer 67-69, 71, 76, 80-82, 87, 116-17, 120, 130, 175, 203, 216
Asia 22n5, 24-25, 27, 66, 76, 85n19, 106, 116, 116n4, 127, 129, 173, 175, 175n2, 179, 181, 199n10, 202n12, 213n20, 215

backwardness 24, 65, 106, 181-182, 234
Barth, Fredrik 35-36, 108, 110, 135, 150, 170
Basu, Paul, and Simon Coleman 24n6, 29, 29n14, 30
Beijing *40*, 44-45, 167, 188, 194, 206, 209, 230-31
belonging 30, 43, 132, 193n6, 231
bitterness and hardship 18-19, 68, 106, 130-31, 200
Bloch, Ernst 182
bodies 29-31, 36, 38, 107-08, *109*, 112, 116, 118, 123, 126, 130, 132-33, 156, 170-71, 173-74, 189, *190*, 192, 204, 216, 231
  feet 79-80, 117, 156
  hands 74, 79, 81, 116, 122, 131, 145, 156
  muscle memory 170
  tactile memory 202, 235
  see also embodiment; senses
Boserup, Ester 25n7, 182
Braverman, Harry 114, 137
Bray, Francesca 25, 25n7, 47n22, 84n18, 85n19-20, 118n6, 119n7, 122, 124, 125n9, 169-70, 174n1, 180n3, 199, 200n11, 230, 234
breeding 69-73, 108, *109*, 119, 124, 127, 129-30; *see also* rice varieties; seeds
Building the new socialist countryside 92

calendar, agricultural 37, 105, 108, *109*, 111, 120-22, 124, 132, 135
  solar terms 50, 106, 117, 120-22, 152
capital 16, 26, 28, 157, 178, 181, 198-99, 201, 216; *see also* resources
capitalism 26, 87n22, 169
care 21, 33, 94, 193n6
  child care 21, 23, 34, 193-94, 198, 203, 220
  elderly care 19, 23, 112, 188, 193-94, 196, 211, 214, 232
  for animals 117
  for fields 33, 49, 157, 188, 191-93, 220, 233
  see also families; health; left-behind people; welfare system
cash 92, 105, 171-72, 178, 180, 194
cash crops 49, 69-70, 77, 91, *190*, 195, 212-19
chaîne opératoire 25, 179
Changsha 19, *42*, 44-45
Chard, Robert L. 46
Chayanov, Aleksandr Vasilevich 214
Chen, Chen 137n13, 208n16-17, 209
Chenzhou City *40*, *42*, 43, 75, 147n1, 149, 216
China, imperial 32n16, 43, 47, 63, 67-69, 74, 76, 79-81, 108, 110, 121-22, 129, 147-49, 163n12, 191, 195, 201, 203, 216, 219
China, northern 44, 89, 175, 177
China, southern 17, 48, 127, 132, 175
China in global perspective 63, 65, 75-76, 63, 91, 96, 149n4, 182, 230
Chinese Communist Party 148-49, 159-62, 216
Chinese language 18n2, 33, 39, 45, 151-54
Chinese New Year 15-16, 18, 45, 50, 62, 132, 135, 147, 167, 188, 203

cities 16, 18-24, 26, 28, 32, 38-41, 43-49, 61-64, 87-88, 90, 93-96, 96n31, 97, 106-07, 111, 126, 130-31, 137-38, 147, 161, 169, 183, 189, 191, 194, 196-97, 199, 205-06, 210, 220, 230-33, 236
climate 37, 41, 68, 71, 117-18, 121-22, 124, 127, 129, 201, 217
  climate change 216, 219, 229
collective 36-37, 48, 62-64, 68, 74, 76, 81, 83, 87-88, 88n24, 90-91, 93, 97, 108, *109*, 110-13, 115, 117, 135-36, 147, 152, 154, 160, 202, 206
communities of practice 23, 28, 31-35, 39, 46, 170, 194-95, 220, 231, 233, 235-36
consumption 30, 93n27, 137
  of consumer goods 92, 208
  of electricity 83
  of farm chemicals 75-78, 80
  of food 87n22, 128, 133, 214, 218
  of grain 133
  of rice 128, 132
  of rice plants 129, 132, 135, 179
cotton 41, 154, 180, 193, 213
Covid-19 pandemic 96
craftwork 30n15, 31, 38, 43, 134, 138, 155, 162, 173, 219
cropping patterns 67-69, 77-78, 110-12, 120-21, 128, 135
  double-cropping 67-69, 77-78, 120-21, 127-28, 193, 199-200, 203
  inter-cropping 129
  multi-cropping 67, 70, 128
  single-cropping 67, 70, 120, *190*, 202-03, 219
crops 17-18, 38, 41, 63, 66-71, 76-77, 80-81, 87, 89, 91-92, 113, 115, 117, 124, 126, 154-55, 157, 172, 179-80, *190*, 195, 201, 211-18, 232
  vegetables 17, 115, 133-34, 138, 212, 213n21
  see also cash crops; cotton; grain; lamp rush; rape seed; rice; soy beans; sweet potatoes; tobacco
Cultural Revolution 64, 84, 87n22, 126, 161

de-collectivization 48, 63, 74, 77, 79, 83, 85-90, 97, 115, 216
decision making 17, 20-22, 24-26, 28, 33, 38, 47, 49, 63-64, 67, 69, 87-88, 90, 111, 122, 128, 156-157, 169-70, 177-79, 182, 189, 191, 192n4, 193, 200-01, 204-06, 211, 214, 217-20, 233
Deng, Xiaoping 64, 71n9
deskilling 26, 48, 80, *109*, 114, 136-37, 177
development 26, 44, 63-66, 69-70, 88, 95, 115, 122, 171, 236
  discourse 47, 49, 64-65, 89, 136, 162, 169, 181-82, 201, 203, 218
  see also modernity
dichotomies
  Cartesian 31
  in migration studies 28-29, 32, 34, 49, 235-36
  inside/outside *109*, 138, 217
diesel 81, 83, 86, 92
differences, rural-urban 40, 63, 90, 92, 94-96; *see also* incomes; migration, reasons; welfare system
drought 74, 119, 129, 158

economies of scale 26, 64, 66, 74, 81-82, 84, 114, 119, 169, 180, 195, 232, 235
economy, diversification of 26-27, 36, 41, 87, 91, 108, 111, 169, 219
Edgerton, David 47n22
education 18-20, 22, 24, 30, 62, 64, 106, 114, 117, 137, 149, 175, 188-89, 200, 203, 214, 219, 230
  of farmers 112, 161-63, 175
  of migrant children 96
  schools 19, 32, 106, 112, 122, 126, 137, 145-46, 169, 175, 188, 197, 214, 230, 234
electricity 66, 73, 75, 81, 83, 86, 91, 105, 117, 133, 169, 171, 200
embodiment 23-24, 29-32, 35, 37-38, 107-08, *109*, 115-18, 132, 137, 155, 236; *see also* bodies; senses; skills
emotions 29, 46, 148, 160
environment 26, 37, 129, 179, 204n14
  farming environment 32, 112, 119, 170, 204
  pollution 75, 79, 81, 127
  protection 80
  socio-material 30-31, 131, 138, 150-51, 178, 231
Europe 43-44, 66, 76, 152, 157, 170, 173, 175, 188, 230, 235-36

families 15-16, 18-19, 21, 33, 44-47, 49, 62, 79, 87n22, 93-94, 105-06, 117, 130, 133, 135-36, 138, 146, 169, 176, 180, 187-89, *190*, 191-93, 195-200, 205, 208, 210-11, 214-15, 217, 220, 232-33, 235-36
  brothers 19-20, 43, 86, 169, 187-88, 210
  children and childhood 15, 18-21, 23, 33-34, 46, 62, 79, 85, 94, 96, 107, 112, 114, 117, 119, 126-27, 136, 145, 163, 169, 175-77, 179, 187-89, 193-94, 196-97, 199, 203, 208, 210, 214, 219-20, 230, 234
  daughters 19-21, 115, 169, 188, 193n6, 194, 198-99, 203, 217
  grandchildren 21, 34, 62, 119, 188, 191, 198, 203, 211, 217
  grandparents 19, 21, 34, 68, 86, 106, 111-12, 114, 193n6, 194
  in-laws 21, 34, 116, 130, 132, 191, 193-94, 208, 220
  mothers 20-21, 34, 46, 74, 138, 167, 191, 193, 197, 203, 217
  parents 18-20, 45, 96, 106, 111, 116, 136, 150, 189, *190*, 191-92, 193n6, 200, 203, 208-10, 214, 230
  son preference 196, 209
  sons 19, 21, 34, 114-15, 126, 130, 172, 188, *190*, 191, 193n6, 196-97, 199, 208-10, 215, 217
  see also households

family planning 94, 115, 196, 209, 229
famines 64-65, 69, 78, 105
farm chemicals 66-67, 75-81, 83, 85, 107, 110, 116-17, 124-25, 127, 129, 134, 137, 163, 172, 199, 201, 216
farmer-state relations 38, 162-63, *190*, 190-91, 195-96, 200, 205, 214, 232, 234
farmers
  farmers' experiments 67, 69, 76, 82, 120, 187
  farmers' projects 22, 38, 49, 170, 189-90, 196, 208, 210, 214, 220, 232
  images of 24, 32n16, 106, 181-82, 231, 234
  landless farmers 63, 96
farmland
  1.8 billion *mu* farmland redline 207
  abandonment 89, 125, 157, 183, *190*, 203-06, 211, 218-19, 232
  allocation 19, 63, 87, 89, 111, 115, 204, 215
  classification and types 17, 74-75, 114-16, 155, 178, 180, 204-05
  fallowing 16-18, 146, 204, 146, 157, 195, 232
  fragmentation 89, 204
  reclamation after return 96, 194-95
  reclamation by the government 88-89, 205
  reclamation of new farmland 67, 197, 115, 121, 197, 212
  see also fields; land, legal situation
farms, experimental *109*, 110-11
Farquhar, Judith 35, 37n17, 38
feminization of agriculture *109*, 112, 116; *see also* gender; women
fertilizers 17, 19, 70, 74, 77, 79, 82, 91, 93, 124, 126, 154, 211, 216
  chemical 70, 75-78, 116, 125, 134, 138, 163, 172
  conventional and organic 76-78, 119, 125-26, 131, 134, 154-55, 172, 181, 210
  fertilizer industry 76
  pit fertilizer 78, 154
fields
  dry fields 17, 89, 115, 157, *190*, 211-12, *213*
  field shapes 124, 180, 212
  field sizes 86, 89, 124, 179-80
  model fields 77, *109*, 110
  wet fields 16-17, 25, 41, 89, 118, 124-25, 211-213, 215
  see also farmland; land areas
financial crises 96, 171
Finnegan, Ruth 46
fishery 78, 83n17, 119, 133, 189n3
Five Year Plans 65
Flitsch, Mareile 30n15, 46, 161, 182
folk literature 46, 147, 149, 160-61
food
  as medicine 132, 188, 214
  cooking 130, 132-135, 214, 231
  poisoning 127
  see also consumption; identities
funerals 45, 135

Gansu Province 18, 209
Geertz, Clifford 25n7, 125n9
gender 37, *109*, 110, 112, 116, 123, 175, 180, 205, 217, 234; *see also* feminization of agriculture; labour, division of; sex ratios; women
genetically modified organisms 26, 70n8, 230
geomancy 44, 120, 211
good life 21, 35, 38-39, 134, 188-89
government
  campaigns and projects 61, 63, 67, 69, 71, 79, 88, 96, 160-62, 175, 216, 236
  central government 67, 71, 75, 79, 87, 90-92, 94, 96, 115, 161, 163, 206, 214, 233
  local governments 64, 67-70, 72, 74, 86, 88-89, 92-93, 120, *190*, 212, 216, 218, 229
  officials and agents 72, 91, 110, 112-13, 120, 160, 181, 203
  promotion of modern agriculture 65, 67-72, 74, 86, 90, 96-97, 110, 120, 169, 182, 206, 213, 218, 229
  quotas 64, 87, 90, 92
  support of farmers 44, 67, 90-94, 180, 203, 206
  see also state; taxes
grain
  for animal feed 189n3
  marketing and supply 90-92, 194, 218
  national sovereignty 66, 91, 204, 219
  output and yields 90, 212
  sufficiency 63, 67, 190, 196, 203, 214, 230
  see also rice
Granet, Marcel 147
Great Famine *see* famines
Great Leap Forward 64, 66-68, 74
Green Revolution 26, 32, 63n1, 64-66, 71, *109*, 111-13, 121, 137, 160, 162, 212, 219
Guangdong Province 18-19, 40, *42*, 43, 188, 196, 199, 202, 210, 220
Guangxi Province *42*, 71
Guizhou Province *42*, 118, 210

Hainan Province 71
Han Chinese 40, 43-44, 67, 138
Han Dynasty 201
harvesting 68-69, 116, 127, 158, 179, 195, 201-02
  harvested fields 78, 126, 167, *168*
  manual 81, 108, 130, 150, 155, 167, *168*, 173-77, 179-80, 182
  mechanical 85-86, 117-18, 128-29, 132, 134, 138, 150, 167, *168*, 173, 175-82, 196-97, 199-200, 232-34
  services 19, 117, 177-78, 196, 198-99
Haudricourt, André-Georges 25n8, 37
He, Congzhi, and Jingzhong Ye 27n11, 192n5, 196n9, 198, 204
health 20-21, 33-34, 62, 93-94, 132-33, 137, 189, 192, 197-98, 205, 219, 232; *see also* welfare system
Heilongjiang Province 219

Heiss, Jan Patrick 158
history of technology-in-use 169, 234
homes 15-16, 18-25, 30, 35, 41, 45, 48, 62, 65, 85, 94-96, 106, 112, 121, 133, 138, 146, 167, 189, 194-95, 203, 205, 208-10, 214, 230-31, 233, 236; *see also* houses; migration, places of origin
Household Responsibility System 36, 50, 69, 108, 111, 115, 120, 135, 182, 195, 235
household strategies 26-27, 33, 35, 48, 76, 91, 183, *190*, 210, 212, 232, 234; *see also* land-arrangement strategies; land-use strategies
householding 33
households
  concept of 33
  model households 79
  sizes 94
  split household pattern 21, 23, 32, 34, 36, 193, 217, 220
  see also families
houses 44, 46, 64, 105-06, 168, 176, 193, 210, 220
  construction 15, 18-20, 44, 134, 138, *190*, 197, 206-11
  costs 19, 208, 210
  floor space 207
  on farmland 206, *207*, 210-11
  underoccupied 106, 207
  urban 130
  with slogans 46, 145
hukou system 24, 32, 63, 95, 111, 206, 229
Hunan Province *40*, *42*
  agriculture 67-68, 71, 73, 75-77, 80, 83-85, 92, 111, 117, 120, 129-30, 175-77, 231
  climate and topography 41
  farming-migration predicament 18, 40-41, 65
  field research 44-45
  history 41, 67-68, 89
  land 43, 89
  land-use strategies 189, 191-92, 194-95, 198-99, 202-08, 213, 218-19
  local economy 41, 43, 111
  oral literature 147n1, 152, 157-58, 162
  population 42-43

identities 29-31, 133, 180
imagination 38-39, 134
incomes
  additional incomes 27, 33, 91, 95, 178, 181, 194, 196, 212, 214-16, 218-19
  from abroad 40
  from migrant work 41, 86, 95, 96n31, 138, 196, 203, 210, 232-33
  from rice farming 19-20, 93, 128
  rural 90-92, 208
  urban 90
Ingold, Tim 31, 150
intensification debate 25, 124, 182
intensification/de-intensification 25, 27, 66, 69, 87, 110, 115, 120, 124, 156, 189-92, 199, 202, 211-12, 217, 246
Internet *109*, 112
irrigation and drainage 67, 73-75, 81, 83n17, 92, 118-21, 126, 153-55, 206, 213
  disintegration of 120, 203, 206
  mechanization of 66, 74-75
  organization of *109*, 110, 119n7, 120, 203
  ponds 44, 119-20, 157, 206
  pumps 41, 75, 119
  reservoirs 74-75, 119-20
  system 118-20, 203

Jacka, Tamara 33, 112n3
Japan 66, 76, 85n19, 179, 181, 202n12, 215
Jiangxi Province *42*, 68, 126, 130, 180, 198, 218

knowledge 23, 25-26, 30-31, 34-35, 37-38, 46, 48, 70, 78, 105, 107-14, 151, 171, 229, 231
  academic 153
  codified 50, 147, 150, 154, 163
  control over 112-13, 119-20, 127, 135-37
  definition 35
  lack 105, 111, 113-14, 132, 137, 151, 220
  moral-political 147-49, 151, 153, 160, 162-63
  negotiation 46, 48, 138, 148, 162
  reinterpretation 160
  repertoires 22, 27, 35, 37-38, 48, 107, 113-14, 136, 147, 157, 163, 170, 189, 195, 200, 206, 212, 218
  scientific *109*, 110, 113-14, 127, 148, 160, 163
  standardization 108
  see also embodiment; skills
knowledge systems 35, 38, 106-08, *109*, 110-14, 125, 136-37, 147-48, 170, 177
  three faces 35-36, 108, *109*, 110-13, 150
    corpus 35, 37, 108, *109*, 110-13, 115-18, 122-27, 129-30, 132-36, 147, 155, 157-58, 162, 175, 202, 204, 211, 231
    media 35, 46, 70, 108, *109*, 110-13, 117-18, 120-21, 128, 132, 135, 147, 149-50, 153-58, 162, 170, 182, 202
    social organization 35, 37, 39, 108, *109*, 110-14, 116, 119-20, 122, 124-25, 127, 130, 132-33, 135, 138-39, 153-54, 156-57, 177, 217, 219, 235
Knutson, Helena 170

labour 22, 26-27, 49, 67, 69, 93, 95, 136, 177, 181, 193, 198, 204
  agricultural 48, 89, 95, 210
  division of 18, 36, 43-44, 64, 110-11, 112n3, 120, 122, 124, 126, 138, 156, 175, 177, 191, 200, 217
  exchange 177, 180, 197-98
  hiring 180, *190*, 197-99
  labour-intensive practices 17-18, 67, 118, 124, 172, 195n8, 211

labour-saving technologies 65, 80, 87, 127, 132, 169, 177-78, 181-82, *190*, 192, 199-200, 216
migrant labour 18, 96-97, 199, 209
physical and manual 69, 131, 198
shortages 18, 26, 36, 49, 68, 120, 122, 124-25, 173, 177-78, 182, 187, *190*, 191, 194, 196, 198, 200, 205, 211-12, 214-15
skilled 18, 25, 192, 199, 202
lamp rush 213-17, *218*
land areas
arable land 63, 69, 74, 89, 92, 175, 206-07
cultivated land 41, 66, 83, 89
land mass 43
loss of 69, 89, 206-07
per capita average 89
scarcity 25, 44, 69, 124, 207, 209-10
sown area 43, 73, 91, 93, 212
see also farmland; fields
land as social security 16, 22, 25, 27, 34, 41, 48, 89, 96-97, 133, 194, 211, 231, 233
land reform 63
land rentals *190*, 192-95, 202, 205, 232, 235
land-arrangement strategies 22-23, 27, 49, 187, 189, *190*, 191-98, 219, 230; *see also* household strategies; land-use strategies
land-labour ratios 25, 89, 193, 203, 210, 219
land, legal situation 87-88, 193-94, 205-06, 209, 230
contracts 87-88, 192-94, 205
entitlement 24, 96, 207
land-use rights 34, 87-88, 111-12, 206, 230
ownership 64, 88, *109*, 110, 115
sales 88, 209-10
transfer 88, 194, 206, 230
land-use strategies 22, 27, 49, 88, 157, 187, 189, *190*, 199-200, 202-03, 206, 211-12, 214, 219, 230; *see also* household strategies; land-arrangement strategies
Lave, Jean, and Etienne Wenger 31-32
left-behind people 16, 23, 27-28, 29n14, 30-35, 39, 45, 49, 179, 189, *190*, 191-93, 198-99, 202, 204-05, 210-11, 220, 231, 233, 236
children 15, 21, 33-34, 46, 106, 230
elderly 21, 34, 106-07, 180, *190*, 198, 215, 232-33
women 34, 49, 62, 116, 118, 167, 177-78, 189, 192, 215, 232-33
Leroi-Gourhan, André 25
Li, Keqiang 152
Li, Liqing 149
Li, Yuyu 137n13, 192n5, 194, 196n9, 198, 199n10, 202n13, 209n18
literacy 149
livelihoods 26-27, 33, 94, 97, 169
local gazetteers 43, 46-47, 67-69, 71, 73, 76-77, 79-82, 85, 87, 89, 116-17, 120, 130, 136, 147, 169, 175, 181, 203, 216
Longshi Township 43-44, 64, 85, 89, 199, 204, *205*, 212, 216-17
Luhmann, Niklas 194
Made in China 2025 152
Mao era 62, 64, 66, 71, 76, 80, 90, 93, 115, 148, 199
Mao, Zedong 64-65, 76, 81, 148, 152-53, 159-60, 163
marketization 62, 87, 111-12, 207, 218
markets 62, 64, 68, 76-77, 81, 87, 93, 111-12, 138, 207, 213, 219, 233
grain 90-91, 218
labour 96
land 194, 205
local 132, 134, 173, 204, 212
seed 72
marriages 20, 23, 43, 45, 112, 188-89, 196, 208-11
Marxism 159, 182n4
material culture and materialization 23, 29-31, 34-39, 49, 120, 131, 170-71, 179, 182, 194, 208, 230-32, 236
Mauss, Marcel 231
mechanization 17-18, 26, 37, 47n21, 49, 66-67, 69-70, 73-75, 79, 81-87, 92-93, 95, 107, 110, 113, 117-18, 125-26, 129-32, 137-38, 162-63, 168-72, 175-81, 197-200, 203-04, 208, 219, 229-30, 233; *see also* harvesting; irrigation and drainage; processing rice; rice cultivation process; technologies, agricultural; tractors; transportation; transplanting
men 18, 84, *109*, 110-12, 116, 122, 133, 136-38, 145, 175, 177, 187-88, 196, 209-10, 217; *see also* families; gender; labour, division of; migrants; sex ratios
Meng, Xiangdan 116, 137n13, 150n8
migrants 16, 18, 20-21, 23-24, 27-30, 32-35, 39-40, 42-46, 49, 61-62, 65, 68, 80, 86, 92, 94-97, 122, 132-34, 167, 169, 173, 177, 188-89, *190*, 191-96, 198, 202-05, 207-11, 218, 220, 229-32, 236
children 96
educational 19-20, 117, 137, 188, 199, 230
female 96, 116, 150
highly-skilled 28, 137, 230, 235
male 112, 114, 187
migrant worlds 28-30, 32, 34, 39, 231, 236
number of 16, 18, 95
occupations 15-16, 18-19, 21, 44-45, 61-62, 68, 79, 95-96, 106, 112-14, 133, 137-38, 167, 187-88, 191, 193, 196, 203, 208, 210, 220, 230
older 96, 137-38, 230
perceptions of 32, 35, 46, 231, 235
skills 30-31, 113, 137-38, 178, 210, 233, 235-36
young 48, 79, 105-07, *109*, 112-13, 137-38, 199, 230
migration
alternatives to 86, 90-91, 95-96, 212, 219
burdens and insecurities 16, 21, 48, 62, 94-97, 188, 230
circular and seasonal migration 23, 44, 84, 176, *190*, 196, 209, 236

decisions 21-22, 25, 28, 47, 156-57, 169, 178, 192, 220, 233
destinations 18-20, 23, 28, 39, 43-45, 188, 233, 235-36
history of migration in Hunan 68, 218
materialities of 23-24, 28-31, 34-39, 49, 120, 131, 170-71, 179, 182, 194, 198, 208, 230-33, 236
opportunities 21, 95, 199, 205, 208-11
overseas migration 40
patterns 18, 23, 28, 34, 40, 42-45, 65, 68, 112, 137, 183, 188, *190*, 192
places of origin 16, 23, 25, 28, 30, 34, 39-40, 42, 65, 236
reasons 18-21, 27, 40, 43, 48, 62-63, 65, 80, 87, 90-92, 94-95, 169, 182, 208-09, 211, 220
return 20-21, 23, 33, 45-46, 62, 96-97, 112, 114, 130, 138, 167, 171, 177, 189, *190*, 191, 195-96, 203, 209-10, 230
urban-to-rural 39, 67, 110, 126, 130, 133, 197
see also decision making; homes; household strategies; *hukou* system; land-arrangement strategies; land-use strategies; left-behind people; mobility
Ming Dynasty 68
mining 18, 20, 43, 188
mnemotechnics 147, 149-51, 170
mobility 23, 28-30, 39; *see also* migration
modernity 24, 38, 46, 49, 63n1, 110, 134, 147, 149, 153, 181-82, 232, 234-35; *see also* development; non-synchronicities
modernization 32n16, 48, 62-63, 65-67, 69, 73-74, 78, 82, 86-87, 93-94, 97, 113-14, 122, 129, 153, 161, 180, *190*, 200
Murphy, Rachel 196n9, 198, 199n10, 209n18
mutual aid 43, *190*, 196-98

networks, social 16, 19, 21-22, 28, 40, 43, 45, 62, 169, 137, 170, 188, *190*, 195-98
New Economics of Labour Migration theory 27
newspapers and radios 79, *109*, 110-12, 216
No. 1 Document 65, 92
non-synchronicities 182, 234
nong, three 65
nongmin 32
nongmin gong 32

objectification 23, 28-29, 34, 231, 236
oil 43, 76n13, 133, 213, 216
On Practice 159
one-child policy *see* family planning
orality 149; *see also* folk literature
Ortner, Sherry B. 38
Oshiro, Kenji K. 179, 181

patrilines 18, 23, 33-34, 43-44, *109*, 110, 120, 136, 138, 189, 194, 196-97, 211, 215, 233
patrilocality 191, 195, 209
Pearl River Delta 43-44
peasants *see* farmers
Peng, Mu 135
people's communes 64, 82, 87, 93, *109*, 111
People's Liberation Army 83, 197
pesticides *see* farm chemicals
pests and diseases 78-79, 151, 155
Pfaffenberger, Bryan 47n22, 36
phenomenological approach 29, 31
Philippines 65, 213n20
ploughing and tilling 69, 81-82, 84, 116-17, 120, 122, 124, 127, 155-58, 168, 171-72, 178-79, 182, 234
harrows 80-81, 84, 131
hoes 80-81, 84, 118, 127, 172
ploughed area 84, 85
see also tractors
policies, rural 38, 47-48, 62-65, 67-68, 73, 76-77, 80, 87-88, 90, 92-97, 106, 113, 115, 149, 152, 195-96, 206-09, 218, 229-30, 232-33
policy makers 63, 67, 80, 114, 137, 152, 180, 207, 229
population 16, 24, 63, 94-95, 97, 132, 149
floating population 95
growth 68-69, 89, 219
pressure 25, 42-43, 68
rural 25, 90, 92-93, 95
size 16, 25, 42-43, 63, 182
structure 94-45, 193, 197, 209
power 33, 37-38
predicament, concept of 20
processing rice
manual 45, 81, 86, 128, 130-32, 155-56, 169, 174, 181, 234
mechanical 70, 82-83, 86, 105, 131-32, 169, 175, 179, 181, 193
propaganda 160, 162
proverbs *see* folk literature

Qing Dynasty 67, 69, 76, 163n12, 195, 203, 216, 219

rape seed 41, 68, 77, 120, 146, 213-14
Rawski, Evelyn Sakakida 219
rebellions 68, 203
reform period 23, 62, 64-65, 71n9, 72-73, 84, 87-91, 93-95, 108, 115, 160-61, 187, 207, 212
remittances 27, 30, 92, 177-78, 182, 199, 208
Republican China 74, 76, 81, 85n19, 108, 135, 159, 216
resistance 71, 232
resources 16, 20, 22-27, 30, 34-38, 41, 44, 47-49, 62, 66-67, 90, 94, 96, 107, 110, 112, 114, 118, 126, 132-34, 136-37, 147, 154, 163-64, 169-70, 181, 189, 191-92, 194-95, 202, 210-11, 214-15, 219-20, 231, 233, 236
rice
prices 90-93, 128, 194
social meaning 132-36

storage 64, 131, 150, 156, 193, 198
taste 71, 128
see also consumption
rice cultivation process
close planting 70, 78
extensive methods 199
gleaning 175
stepping into the fields 76, 80, 117, 126
see also harvesting; ploughing and tilling; processing rice; sowing; timing of tasks; transplanting
rice economies 25
rice plants, parts of
chaff 78
husks 131, 133-34
seedlings 82, 120, 123-24, 134, 151, 154-55, 172, 195n8, 203
straw 78, 131, 133-35, 168, 174-75, 178-79, 194
stubbles 68, 86-87, 126
see also consumption
rice production
geographic shifts 219
inputs and costs 19, 25, 66-67, 69, 85, 91-93, 124, 128, 138, 171, 176-78, 194, 195n8, 197-98, 200, 211, 216, 232
output 41, 89-91, 203, 216, 219
yields 17, 66-67, 69-71, 73-74, 77, 110, 118, 124, 128, 163, *190*, 201, 211-12, 219, 229
rice varieties 67-73, 111, 122, 124, 127-29, 132, 154-55, 159
conventional 17, 69-70, 78, 127-28
high-yielding varieties 66, 69-71, 73-74, 77-78, 92, 129, 137, 163
hybrid 65-66, 69-73, 77, 79, 107, 110, 128-30, 136-37, 201, 234
Richards, Paul 25
risk reduction 27, 129, 159, 194, 212
rituals 25, 36, 108, *109*, 112, 117, 120-21, 127, 135-36
roads 16, 44, 86, 145, 167

Santao Jicheng project 161-62; *see also* folk literature
Santos, Gonçalo 63n2, 181
Sargeson, Sally 89n25, 207n15, 208n17, 209n18-19, 211
Schäfer, Ingo 147-48, 152, 159-60, 163
Schippers, Thomas K. 35-37, 157, 189, 202
Schmalzer, Sigrid 66, 71n9, 72n11, 136, 150n8, 154, 160
science 25, 63n1, 66-67, 71-74, *109*, 110, 113-14, 117, 127, 129, 148-50, 160, 162-63
scientists 66-67, 70-71, 73, *109*, 110-13, 130, 137
Scott, James C. 232
seasons 15, 41, 61, 71-72, 78, 84, 106, 117-18, 120-22, 126-29, 145-46, 153, 155, 158-59, 176, *190*, 195-96, 198, 201-04, 212, 214-16, 219
peak seasons 116, 122, 124, 129, 157, 173, 177, *190*, 194-98, 200, 214, 217
seeds 67, 69-73, 82, 92-93, 107-08, *109*, 110, 121-25, 127-28, 130, 137, 151, 155, 158
seed companies, shops, and stations 70, 72, 79, *109*, 112, 128, 130
see also breeding; genetically modified organisms; hybrid rice; rice varieties; sowing
senses 29-30, 71, 116, 132, 150-51, 193, 202, 235
sent-down youth 39, 67, 110, 126, 130, 133
sex ratios 209; *see also* families, son preference
Shakespeare, Tom 20
Shanghai 15-16, 28, 39, 44-45, 61, 90, 96, 106, 126, 130, 132-33, 138, 191, 193, 206
Shennong, Divine Farmer 121, 135
Shiji 153
Sichuan Province 40, 163n12, 210
Sigaut, François 25n7, 31-32, 47n22, 173-75, 179, 235
Sima, Qian 153
skill perspective on migration 23-24, 30, 34, 49, 234-36
skill-producing group 31-32, 110, 235
skills
concept of 22, 30-32, 107, 131
enskilment 30-32, 37, 71, 73, 79, 106-07, 113-14, 116-17, 119, 126, 131, 133, 136-38, 150-51, 153, 155, 170, 175, 188, 193, 210, 233
evaluations of 114, 136-37, 160, 234-35
new skills 117-18, 137-38, 177, 210, 232
see also deskilling; embodiment; knowledge; migrants, skills; senses; techniques, concept of
smallholder model 24
smart farming 203, 230
smartphones 46, 112, 122, 229, 234
socio-technical systems 36, 134, 181, 235
soil 17, 27, 37, 66, 74-79, 82, 84, 97, 114-18, 125, 127, 137, 146, 154, 156-58
Song Dynasty 43, 68, 74, 80, 129
Soviet Union 66, 153
sowing 43, 69, 73, 93, 120, 122-25, 128-29, 155, 157, 180, 212
direct seeding 80, 125, 182, *190*, 200-02
soy beans 68, 105
Spring Festival *see* Chinese New Year
state 21, 32, 46-49, 62-64, 66, 83-84, 90-93, *109*, 110-13, 148, 161-63, 181
interests 67, 69, 78, 81, 108, 112, 136, 147, 154, 161-63, 175, *190*, 191, 195-96, 214, 232, 234
see also farmer-state relations; government; grain, national sovereignty; taxes
statistics 46n21, 71, 83, 169, 203, 212-13
Stavis, Benedict 71
Strategic Plan for Rural Revitalization 2018-2022 92
subsidies 67, 91-94, 180, 203, 206
superstitions 65, 111, 135, 216

suzhi discourse 106
sweet potatoes 68, 133, 211-12, 214, 218

taxes 88, 91-92, *190*, 194, 209, 218
technical lineage 173
technical linkage 179, 181, 201
technicians 71-73, 79, 132
techniques, concept of 171
techniques of the body 231
technological choices, concept of 25, 171
technologies, competitive edges 178
technologies, agricultural
  farmers' attitudes towards 68, 71, 77, 79-80, 82-83, 86, 169, 181-82, 214, 216-17, 229
  foreign 66, 77, 82, 85n19, 129
  local production 71, 77, 83-84, 173
  problems with new technologies 68, 71-72, 76, 78, 81, 83-84, 127
  simultaneous use 129, 168-70, 178, 182, 212
technology, understandings of 47, 169-70, 234
telephones 15, 46, 112, 122, 188, 229, 234
television 79, 106, *109*, 112, 147, 208
temperatures 15, 41, 71, 118, 122, 129, 131, 145, 155, 195n8, 198, 201
tenure system *see* land, legal situation
threshing 81-85, 130-31, 155, 168-69, 174-76, 179, 198, 200
timing of tasks 37, 79, 107, 119-22, 124-26, 129-30, 151, 155, 157-59, 172-73, 177, 179, 181, 195n8, 198, 200-01, 204, 211, 217
  ripening times 70-71, 118, 124-26, 128-29, 180, 195n8, 201
tobacco 43, 69, 213, 218
toilets 78, 181, 210, 231
tool reform 82
topography 41, 68
Township and Village Enterprises 90-91
tractors 84, 92, 116-17, 200, 234
trade 63, 93, 138, 199, 218
  going out strategy 230
  import and export 42, 70, 93, 96, 218
  trade war 230
  see also markets
transplanting 80-83, 86, 110, 122-26, 128, 154-55, 180, 182, 195-97, 200-02
  machines 82-83, 86, 113, 125, 203, 229
transportation, agricultural 81, 86, 131, 174-75
treatises, agricultural 77n14, *109*, 110, 121
trust 194, 233

uncertainty 21, 87, 211, 231
urbanization 24, 88, 95, 230

Vietnam 116n4, 129, 199n10, 202n12, 213n20
villages
  administrative 43, 83, 88n24
  native (*niangjia*) 188
  natural 43, 64, 86, 88n24, 119
  village groups 43, 88n24, 197

Wang, Cangbai 24n6, 28-30
water *see* irrigation and drainage
weeds and weeding 17, 78-80, 108, 118, 124-27, 131, 146, 151-52, 155, 157, 201, 214, 230
welfare system
  abolition of the collective 93-94, 97
  collective 48, 62, 64, 87, 93
  rural 21, 94, 89, 172, 189, 211
  urban 24, 94-95
  see also care; families; health
White, Ben 137
Whittaker, John C. 25n8, 131, 178-79
Whyte, Susan R. 21
Wilken, Gene 114
women 18, 20-21, 34, 49, 79, 84-85, 96, 106, *109*, 110-12, 116, 122-23, 130, 132, 134-35, 137-38, 149n4, 149-50, 163, 168, 175, 177-78, 189, 193n6, 197, 209-10, 215-17, 232; *see also* families; feminization of agriculture; gender; labour, division of; left-behind people, women; migrants, female; sex ratios
Wu, Chunyuan 133, 135-36

Xiang, Biao 33

Yan Emperor 44
Yangtze 40-41
Yuan, Juanwen, and Anke Niehof 118, 137n13, 199n10
Yuan, Longping 71

Zhejiang Province 175
Zhi, Fujing 216
Zhou Dynasty 121

For Product Safety Concerns and Information please contact our EU representative GPSR@taylorandfrancis.com Taylor & Francis Verlag GmbH, Kaufingerstraße 24, 80331 München, Germany